To a wise man, the whole earth is open because the true country of the soul is the entire universe.

Democritus

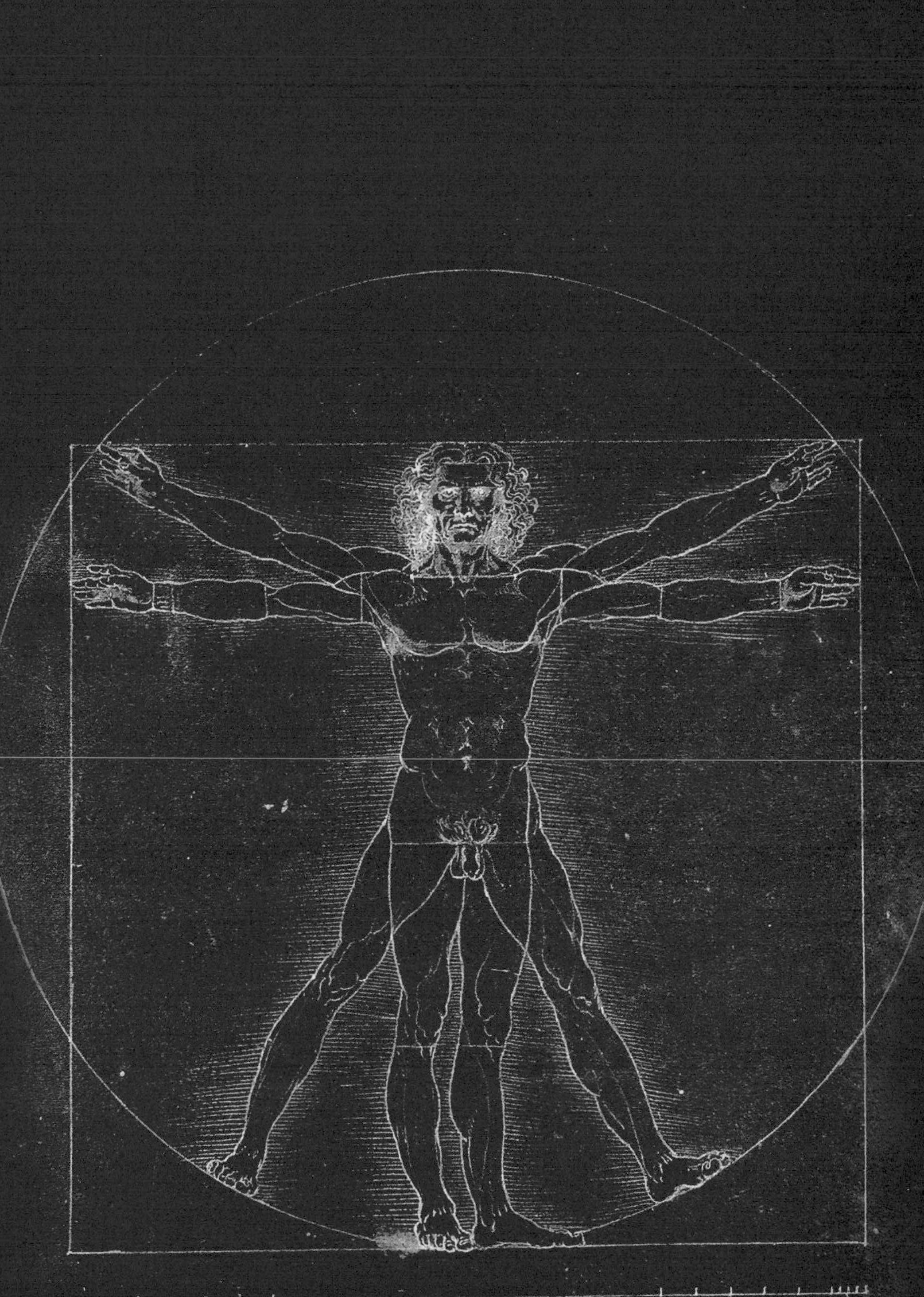

A NEW MAP OF WONDERS

BY THE SAME AUTHOR

The Book of Barely Imagined Beings

CASPAR HENDERSON

A NEW MAP OF WONDERS

A JOURNEY IN SEARCH OF MODERN MARVELS

GRANTA

Granta Publications, 12 Addison Avenue, London W11 4QR

First published in Great Britain by Granta Books in 2017

A CIP catalogue record is available
from the British Library

1 3 5 7 9 10 8 6 4 2

ISBN 978 1 78378 133 1 (hardback)
ISBN 978 1 78378 136 2 (ebook)

Text designed and typeset in Dante and La Gioconda by M Rules
Printed and bound by CPI Group (UK) Ltd, Croydon, CR0 4YY

www.grantabooks.com

We carry with us the wonders we seek without us.

Thomas Browne

Still I felt no fear my wonder seeking happiness had no room for it.

John Clare

Nobody knows how [nature] can be like that.

Richard Feynman

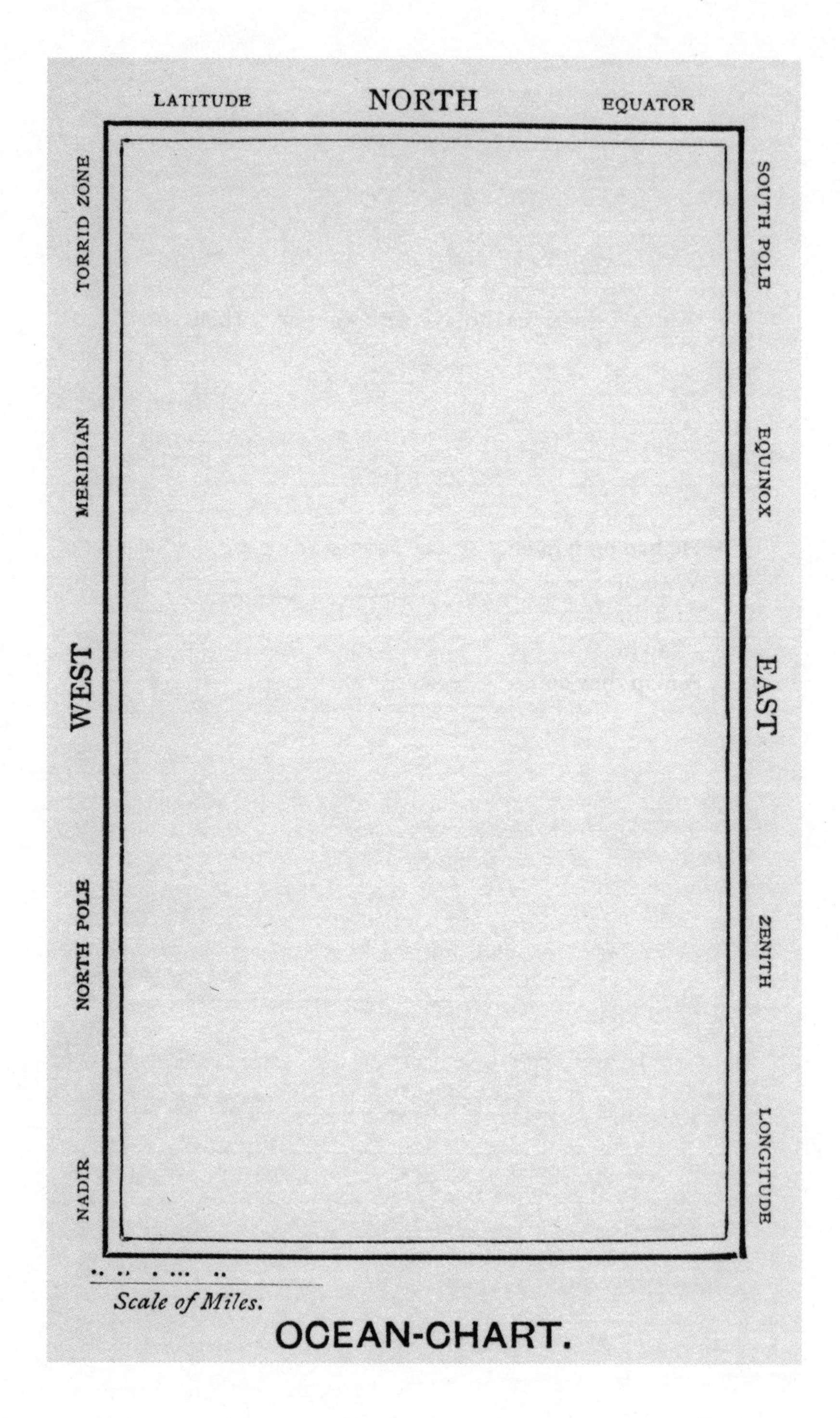
LATITUDE
NORTH
EQUATOR
TORRID ZONE
MERIDIAN
WEST
NORTH POLE
NADIR
SOUTH POLE
EQUINOX
EAST
ZENITH
LONGITUDE
Scale of Miles.

OCEAN-CHART.

He had bought a large map representing the sea,
Without the least vestige of land:
And the crew were much pleased when they found it
to be
A map they could all understand.

The Hunting of the Snark, Lewis Carroll

Contents

Introduction

Celebrate. Yes, but what?

Friedrich Hölderlin

Why, why do we feel . . . this sweet sensation of joy?

Elizabeth Bishop

One morning in early spring I came downstairs with my young daughter to find a brilliant pool of light on the kitchen ceiling. At first I couldn't account for this strange thing, which wobbled, reformed and was momentarily darkened by shadows. Slowly, I worked out what was going on. The Sun, which had been hidden by clouds on many previous days, had broken free and risen high enough to illuminate the windows of a building facing towards it. And those windows reflected the light through the undulating branches of a tree down onto another reflective surface that happened to be angled in just such a way as to bounce the branch-shadowed light up through our kitchen window and onto the ceiling.

Sometimes it takes extreme or unusual circumstances to make ordinary things seem wonderful. In the case of the poet Ko Un, for instance, a postage stamp-sized patch of sunlight on the wall of his cell in the Korean military prison in which he was being held was enough to rekindle a sense of wonder and hope, even as he feared for his life. But there was nothing extreme about my circumstances that morning. I didn't fear for my life. I wasn't in some breathtakingly beautiful or exotic location. It was a bog-standard working Tuesday. Or Wednesday. Or some other day. I forget. At any rate, there was nothing extraordinary about the time and place or, it might seem, the phenomenon. Who hasn't seen sunlight dappling on a wall and wondered how the effect occurs? Who, in the climate in which I live, hasn't felt elated when the sun finally appears after many dark days?

Still, my sense of wonder – of being completely awake – was exceptional. Being a bit geeky, I knew that the patch of light and shadow-play, so gentle and so alive, was created by trillions of photons (particles of light) flowing from a stupendous thermonuclear explosion tens of millions of miles away. And I knew that those photons were a tiny proportion of a vastly larger number pouring silently onto the planet every second at a speed far beyond anyone's power to visualize. As Ko Un writes in another poem, 'I am gazing at the invisible movements of all things.'

The presence of my daughter on that morning made the moment especially joyous for me. She was five at the time, and the patch of light was probably no more and no less remarkable to her than many things that a five-year-old sees in any given week, from postmen to fish fingers. But she saw that her father was laughing, decided that something must be funny, and laughed too. So love was there, and that was wonderful. But it was not the whole story.

Wondering about wonder

The experience in the kitchen set me thinking about wonder itself – about what prompts it and how we experience it, about how it elusive it can be, about how there are so many ways in which it can be closed down and destroyed, but also how it can impart a sense of meaning and be constitutive of a life well lived. I decided that all this was worth exploration. *A New Map of Wonders* is the result.

This book looks into philosophy, history, art, religion, science and technology in search of a better appreciation of both the things we wonder at and the nature of wonder itself. I have no particular expertise, and no qualifications beyond curiosity and stubbornness. I do agree, however, with Samuel Johnson: 'Nothing will ever be attempted if all objections must first be overcome.' And, although have I left out a lot (verging, in fact, on everything), I have tried to keep the account as grounded and coherent as I can. Grounded in that the various and diverse wonders explored in the book are already present to some extent in apparently simple and mundane moments such as the one in my kitchen. Coherent in that these

Two good places to start are *Wonders and the Order of Nature: 1150–1750* by Katherine Park and Lorraine Daston, and *The Age of Wonder* by Richard Holmes.

various wonders are linked through the phenomenon of emergence.

I will say more about emergence later in this introduction, but first here are a few observations on the meaning of the word 'wonder', on its possible history, and on how wonder relates to the great project of trying to understand and be in the world.

The Oxford Companion to Consciousness doesn't contain an entry on wonder – though it does have one on wine, and maybe we should take the hint. A standard dictionary is only a little more helpful. A typical definition describes wonder as something that causes astonishment or profound admiration, and wonder as the state of the person contemplating it. 'Astonishment' comes from the Latin for thunderstruck, while 'admiration' and 'marvel' derive from terms that simply mean to look at. (The derivation is unproven, but the word 'miracle', which is from the Latin *mirari* – 'to wonder at, marvel, be astonished' – may ultimately stem from a proto-Indo-European word meaning to smile or laugh.) 'Wonder' is from Old English *wundor*, but the origin of this word (and its old Germanic root, *Wundran*) is unclear. Henry David Thoreau suggested a shared root with 'wander,' and others have suggested 'wound', but such derivations are entirely speculative.

In the short story 'Undr', Jorge Luis Borges makes wonder an ur-word – one that precedes and supersedes all others. Ralph Waldo Emerson writes that, 'though the origin of most of our words is forgotten, each word was at first a stroke of genius, and obtained currency because for the moment it symbolized the world to the first speaker and to the hearer.'

Here is part of a definition that goes a little further. Recalling his experience of an ash tree in the evening sunlight, the philosopher Martyn Evans describes wonder as:

> an attitude of altered, compellingly intensified attention towards something that we immediately acknowledge as somehow important – something whose appearance engages our imagination before our understanding but which we will probably want to understand more fully with time.

'Imagination is a tool for making sense of a world with infinite possibilities, by reducing them.' Michael Lewis

This does, I think, capture an important part of what is often going on when we are filled with wonder. At least it does in my case. As Evans puts it, we recognize or intuit something essential and beautiful (perhaps an underlying structure or order), and we become highly attentive.

Where did wonder start? Has it been part of our experience since the very beginnings of human history? Or does it go back even further than that? A few years ago at Gombe Stream National Park in Tanzania, two chimpanzees were observed to climb separately to the top of a ridge at sunset. There they greeted each other, clasped hands, sat down together and watched the Sun go down, staring for a long time at its fading light. What are we to make of such a report? The primatologist Jane Goodall has few doubts. Not far from that spot she has observed other chimps watch a waterfall and then display and dance extravagantly. She says:

> I can't help feeling that [such behaviour] is triggered by feelings of awe, wonder that we feel. The Chimpanzees' brains are so like ours. They have emotions that are similar to or the same as [ours] ... and incredible intellectual abilities that we used to think unique to us. So why wouldn't they also have ... some kind of spiritual [life], which is really being amazed at things outside yourself ... I think chimps are as spiritual as we are but they can't analyse it, they don't talk about it ... It's all locked up inside them, and the only way they can express it is through this fantastic rhythmic dance ...

'Unjustified linguistic barriers fragment the unity with which nature presents us. Apes and humans did not have enough time to independently evolve strikingly similar behaviours.' Frans de Waal

If Goodall and other researchers are right then the common ancestors we share with chimps, who lived more than five million years ago, may also have felt a sense of wonder.

By fifty to a hundred thousand years ago anatomically modern humans were making sophisticated tools and trading over long distances. To do these things they would have needed language, and we may therefore presume they had stories. By no later than about forty thousand years ago people were creating sculptures and murals of animals and other beings which are widely regarded today as great art. Nobody disputes the skill of these early creators, but what can we say about their emotions and beliefs? Take a thumb-length sculpture of a water bird, perhaps a

Profiled hands are widespread in Paleolithic art, and may be the first universally recognized symbol of the human form. These, found in Santa Cruz Province in Argentina, are between 13,000 and 9,000 years old.

cormorant, which was found in a cave in southern Germany and is more than thirty thousand years old. It is perfectly streamlined, as if caught in the act of diving. Jill Cook, a senior curator in prehistory at the British Museum, says it may be a 'spiritual symbol connecting the upper, middle and lower worlds of the cosmos . . . Alternatively, it may be an image of a small meal and a bag of useful feathers.' But consider too some of the surviving representations of human and half-human forms, from highly stylized female nudes to the supposed half-deer, half-man 'sorcerer' of the Chauvet cave in France and the Lion-Man of Hohlenstein-Stadel in Germany. The makers of these objects were not only observing and copying with what many now regard as exquisite precision; they were also creating. Could their sensibility really have been so distant from that of an artist such as Paul Klee, who in 1920 wrote that 'art does not reproduce the visible; rather, it makes visible'. Could their emotions really not have included wonder?

'The thrust of an animal's neck or the set of its mouth or the energy of its haunches was observed and recreated [in the Chauvet cave paintings] with a nervousness and control comparable to what we find in the works of a Fra Lippo Lippi, a Velasquez or a Brancusi.'
John Berger

*

Around ten thousand years ago, at the dawn of agriculture, a society capable of building monumentally in stone thrived in Anatolia in modern-day southeast Turkey. On a mountain ridge at Göbekli Tepe, tall pillars were decorated with pictograms, which are thought to be sacred symbols, and with reliefs depicting various creatures. Not far from Göbekli Tepe, at Nevali Çori, a site on a riverbank that was excavated just before it was flooded behind a giant dam in the 1990s, an amphitheatre was surrounded by giant stone figures. In one sculpture a snake writhed across a man's head. Another depicted a bird of prey landing on embracing twins. Huge, T-shaped megaliths had faceless, oblong heads and human arms engraved on their sides. As people sat on benches around the walls of the buildings, these forms would have loomed over them. Did those who saw these forms feel wonder, dread or something else? It's unlikely we'll ever be able to do more than speculate, but a context of ritual and religion seems plausible: we know that today places like this often facilitate emotionally charged states of reverence, awe and wonder.

Wonder, wrote the scholar Philip Fisher in 1999, 'is a feature of the middle distance of explanation, outside the ordinary, short of the irrational or unsolvable'. It is a horizon, both personally and historically, 'between what is so well known that it seems commonplace and what is too far out in the sea of truth even to have been sighted except as something unmentionable.' Wonder, then, is linked to the love of knowledge and wisdom. Of course, this should come as no surprise. For both Plato and Aristotle root philosophy in wonder, or *thaumazein*. Plato writes that 'wonder [is] where philosophy begins'. Aristotle says that 'it is owing to their wonder that men both now begin and at first began to philosophize'.

Many traditions honour something that we in modern industrialized countries take to be a sense of wonder, though it may be wonder of a subtly different kind. While not necessarily hostile to further knowledge, that sensibility has often been less hungry and restless than in our own societies. The peoples who we term 'animist' for example, may be united in what the anthropologist Tim Ingold calls 'a way of being that is alive and open to a world in continuous birth.' For them, the world is a perpetual source of 'astonishment but not surprise.' In Yogic philosophy the wonder felt upon recognizing one's ignorance of the world is occasion for liberation. And in the work of classical Chinese poets such as Li Po and Tu Fu, close attention to the marvels of existence is not followed by a restless seeking after more facts but by wonder (though it may often also arouse feelings of melancholy and separation). And in Europe, the twentieth-century philosopher Martin Heidegger contrasted what he saw as industrial civilization's instrumental attitude, in which everything that is not us becomes part of a standing reserve that can be consumed, with what he saw as true wonder.

Wonder as Heidegger envisaged it (which he called *Erstaunen*, and associated with restraint, or *Verhaltenheit*) reveals people and things as they simply are, and moves us to want to safeguard the beauty and complexity of the world. This is, I think, an attractive thought, and one that can contribute to (but not define) an ethic that is adequate to our times. But one doesn't need to follow Heidegger very far – and especially not towards the catastrophic political choices he made – to make use of some distinctions he drew between true wonder and states of mind that are close but not the same.

Start with astonishment and awe (for Heidegger, *Staunen* and *Bestaunen*). These are particularly prevalent in pre-modern and religious thought, where the world is filled with mysteries, and the gods or God can be terrifying. Two striking examples, which were probably both first written down in about the fourth century BC, come from the *Bhagavad Gita* and the Hebrew Bible. In Chapter 11 of the *Gita*, Krishna grants Arjuna a vision of his divine form. Before the wonder of this majesty, which is brighter than the light of a thousand suns, heaven, Earth and all the infinite spaces tremble in fear. In verses 38 to 41 of the Book of Job, Yahweh (Jehovah) speaks to Job out of the whirlwind, asking, 'Where wast thou when I laid the foundations of the earth? . . . When the morning

The physicist Robert Oppenheimer famously quoted from the *Gita* when he recalled his feelings at the explosion of *Trinity*, the first atomic bomb: 'I am become death, the destroyer of worlds.' A more accurate translation is 'I am all-powerful Time, which destroys all things.' The ultimate message of the *Gita*, however, is joy.

stars sang together, and all the sons of God shouted for joy? Or who shut up the sea with doors, when it brake forth, as if it had issued out of the womb?' (Job, of course, has no answer.) Similarly, the Psalmist gives a special place to yir'ah – the awe, dread and reverence felt by those who have witnessed the signs and portents of God's works in the world – and portrays it as the beginning of wisdom.

'Let all the earth fear the LORD: let all the inhabitants of the world stand in awe of him.' Psalm 33

Astonishment and awe are also present in the sublime and romantic sensibility that arose in eighteenth- and nineteenth-century Europe. But the emotions were deflected and altered in this new world. A sense of awe before the great works of nature such as a mountain or a mighty waterfall was an aesthetic as much as or more than it was a religious experience. An influential expression of this perspective appeared in 1756 in Edmund Burke's *Philosophical Enquiry into the Origin of Our Ideas of the Sublime and the Beautiful*. Burke describes the sublime as a state of astonishment 'in which all [thought is] suspended with some degree of horror'. But, he notes, we also feel a sense of delight in the presence of the sublime in spite of the danger. The fact that a huge waterfall could swallow you is at some level part of its appeal, because facing it enables you to face and to some extent master your fear, and thereby makes you vividly aware that you are alive.

Quite different from a sense of the sublime is the typical human reaction to curiosities. A key factor in this sensation (which Heidegger calls *Verwunderung*) is novelty. The object of curiosity will vary greatly. It may be a feat of extraordinary skill or daring, such as juggling with a chainsaw. It may be a matter of sheer strangeness and unfamiliarity: think of the scene in *The Tempest* where Trinculo's first idea upon encountering Caliban is to try and think of a way to get him back to England and make money by putting him on display. It may be something lurid or abnormal, verging on or well into what many people find grotesque, such as the Elephant Man or the microcephalics in Tod Browning's 1932 film *Freaks* – or, in the age of the Internet, an image of a harlequin baby, or as a penis-like appendage on the face of a piglet. Or it may be something incongruous and charming: 'This baby

'[In England] they will not give a [penny] to relieve a lame beggar [but] they will lay out ten to see a dead Indian.'

hippo got separated from his family by a tsunami and a 103-year-old tortoise became his best friend' is among my favourites. But whatever the objects of such curiosity, no one thing holds the attention for long. There is always hunger for something new, and seldom much interest in deep explanation or meaning.

Different and distinct from the fascination and craving for novelties is a cooler kind of admiration (in Heidegger's categorization, *Bewunderung*), in which the intellect remains active and the wonderer maintains some emotional distance from the object of wonder. This form of wonder may be closest to what Plato and Aristotle had in mind. For them, wonder is the beginning of philosophy but not its goal, which is to use reason to improve the human condition. And following this, the mathematician and philosopher René Descartes, writing in 1649, describes wonder (*l'admiration*) as the first of the passions – and a uniquely mental one, unaccompanied by fluttering pulse or pounding heart. Wonder, according to Descartes, is 'a sudden surprise of the soul which causes it to apply itself to consider with attention the objects which seem to it rare and extraordinary'. It is to be welcomed because it helps us to focus on objects for what they are, instead of for what they are for us, and in this way it disposes us to the acquisition of sciences. And, he says, once we've been inspired to pursue this higher knowledge, we have no further need of wonder. So it is instrumental, not a place to dwell. This form of wonder is, broadly speaking, the spirit of the scientific revolution of which Heidegger was so suspicious. (In a more hopeful view, championed by Steven Johnson in his 2017 book *Wonderland*, delight in novelty and the pursuit of fun have, when combined with cooler analysis, frequently given rise to technological and social advances of great significance and benefit. Programmable computers spring from self-playing musical instruments; democracy from the meeting of all people as equals in taverns and coffee houses.)

In Descartes' schema, besides wonder, the passions are love, hatred, desire, joy and sadness.

The rise of science

The flowering of enquiry in early modern Europe changed profoundly what people wondered at and about. The historian David Wootton illustrates the nature of the change with a comparison between well-educated Englishmen before and after science, then known as natural philosophy, came to play a major role in his country's culture. In 1600, a decade before Galileo's discoveries with a telescope prompted John Donne to write that 'new philosophy puts all in doubt', a well-educated Englishman believes that magicians and witches actually exist, and that witches can summon up storms that sink ships at sea. He believes mice are spontaneously generated in piles of straw. He has seen a unicorn's horn but not a unicorn. He believes that there is an ointment which, if rubbed on a dagger which has caused a wound, will cure the wound, and that a murdered body will bleed in the presence of the murderer. He believes that the shape, colour and texture of a plant reveal its medicinal properties. He believes it is possible to turn base metal into gold. He believes a rainbow is a sign from God and that comets portend evil. He believes in astrology, and, although it is nearly sixty years since Nicolaus Copernicus published his argument to the contrary, he believes that the Sun revolves around the Earth. He believes that Aristotle is the greatest philosopher and that Pliny, Galen and Ptolemy, all ancient Romans, are the greatest authorities on natural history, medicine and astronomy.

By the 1730s a well-educated Englishman has looked through a telescope and a microscope. He owns a pendulum clock and a barometer (and he knows there is a vacuum at the end of the tube). He does not know anyone educated and reasonably sophisticated who believes in magic, witches, alchemy or astrology. He knows that the unicorn is a mythical beast. He does not believe that the shape or colour of a plant reveals anything about how it may work as a medicine. He does not believe that any creature large enough to be seen with the naked eye is generated spontaneously. He does not believe in the weapon salve or that murdered bodies bleed in the presence of the murderer. He knows that a rainbow is produced by refracted light, that the Earth goes round the Sun, and that comets have no significance for our lives on Earth. He knows that the heart is a pump, and he may even have seen a steam engine at work. He believes that natural philosophy is going to transform the world and that the moderns have outstripped the ancients in every respect.

Thomas Newcomen's atmospheric engine – harbinger of the Anthropocene – started pumping water out of coal mines in 1712.

This new way of thinking – in Wootton's phrase 'a new kind of engagement with sensory reality' – vastly increased human power and choice in the face of nature, and diminished the amount of fear in daily life. And this perspective continues largely unchanged to this day, and it can on occasion be very phlegmatic. As the twentieth-century physicist Richard Feynman put it, 'People say to me, "Are you looking for the ultimate laws of physics?" No, I'm not. I'm just looking to find out more about the world and if it turns out there is a simple ultimate law which explains everything, so be it; that would be very nice to discover. If it turns out it's like an onion with millions of layers and we're just sick and tired of looking at the layers, then that's the way it is . . . My interest in science is to simply find out more about the world.'

But for all that Feynman talks it down here, the process of scientific discovery can lead to great joy, in both the discoverer and those who follow in his or her footsteps. Take the theory of relativity, which Albert Einstein arrived at through what he called 'combinatory play', in which two previously unrelated ideas are brought together. What, Einstein asked, if gravity is not some mysterious force acting on objects at a distance but is more like the electromagnetic field, and *is* space? The contemporary physicist Carlo Rovelli describes the emotional impact on him as a student of coming to understand Einstein's breakthrough:

> Every so often I would raise my eyes from the book and look at the glittering sea: it seemed to me that I was actually seeing the curvature of space and time imagined by Einstein. As if by magic: as if a friend were whispering into my ear an

extraordinary hidden truth, suddenly raising the veil of reality to disclose a simpler, deeper order. Ever since we discovered that Earth is round and turns like a mad spinning-top, we have understood that reality is not as it appears to us: every time we glimpse a new aspect of it, it is a deeply emotional experience. Another veil has fallen.

The shadow

Science and technology are strongly associated in our culture with the idea of progress. Most discoveries and developments are greeted with enthusiasm because they are fascinating or exciting or because they increase the range of possibilities and diminish the sway of misery. Science is, supposedly, disinterested. But there is a shadow, for the new knowledge brings power and the possibility to abuse that power. As an ancient Greek myth foretells, one of the daughters of Thaumas, the god of wonder, is Iris, the beautiful goddess of the rainbow, but the others are the harpies: cruel harbingers of disruption and of death.

'Wonder, therefore, and not any expectation of advantage from its discoveries, is the first principle which prompts mankind to the study of Philosophy, of that science which [strives] to lay open the concealed connections that unite the various appearance of nature; and they pursue this study for its own sake, as an original pleasure or good in itself, without regarding its tendency to procure them the means of many other pleasures.' Adam Smith

For early modern Europeans, the discovery of the Americas, which arguably sparked the scientific revolution in the first place, was astonishing – not least because it showed that there were genuinely new things, unknown to the ancients, to be found. But, as the more sensitive and thoughtful among them realized, these discoveries were often followed by catastrophes. 'The marvellous discovery of the Americas . . . silence[s] all talk of other wonders,' wrote Bartolomé de las Casas in 1542. But this single sentence is followed by an entire book documenting the genocide of the native peoples. The US founding story is undergirded with many genocides, some of them only glimpsed through individual incidents such as the torture to death of Native American women and children 'for public amusement' and the high-spirited use of some eighty heads as footballs in the streets of Manhattan.

Today, science and technology have increased the

scope for human wellbeing far beyond the dreams of our ancestors. But they have also made possible weapons that can kill hundreds of millions of human beings in seconds. 'Our entire much-praised technological progress, and civilization generally,' wrote Albert Einstein in the mid-twentieth century, 'could be compared to an axe in the hand of a pathological criminal.' We may be a little less concerned about nuclear war than during the Cold War, but other spectres haunt us. One of them is that science and technology enable our societies to perturb and pollute the ecosystems on which we depend with profoundly destabilizing consequences. Another is the fear that, for all their great promise, our freedom may actually be diminished by new technologies. In the 1950s Aldous Huxley warned that pharmacology and brainwashing might one day cause people to love their servitude. If we substitute sugar, smartphones and mass disinformation then maybe he was not so far wrong, and this is before we start to see the full impact of intelligent systems that can read our emotions and anticipate our desires better than we can ourselves.

The average weekly screen time for a US adult is seventy-four hours, and rising. Many millions of un-working young men – out of work and not looking for jobs – sit in front of screens all day, stoned.

Melancholy

Even with a relatively healthy and secure existence, a bad mood can spoil almost everything. In my case, the opposite of wonder sometimes overcomes me when I think of some of the choices I have made (if that is what they really were), or when I reflect on a political and economic system that squcks our thrugs till all we can whupple is geep. In these moments, I am, if not in the Slough of Despond, then certainly in one of its Basingstokes. For the most part, however, I soon realize that mine are First World problems and things could be a lot worse: as the joke goes, terrible food, and such small portions. The mood eventually passes, and I try to do something positive. 'Dream delivers us to dream, and there is no end to illusion,' wrote Ralph Waldo Emerson. 'Life is a train of moods like a string of beads, and, as we pass through them, they prove to be many-colored lenses which paint the

'There is a destination, but no path to it; what we call a path is hesitation.'
Franz Kafka

world their own hue, and each shows only what lies in its focus.'

'It seems to me sometimes that we never get used to being on this earth and life is just one great, ongoing, incomprehensible blunder.'
W. G. Sebald

There are many reasons for feeling despondent, and other people fall into deeper darkness than I have ever done, and for different reasons. Some trajectories have been portrayed brilliantly in literature. 'I tell you solemnly that I have wanted to make an insect of myself many times,' says Fyodor Dostoyevsky's Underground Man. 'Secretly, in my heart, I would gnaw and nibble and probe and suck away at myself until the bitter taste turned at last into a kind of shameful, devilish sweetness and, finally, downright definite pleasure.' Even darker is Franz Kafka's novella *Metamorphosis*, in which the protagonist Gregor Samsa turns into a dirty giant bug: a metaphorical or fable-like representation of a psychological state.

Real lives can feel just as painful or ugly as the worst experiences described in books. The mere realization that we exist at all in a roiling and ever-changing world, and are going to die, can be dreadful. Ressentiment and anger at real or imagined slights and injustices may be expressed in hate-thought and hate-speech, and in acts of violence. But attempts to overcome a sense of emptiness or to numb oneself against feeling altogether – through alcohol abuse, drug abuse, self-harm or other behaviours – are also widespread.

Ressentiment, a term introduced by the philosopher Søren Kierkegaard, is a psychological state resulting from suppressed feelings of envy and hatred which cannot be satisfied. The writer Albert Camus framed it as 'autointoxication – the malignant secretion of one's preconceived impotence inside the enclosure of the self'.

In early modern Europe, dark moods – sleeplessness, irritability, anxiety and despair – were seen as symptoms of melancholy, which was believed to result from excessive concentration of black bile (one of the four supposed 'humours', along with yellow bile, phlegm and blood). Melancholy is what causes Hamlet to see the firmament not as a majestical roof fretted with golden fire but as a foul and pestilent congregation of vapours. It is the beast that looms over Robert Burton's gargantuan work of 1621, *The Anatomy of Melancholy*. But for Renaissance humanists there was also a positive side to the condition. According to Marsilio Ficino, melancholy was associated with 'genius' and therefore with the potential for creativity and change. In this instance, and for this book, Albrecht Dürer's *Melencolia I* helps point the way.

'A mere temptation is our life. Who can endure the miseries of it? In prosperity we are insolent and intolerable, dejected in adversity, in all fortunes foolish and miserable.'
Robert Burton

For Freud, melancholy was 'a profoundly painful dejection, cessation of interest in the outside world, loss of the capacity to love, inhibition of all activity, and a lowering of the self-regarding feelings to a degree that finds utterance in self-reproaches and self-revilings, and culminates in a delusional expectation of punishment.'

In Alan Moore's novel *Jerusalem*, a gang of archangels plays billiards for human souls in a dingy snooker hall in Northampton.

The 'I' in *Melencolia I* may indicate that the image was intended to be the first of a series. If so, the others were never created or have not survived. One possibility is that it refers to imagination as the first and lowest of the three categories of genius, the next being reason, and the highest the spirit. Another is that the 'I' is for *Ite* – Latin for 'go', in which case the title would mean something like 'Go away, Melancholy'.

Ever since its creation in 1514, Dürer's engraving has fascinated people who have, in various ways, expanded the realm of wonder or its shadow. William Blake kept hold of his copy even when poverty forced him to sell almost everything else he owned. Albert Einstein and Sigmund Freud had reproductions on the walls of their studies.

According to a classic account by the mid-twentieth-century scholar Erwin Panofsky, the angel in the picture is a personification of two distinct ideas: Melancholy as one of the four humours, and Geometry as one of the Seven Liberal Arts. She represents the spirit of the Renaissance artist, who respects practical skill but longs for the beauty and abstraction of mathematical theory, and who is inspired by celestial influences and eternal ideas but suffers all the more deeply from his human frailty and intellectual limits. So far, so respectable and well grounded in scholarship and art history. But one of the things that makes *Melencolia I* such a fascinating work is that it seems bigger than any one interpretation – and always open to new readings. Here is mine.

Angels were (and by some people still are) believed to exist in a realm between God and Man, serving the former and sometimes carrying messages to the latter. Dürer's angel, physically robust, seems to exist solidly in the material realm, or at least be fully engaged in it. There is darkness in the scene, which is illuminated by moonlight, but we do not seem to be in the nightmarish world created by the demiurge of Gnostic belief. The strange bat carrying the title of the print is puny rather than terrifying, and in any case may be flying away. The unorthodox spelling of the word it carries on its banner makes sense if we read it as a jumbled anagram, *Limen Caelo*, or gateway to heaven.

And whatever this angel is doing, she has not given up. Her eyes are not cast down, as they would be if she were dejected, but gaze upwards. Perhaps she has stopped for a moment, having seen something in her mind's eye, before continuing with whatever it is she is drafting with the compass on her lap. The woodworking tools around her feet are disordered but not broken; this looks like the workshop of someone busy

MELENCOLIA§I
16 3 2 13
5 10 11 8
9 6 7 12
4 15 14 1

Leonardo's outline for a treatise entitled 'On the Nature, Weight and Movement of Water', for example, proposes fifteen separate books. It contains dozens of notes on phenomena each demanding meticulous observation but ultimately goes nowhere, because Leonardo does not pursue the issues systematically.

creating, not someone whose inner world is breaking down. (If the outpourings in his notebooks are any indication, the studio of Dürer's contemporary Leonardo da Vinci must have looked a bit like this, and Leonardo was no more subject to lethargy than is an active volcano.) 'Melancholy shares nothing with the desire for death,' writes W. G. Sebald. 'It is a form of resistance.'

A large polyhedron at centre-left of the image and on the periphery of the angel's gaze seems to be a product of the workshop. The parameters of this object, which is a truncated rhombohedron and is known today as Dürer's solid, have been much debated. It may be an attempt to construct a new Archimedean solid (a symmetric polyhedron made of two or more types of regular polygon), the vertices of which would all touch the inside of a sphere. If so, the angel, like the artist, has failed because the geometric problem posed has no mathematical solution.

Modern physics says that fundamental particles can be divided into two groups: particles of spin ½, which make up the matter of the universe, and particles of spin 0, 1 and 2, which give rise to the forces between matter. Two particles from the former group can never be in the same place at once but an indefinite number from the latter group can. Fundamental particles are emergent phenomena of quantum fields, which in turn are thought to arise, together with space-time, from covariant quantum fields.

Geometers – cutting-edge natural scientists in the early 1500s – followed Plato in believing that regular polyhedrons made of one type of polygon, and known as Platonic solids, were the building blocks of matter. The tetrahedron made fire, the cube made earth, the octahedron air, the icosahedron water and the dodecahedron aether (the mysterious substance that, supposedly, filled the heavens). They were wrong, of course, but, as the physicist Frank Wilczek comments, they were usefully wrong in that they helped set people thinking about the possibility of a limited number of discrete entities at the foundation of the material world. The Standard Model of particle physics, or Core Theory, also identifies a limited number of fundamental particles (seventeen so far), which combine in various ways to make all that is. Is the angel a forerunner of today's physicists – seeking to understand the basic building blocks of the world, and in doing so, opening doors to its manipulation at the most profound levels?

Melencolia I contains much else, including a cherub, a dog, an hourglass, a pair of scales, a bell and a bunch of keys hanging from the angel's waist. Notable too are alchemical symbols: a crucible sits on the ledge to the left of the polyhedron, and a ladder with seven rungs – which is conventionally interpreted as representing the seven metals and planets of the alchemical system – rises to an unseen tower or lookout point above the frame. And then there is the set of numbers behind and to the right of the angel's head. These form a magic square – an array in which the sum in any horizontal, vertical or main diagonal is always the same. (In this case the sum is 34, and the square has the additional property that the sums in any of the four quadrants, as well as the sum of the middle four numbers, also equal 34.) The central two numbers at the bottom comprise the date of the engraving, 1514, while the 1 and 4 on either side of them correspond to the first and fourth letters of the alphabet, A and D, the two letters with which Albrecht Dürer always signed his prints and drawings.

What does it all mean? Maybe Dürer was wrestling with the idea (first expressed, as far as we know, by Pythagoras) that all things are number. Perhaps he also wanted the work to be an enduring puzzle, and hoped that in engaging with it, the viewer might unlock his or her own mind. For as well as drawing us into the detailed, almost obsessional, symbolic world of the angel's studio, Dürer (whose name means 'maker of doors') allows us to see and think beyond it. From Paracelsus to Jung, alchemy has been associated with the growth of the individual, but I like to see in the ladder a predawn intuition of the scientific method, in which a larger view can be obtained by careful step-by-step progress. I also like to see in Dürer's magic square an expression of an enduring, real-world riddle: there seems to be (as the twentieth-century physicist Eugene Wigner put it) something 'unreasonably effective' about mathematics in that its concepts often apply far beyond the context in which they were originally developed. Why, for example, does π, the ratio of the circumference of a circle to its diameter, appear in a statistical analysis of population trends, not to mention definitions of the fine structure constant, the Einstein field equations and the definition of the Planck length, the smallest meaningful measure? Almost uncannily, the universe appears to

Albert Einstein was not an especially gifted mathematician but, like Galileo, he recognized the power of maths to explain the world. Still the puzzle remained as to why. 'The eternally incomprehensible thing about the world,' he wrote, 'is its comprehensibility.' The physicist James Hartle has a simple answer: 'The world must be comprehensible in order for information-gathering and -utilizing systems, including human beings like us, to exist [in the first place].'

'No question is more sublime than why there is a Universe.' Derek Parfit

be written in some kind of code – and one that, piece by piece, can be deciphered. We may have little or no idea as to what breathes fire into the equations, and yet somehow we find ourselves situated in the world they describe.

Melencolia I has other resonances. The sphere at the angel's feet is a 'perfect' symmetrical shape beloved of the ancient Greeks, but in 1514 it was also acquiring new significance. A little over twenty years before, Columbus's first voyage to the Americas had seemed to confirm that the Earth was a sphere, but also suggested that there was more to discover on its surface than the ancients had ever imagined. The Behaim globe made in Dürer's home city of Nuremberg in 1493 had depicted all the world's landmasses in confident detail but had left no place on it for the Americas: Japan and China were a straight sail west from Lisbon. The Salviati map of 1525, by contrast, tentatively depicts parts of the coast of the newly discovered Americas but, confident in its ignorance, also leaves large areas of the planet blank and unexplored. *Melencolia I* also appeared shortly after Copernicus displaced the globe from the centre of the universe in the *Commentariolus*, or *Little Commentary*, of 1510, and it is possible that Dürer, who was one of the best-informed artists of the age and a keen amateur astronomer, had heard tell of its sensational thesis.

Working with two professional astronomers, Dürer made the first printed star map in 1515. It depicted the sky of the southern hemisphere, and left large areas blank . . . for future exploration.

And there, in the top left of *Melencolia I*, is a glimpse of a *Weltlandschaft* (or, more precisely, a *Seelandschaft*) – a wide world beyond the angel's studio. A beautiful town stands on the shore of a great lake or a calm sea, which reaches off into untold distance – an ancestor of Paul Klee's *Lagunenstadt*. Arcing above the water is a moonbow, the rare night-time counterpart of a rainbow. In exceptional conditions, a moonbow reveals to the most sensitive eyes a hint of the colour spectrum within its ostensibly white light. (Usually, a camera more sensitive than the human eye is needed.) But whether or not one see the colours in a moonbow, those who see them experience light in a new and rare way: sunlight must reflect off the Moon before refracting and reflecting in water droplets to produce something at the very edge of visibility, and there is great wonder in this. Also in the sky is a comet resembling the one recorded in the great chronicle printed in Nuremberg in 1493. For Dürer, perhaps, the significance of this comet is ambiguous rather than necessarily a portent of evil. After all, the comet that passed the Earth in Dürer's lifetime did not seem to have brought the disaster some had feared. Maybe it was OK to question a little more and fear a little less – to recognize one's ignorance and not to rely on authority for answers.

For me, then, *Melencolia I* is an invitation to wonder. The angel may be preoccupied, but a deeper and wider world beckons, with ever more to explore and appreciate. ('The whole of matter,' wrote the twentieth-century author and artist Bruno Schulz some years before he was brutally murdered, 'pulsates with infinite possibilities.') With the compass on her lap the angel is making a map that may somehow link the mysteries in front of her with the larger world beyond, and the artist is encouraging us to do something similar.

An invitation to wonder

This book is an attempt to inspire and share curiosity and wonder, but how to select and organize its objects of attention? Some works from the Islamic golden age are stunning and compendious, and offer one possible model. The *Book of Curiosities of the Sciences and Marvels for the Eyes*, for example, brings together an astonishing series of diagrams of the heavens and maps of the earth, combining material from astronomers, historians, scholars and travellers of the ninth to eleventh centuries AD. Then there are splendid if unreliable examples such as *The Ultimate Ambition in the Arts of Erudition*, a work published in 1314 which finds space for vital insights such as that if a man urinates on a rhinoceros's ear the animal will run away. European volumes such as *The Book of Miraculous Signs*, published in Augsburg in about 1550, contain gorgeous illustrations of strange meteorological events and bizarre phenomena believed to be portents of apocalypse. Books like these are all well and good if you have decades, or a team of illuminators, or – ignoring the arguments against miracles made by David Hume – are given to magical thinking. In the absence of any of those, however, some relatively simple organizing principles will have to do.

Hume argued, principally, that the laws of nature are invariant, and human testimony is unreliable.

One possible model is *De Rerum Varietate*, or *On the Variety of Things*, published in 1557 by the mathematician and gambler Girolamo Cardano. This describes wonders according to place ('wonders of the earth', 'wonders of water' and so on) and according to what Cardano judges to be their magnitude. So, for example, in the category of the truly wonderful are the 'blue clouds' sighted above the Straits of Magellan, while 'worthy of wonder, but not great wonder' are the foot jugglers of Mexico.

Cardano was a brilliant polymath but often short of money, and his disputes over gambling debts are said to sometimes have ended in knife fights. His *Liber de ludo aleae* (*Book on Games of Chance*) of 1564 is the first systematic treatment of probability in European mathematics. He is said to have predicted the date of his own death and made sure he won by killing himself when the day arrived.

I have done something like this, using two organizing principles. First, I have used the old idea of seven wonders. It may be a familiar approach but it does have the merit of being manageable, and it has been applied well at least once, in a brief essay by the physician Lewis Thomas. Second, I have ordered and linked these seven wonders through the principle of emergence – the process whereby novel properties and behaviours arise from the combination of simpler parts. In this book, the relatively simple phenomenon in the first chapter is implicated in the emergence of the one in the second. And so on.

Writing in the early 1980s, Thomas gave the seven wonders, leaving the greatest until last, as:

(1) bacteria that thrive on deep-ocean hydrothermal vents;
(2) oncideres, a beetle that lives in mimosa trees;
(3) the scrapie virus;
(4) olfactory receptor cells;
(5) termites;
(6) any human child;
(7) the world, seen as a living system.

If I could have, I would have travelled as far and wide to make a map of wonders as does the engineer Mabouloff of the Institute of Incoherent Geography in Georges Méliès's 1904 film *The Impossible Voyage*. My budget, however, seldom took me further than my shed, and the furthest it stretched was to a week in a different shed. So I was largely constrained to explore wonders of the nearby – the library, the laboratory and the occasional wood or shoreline within striking distance rather than the distant places I dream of. But present in my undiscovered nearby I sometimes glimpsed what the mystic Thomas Traherne calls 'things strange yet common; incredible, yet known; most high, yet plain; infinitely profitable, but not esteemed'. Even the greatest of wonders is implicated in an incident as tiny as the one in my kitchen – and in that sense it took a small thought to fill this whole book.

Ludwig Wittgenstein's observation – 'How small a thought it takes to fill a whole life' – is text enough for 'Proverb', a remarkable musical composition by Steve Reich.

Here's how it goes. Chapter One is about light – one of the fundamental phenomena of the universe, and a starting point for much else. It describes what light is and how we see, and it takes the rainbow, the Sun and kinds of darkness as instances. Chapter Two explores the origin of life itself and the material world from which it arose, and briefly considers a tiny proportion of the vast number of things that make life astonishing. Chapter Three focuses on the human heart, the beating of which is an inescapable reminder of our short but remarkably resilient physical existence. Chapter Four is about the brain, and explores seven simple things you can say about what may be the most complex single object in the universe. Chapter Five traces the arc of a human life – a phenomenon that emerges from the sustained functioning of (among other things) heart and brain. Dividing a life into three ages of youth, maturity and old age as the ancients did, rather than the seven ages of medieval and early modern lore, it explores ways in which wonder is experienced over a lifetime and how that changes. Chapter Six is about wonder at the world in which we find ourselves. It considers how that world shapes us, and how we shape it through maps and dreams. Chapter Seven touches on some of the technologies that are transforming both human experience of the world and the world itself, and thereby changing what we find wonderful. It looks at the future of wonder in a world increasingly manipulated, or abused, by humans, and asks what wonders may be in store. An afterword reflects on the territory crossed in the book.

A New Map of Wonders is filled with findings from science. This is not because I believe this is all we need. I don't. To adapt and distort a famous passage from Immanuel Kant's *Critique of Practical Reason* of 1788, two things fill me with wonder and apprehension: the natural world that is the ground of our being, and what humans do to it and to each other. There are important matters on which, at present, science offers little guidance, including many but not all questions

'Two things fill the mind with ever new and increasing admiration and awe . . . the starry sky above me and the moral law within. The former view of a countless multitude of [stars] annihilates . . . my importance as an animal creature, which after . . . a short time must again give back the matter of which it was formed to the planet it inhabits (a mere speck in the universe). The second, on the contrary, infinitely elevates my worth as an intelligence . . . reaching into the infinite.'
Immanuel Kant

about what we should value and why. As for new technologies (which are both an expression and a source of scientific breakthroughs), the warning that they can diminish existence should be taken seriously. But it is often the use to which new technologies are put rather than the technologies themselves that is most to blame. When used wisely, science and technology can clear away many kinds of error and expand the realms of possibility, including the kinds of wonder and wonderful experience we encounter. If we are serious about helping all sentient beings to flourish, science and technology can help.

Consider what the historian Sven Beckert calls 'war capitalism': 'If our allegedly new global age is truly a revolutionary departure from the past, the departure is not in the degree of global connection but the fact that capitalists are for the first time able to emancipate themselves from particular nation states.'

There is always more. Knowledge is provisional, open-ended and insufficient, and likely to remain so. The provable statements of mathematics, for example, are infinite in number. 'What we observe', said Werner Heisenberg, 'is not nature itself, but nature exposed to our method of questioning.' And, as usual, Thoreau got to the heart of the matter. Higher knowledge, he wrote, is nothing more than 'a novel and grand surprise on a sudden revelation of the insufficiency of all that we called Knowledge before'. The highest that we can attain to therefore 'is not knowledge, but sympathy with intelligence'. I take that to mean that the greatest wonder comes with love and the wisdom to recognize what is front of us. 'It's not about understanding,' says the poet W. S. Merwin. 'It's about our one life, our one and only life.'

'We will always be at the beginning of infinity, alike in our infinite ignorance.' David Deutsch

This book is not a literal map, although it contains a few. As a thaumatologue, or text of marvels, it is not remotely comprehensive. The aim is simply to witness some wonders as far as I can – and to get a better sense of the size of those wonders and where they are: to recognize them. I hope the reader will find more wonder by looking in some of the directions I suggest.

In her memoir of the writing life, Annie Dillard also quotes Thoreau: 'The youth gets together his materials to build a bridge to the Moon, or perchance a temple or palace on Earth, and at length the middle-aged man concludes to build a wood-shed with them.'

Wonder occurs when we are firing on all cylinders intellectually and emotionally. It can take us to the edge of terror, or it can bring the unpredictable and the strange right up to our noses without imparting a sense of threat. Like levity – a mythical force that Aristotle supposed to be in opposition to gravity – it can be exhilarating. Like a good joke, it can say what needs to be said . . . in too few words. Wonder opens up new possibilities – the psychological equivalent of the pair of bolt cutters that the director Werner Herzog advises filmmakers always to carry. It can feel like the apprehension of something bigger and better of which we are momentarily a part. It can feel like discovery, or at least the first step on a journey towards one. And it can feel like return or recovery – a sense that something is being put right. Wonder can feel like enough, or like a good point from which to start. It is a state of mind in which we can accept a gift, and be aware of its importance, if not necessarily its meaning. It is a kind of grace.

1

THE RAINBOW AND THE STAR

Light

The changing of bodies into light, and light into bodies, is very conformable to the course of nature, which seems delighted with transmutations.

Isaac Newton

'Yes, I have a pair of eyes,' replied Sam, 'and that's just it. If they wos a pair o' patent double million magnifyin' gas microscopes of hextra power, p'raps I might be able to see through a flight o'stairs and a deal door; but bein' only eyes, you see, my wision's limited.'

Charles Dickens

On a February morning in 1962, US Marine Colonel John Glenn squeezed into a capsule on top of a rocket and was hurled more than a hundred miles up into space. He orbited the Earth three times in just under five hours before gravity's rainbow plunged him safely into the Atlantic Ocean.

In those five hours Glenn flew through three days and three nights. His first day, lasting some forty-five minutes from launch at Cape Canaveral, took him over the Canary Islands and Kano in Nigeria before he saw the Sun set over the Indian Ocean on the other side of Africa. Twilight, he said, was beautiful. The sky in space was very black with a thin band of blue along the horizon. The Sun set fast, though not as fast as he had expected. Brilliant orange and blue layers spread out on either side of it, tapering towards the horizon. It was night by the time he flew over the Australian coast near Perth. Over the Pacific he was preparing for his first dawn.

As the Sun rose over Kanton, an atoll in the Phoenix Islands about halfway between Fiji and Hawaii, Glenn reported seeing thousands of tiny glowing orbs outside the capsule. 'They're brilliantly lit up like they're luminescent. I never saw anything like it ... they're coming by the capsule and they look like little stars. A whole shower of them coming by. They swirl around the capsule and go in front of the window and they're all brilliantly lighted.' For Glenn, the sight of these orbs, which disappeared as his craft moved into sunlight, was one of the most moving experiences of his flight. He was a deeply religious man, and their angelic appearance stayed with him for a long time afterwards.

NASA later determined that the orbs were Glenn's urine, frozen into perfect spherical droplets as it vented from the spacecraft. It's easy to laugh at the bathos, or deflation, in this discovery. The whole enterprise of the US space programme in its first years – employing hundreds of thousands of people and soaking up a sizeable portion of the federal budget – was, after all, in large part a giant pissing contest with the Soviets. But wonder and humour are not mutually exclusive, and the orbs were evidently a marvellous sight, however lowly their origin. They even have a kind purity compared to the haze of manmade junk that now orbits the Earth, and the wonders of Glenn's flight, arcing briefly above the Earth at 28,000 kilometres per hour (or 7,843 metres per second), are no less great for it.

Speed

Whether in low Earth orbit or in dappled shadows on a wall, many phenomena associated with light arouse profound wonder. Discussions of its nature are often freighted with mystical and religious associations, not least in Western culture – see, for example, the opening of Genesis. But I will start with a brute fact – which is also a mystery – the speed of light.

Unlike a gust of wind or the swiftest arrow, light appears to already be everywhere that it is – to travel everywhere in no time at all. Common sense might suggest that its speed is therefore infinite. From ancient times, a few thinkers challenged this view, but only from the seventeenth century onwards are there records of attempts to actually prove otherwise. Galileo Galilei suggested placing a lantern on a distant hillside at night, uncovering it at a given moment and attempting to measure the time before this uncovering was observed some distance away. This only showed that light must be extraordinarily quick. A way to measure the speed of light was found, however, by using observations undertaken for an entirely different purpose.

If the speed of light really were infinite, particles and the information they carry would move from A to B instantaneously, cause would sit on top of effect and everything would happen at once. The universe would have no history and no future, and time as we understand it would disappear.

In 1610 Galileo pointed a telescope at Jupiter and found four bright moons, hanging like ship's lanterns on a calm night at sea, orbiting it. It was a momentous discovery: compelling evidence that not all heavenly

In reality not all the moons of Jupiter are impassive orbs. Io, which is about the same size as our Moon, is dotted with volcanoes the size of Everest and is the most volcanically active body in the solar system. Enceladus, a small moon unknown to Galileo, spouts water though its icy crust.

bodies go around the Earth, and a challenge to the teachings of the Church. But Galileo also thought he saw a practical application. The regular motion of Jupiter's moons could, he suggested, be used as a kind of clock in the sky. Navigators and mapmakers anywhere in the world should be able to observe when the moons appeared or disappeared behind the planet and compare the local solar time of these eclipses to standard time at a place of known longitude. From the time difference they should be able calculate their relative longitude.

The idea made perfect sense. Footage captured as the *Juno* space probe approached Jupiter in 2016 does indeed show its moons orbiting the planet like the hands of a heavenly clock. But turning the idea into a reality proved to be beyond the reach of seventeenth-century measurement techniques. In the attempt, however, Ole Rømer, an astronomer working at the Paris observatory a generation after Galileo, compiled extensive data on the motion of Io, Jupiter's innermost moon, and found a strange anomaly. When the Earth was nearest to Jupiter, the eclipses of Io (which goes around it once every forty-two-and-a-half hours) occurred about eleven minutes earlier than predicted. Six months later, when the Earth was farthest from Jupiter, the eclipses occurred about eleven minutes later than predicted. Rømer knew that the amount of time it took Io to travel around Jupiter could have nothing to do with the relative positions of the Earth and Jupiter, and realized that the time difference must be because light travelled at a finite speed: light was taking about twenty-two minutes longer to reach the Earth from Jupiter when the two planets were on opposite sides of the Sun than when they were on the same side. Determining the speed of light was simply a matter of dividing the diameter of the Earth's orbit by this time difference.

Rømer did just that, and in 1676 he calculated the speed of light to be about 210,000 kilometres per second. We now know that he underestimated its true value (which is nearly 300,000 kilometres per second) because he mistook the maximum time delay between eclipses of Io and the diameter of the Earth's

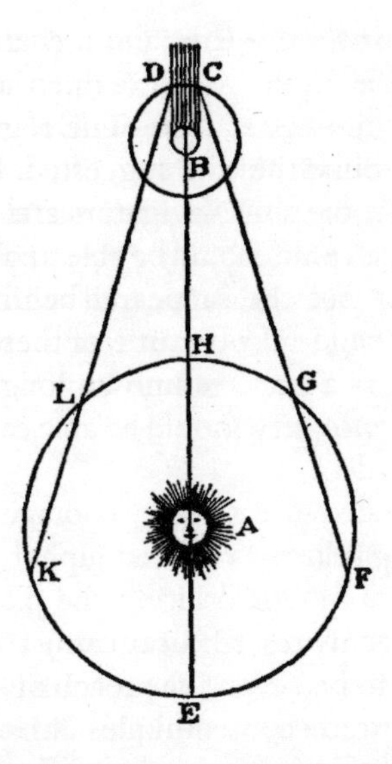

In Rømer's schematic representation, the sunlight reflected off Io as it passes into and out of the shade of Jupiter at B at points C and D takes less time to reach the Earth when the Earth is at G or L on its orbit around the Sun at A than it does when the Earth is at F or K..

orbit around the Sun. But it was a stunning result: powerful evidence that not only is the speed of light finite (though astonishingly fast) but that it can be measured by experiment.

Even now the speed of light remains hard to conceive. In a 1982 essay, Annie Dillard describes witnessing a solar eclipse. In the second before the Sun goes out, a wall of shadow races towards the hill on which she and her companions stand. It roars up the valley, slams their hill and knocks them out: 'the monstrous swift shadow of the Moon.' Dillard learns later that the Moon's shadow was moving at 1,800 miles an hour (or about 2,900 kilometres per hour), and says that language can give no sense of this sort of speed. And yet this terrific rate – nearly two and half times the speed of sound in air – is in the region of a million times slower than the speed of light.

Consider an athlete on the blocks at the start of a hundred-metre sprint. He or she may take almost a seventh of a second to react to the starting gun. By that time, the 'b' of the bang will already be about halfway to the finishing line. The light from the flash of the gun, meanwhile, will already have travelled about 100,000 kilometres – the equivalent to two and a half times around the world. If, like another kind of strange particle we will come to later, the particles of light that we call photons could pass straight through the Earth and out the other side without slowing down, they would take about four-hundredths of a second to do so – barely enough time for a hummingbird to flap its wings twice.

'A photon is a minimal disturbance in an electromagnetic field . . . Quantum theory [states] that energy comes in discrete units or quanta. Because these units cannot be broken down further they have the sort of integrity we associate with particles, and in some circumstances it is helpful to think about them that way. In that sense, photons are particles of light.'
Frank Wilczek

What is light?

For what it is worth, the nature of light can be described in a few words. The problem is that many of those words and the concepts behind them are remote from everyday experience and stretch the powers of comprehension of those of us who are not specialists. One can say, for example, that visible light is a tiny part of the electromagnetic spectrum, and as such a self-propagating transverse oscillating wave of electric and magnetic fields able to travel through a complete vacuum. One can also say that light is made of particles called photons, which are the smallest quantity of energy that can be transported, and that it is the force carrier for electromagnetism which (along with gravity and the strong and weak forces) is one of four known fundamental forces in the universe.

The strong nuclear force binds protons and neutrons together into atomic nuclei. The weak nuclear force is responsible for radioactive decay and nuclear fission.

If little of that is especially helpful, the following may not be either. Photons (the particles of light) have no mass and no charge, but they are exchanged between charged particles such as electrons and protons whenever they interact. They cannot come to rest, but only transform; and, for reasons that are not yet fully understood, they travel at the speed limit for all that is.

When you first hear that light is made of particles, it seems reasonable to ask how big they are. But there is no good answer to this question, because the 'particle' label is a half truth. In physics, common-sense notions based on our perceptions of the world don't always apply. Actually, many things that are easily visible to

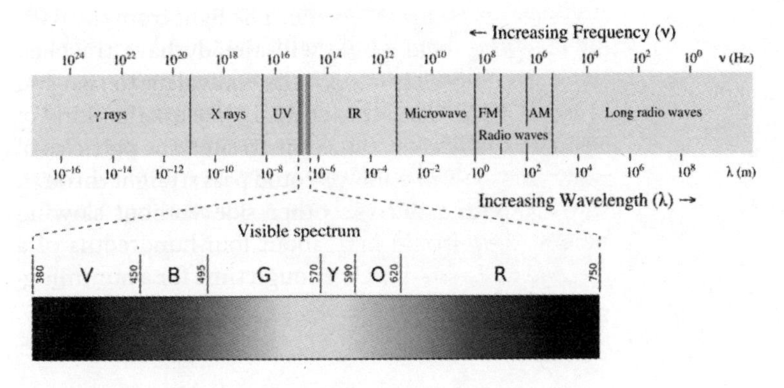

Visible light as a proportion of the electromagnetic spectrum.

the naked eye don't obey common sense either. The fact that the Moon is a solid rock weighing more than 73 billion billion tonnes but shows no sign of falling out of the sky is one example. The persistence of separate taps for hot and cold water in public washrooms in Britain is another. And unlike anything we can easily conceive, photons sometimes behave like particles and sometimes like waves.

If pressed, however, some physicists (maybe just because they want you to go away) will tell you that one way to think about the size of a photon is to consider its wavelength. But how big is *that*? The answer is that for visible light it ranges from about 400 nanometres for violet light to 700 nanometres for red. A nanometre (abbreviated nm), is a billionth of a metre, and this too is hard to comprehend. But try the following. Imagine the last joint of your little finger expanded to the size of a typical room in a house. Each of the billion or so cells in it would be about the size of a grain of rice. Green light in the middle of the visible spectrum (at around 550nm) would have a wavelength about a twentieth of the length of each grain.

But the wavelengths of the electromagnetic spectrum extend over a much larger range than visible light, so photons can be a lot smaller or bigger than this. Those of gamma rays are measured in picometres – trillionths of a metre. This makes them around a thousand times smaller than an atom. (To envisage the size of an atom, consider that a grain of salt contains about a billion billion, or 10^{18}, of them – about 10 million times the number of stars in the Milky Way galaxy.) By contrast, microwaves used to heat food approach the dimensions of a grapefruit (note to self: do not microwave grapefruit), and radio waves can be anything from a few metres to many kilometres long.

Light and sight

The existence of light is a profound wonder. So too is the fact that we can see it. Photons in the range of visible light are the only elementary particles that our ancestors evolved to be able to detect directly. The reason we see light in the wavelengths we do – the wavelengths that create all the colours and shades in our world – has a lot to do with the fact that most of the sunlight that reaches the surface of the Earth (and also travels through water) is in these frequencies. If we could see electromagnetic waves from other parts of the spectrum, the world would look completely different. Gamma rays alone would show the Moon as brighter than the Sun. Seen with X-rays, everyone's birthday suit would be transparent and we would all be dancing around in our bones, like the Mexican Day of the Dead.

The science of how we see is incredibly complicated and only partly understood. The human brain, which is probably the most complex single thing in the known universe, allocates a significant proportion of its circuits to processing the light that enters our eyes. But some of the basics of how the eye works can be outlined in a few short paragraphs.

Light passing through the pupil and lens is projected onto the retina, which is lined with light-sensitive cells called rods and cones. The cells are packed next to each other like pencils in a box, with their ends (flattish in the case of the rods, somewhat pointed in the case of the cones) facing the light. Each retina has about 120 million rods and 6 million to 7 million cones. Rods, which are sensitive to light intensity, are mostly distributed around the outer regions of the retina, while cones, which are sensitive to colour, are mostly concentrated in the middle.

Each rod contains about a billion molecules of rhodopsin, a protein sensitive to light, which are stacked on transparent plates facing towards the light. A single photon strike is enough to tweak the shape of a central part of these amazing molecules, which are called chromophores, and this gives rise to a cascade of effects via intermediary molecules which amplify the original signal hundreds and thousands of times. Until recently it was thought that it took a minimum of about seven simultaneous photon strikes to create a signal strong enough for the rod to tell the brain that light is present, but experiments have now shown that some people can detect a single photon. Test subjects have described the sensation as 'almost a feeling, at

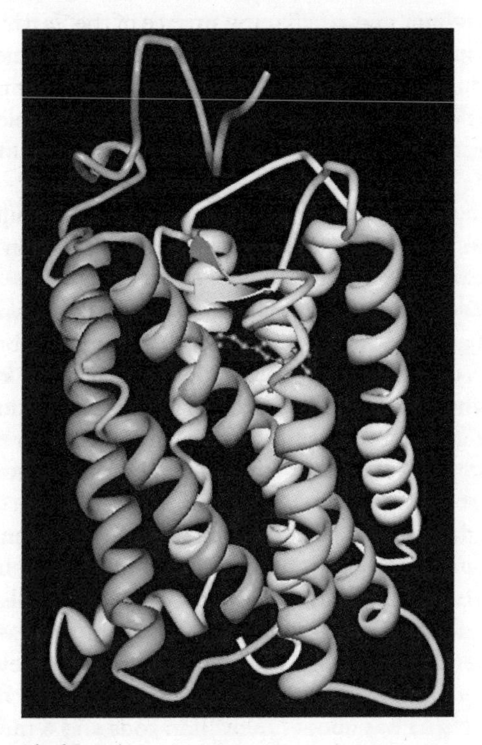

Rhodopsin

the threshold of imagination'. At any rate, our eyes can be exquisitely sensitive and, once they have adjusted, enable us make our way by nothing more than starlight. Yet our visual system is also robust enough to allow us to discriminate objects clearly in daylight more than 10 billion times as bright. Dappled sunlight reflected on a wall lies somewhere in the mid-range of our vision.

We never see the world as our retina sees it, says the neuroscientist Stanislas Dehaene. 'In fact, it would be a pretty horrible sight: a highly distorted set of light and dark pixels, blown up toward the centre of the retina, masked by blood vessels, with a massive hole at the location of the 'blind spot' where cables leave for the brain; the image would constantly blur and change as our gaze moved around. What we see, instead, is a three-dimensional scene, corrected for retinal defects, mended at the blind spot, stabilized for our eye and head movement, and massively reinterpreted based on our previous experience of similar visual scenes.

At the centre of the retina and comprising about one per cent of its total area is the fovea: a cup-shaped depression about the width of a poppy seed. The fovea is lined with cone cells, which are thinner than the rods – just one-millionth of a metre across compared with about five-millionths – and more densely packed. This allows them, together, to achieve far greater spatial resolution. Indeed, about half of the information reaching the brain from the eye comes from here. It is the fovea that, in good light, allows you to focus on the fine details of a spot directly in front of you. Even in healthy young eyes that spot is just two degrees across, or about the width of your thumbnail at arm's length. It's up to your brain to put together an impression of a whole scene from a torrent of narrow snapshots taken largely with the fovea many times a second.

The physician Thomas Young laid the foundations for our modern understanding of colour vision in 1802, when he noticed that all colour sensations could be produced by combinations of red, green and blue, and proposed that there are three types of nerve in the eye, each one sensitive to one of these colours. In the twentieth century the nerves were identified as the cone cells in the retina and they did indeed come in three varieties, with each kind containing one of three different kinds of opsins (light-sensitive proteins) that absorb light at different frequencies. The first variety absorbs more light towards the longer end of the visible spectrum, which we see as red, while the second peaks in the middle of the spectrum, which we see as green, and the third towards the shorter end, which we see as blueish. Fifty years later, the physicists Hermann von Helmholtz and James Clerk Maxwell refined Young's work, and Maxwell applied

To be more specific, the peak absorption frequencies for the different kinds of cone cells in our eyes are: 564nm, which is towards the redder end of the spectrum but is actually yellowish-green; 534nm, blueish-green; and 420nm, blueish-violet.

The artist Ben Conrad wearing his pinhole suit and helmet with 135 pinhole cameras (1994). Unlike a pinhole camera, which captures a wide area in equal focus, our eyes hold only a 'pinhole' area – a tiny sparkling fragment of reality – in focus at any one moment.

what we now call the trichromatic theory to create the first colour photograph. Almost every colour image you see today is based on a distant descendant of his technique.

Cones need a lot more light than rods to fire, which is why we don't see much colour in low light. And the colours we do perceive depend on how many of each type of cone are stimulated and how strongly at a given moment, with the eye and the brain averaging the signals from many cones. So, for example, yellow is perceived when cones sensitive to the reddish part of the spectrum are stimulated slightly more than those sensitive to the blueish part. You may have noticed that you are able to distinguish more shades of green than any other colour, and that those greens tend to remain visible for longer as the light fades. This is because human eyes are especially sensitive to green light, which, with a wavelength of around

555nm, stimulates two of the three kinds of cones: the ones sensitive to longer- and medium-frequency visible light.

It is hard to say why certain combinations of wavelengths produce particular colour sensations. The philosopher Alva Noë points to the case of yellow. You can get yellow by mixing red and green light, he observes, just as you get purple by mixing red and blue. But yellow isn't reddish-green or greenish-red in the way that purple is reddish-blue. In fact, Noe says, there is no such thing as reddish-green. Moreover, 'you don't see red or green in yellow the way you see blue and red in purple. Yellow, like blue and red, but not like purple, is unary, not binary.' The psychologist and neuroscientist Michael Graziano suggests our sensation of white is a puzzle. White, he says, should be 'the dirtiest, muddiest colour possible' because it contains a mixture of all wavelengths in the visible spectrum. But the brain's model of white light is a high value of brightness and a low value of colour: a purity of luminance – and a physical impossibility.

'Rays [of light themselves] are not coloured,' wrote Isaac Newton in 1704, '[and] in them there is nothing else than a certain Power and Disposition to stir up a Sensation of this or that Colour.' Three hundred years later this is still largely the accepted wisdom: colour is described as a part of how the brain interprets information rather than as a representation of some absolute, objective reality. It can tell us about motion, for instance: a black-and-white wheel set spinning can reveal the rainbow. And it can tell us about depth: distant hills typically appear blue in bright daylight because – just as happens when we look at the sky above our heads – the volume of air between us and them scatters blue light more than it does the longer wavelengths. According to the philosopher Mazviita Chirimuuta, 'colours are not properties of minds, objects or lights, but of perceptual processes – interactions that involve all three terms'. In what she calls 'colour adverbialism', colours are not properties of things but ways that stimuli appear to certain kinds of individuals or even cultures. 'Instead of treating colour words as adjectives, we should treat them as

'The word is blue at its edges and in its depths. This blue light is the light that got lost'
Rebecca Solnit

adverbs. I eat hurriedly, walk gracelessly, and on a fine day I see the sky bluely!'

If human colour vision sounds like a bit of a kludge that's because it is. But it's a pretty good kludge. Depending on whom you talk to, we can distinguish anything from tens of thousands to millions of different colours. In reality, our eyes distinguish only about a hundred and fifty hues, but our visual system takes account of at least two other factors to create the experience we call 'colour'. First, there is brightness – the amount of light emitted or reflected by an object. Second, there is saturation – the intensity of a hue, where a less saturated colour is duller and a highly saturated one is more vivid. Combining hue, brightness and saturation in varying degrees produces the enormous range we experience. There are other factors at work too: context also influences colour perception. In the checker shadow illusion, for instance, squares that are actually the same colour appear to be totally different because the brain interprets one of them as being in shadow.

How does our colour vision compare to that of other animals? While it is inferior in some respects to that of birds and bees, which can see ultraviolet, it turns out that it is actually superior to that of the mantis shrimp, which became famous, as obscure and weird crustaceans go, for having twelve or more different kinds of colour-sensitive opsins in the cone cells of its eyes compared to the three in ours. In the deep and distant past, also known as 2013, researchers assumed that such a range of opsins would give these animals superbly refined colour discrimination. It turns out that the opposite is true: mantis shrimps are actually terrible at telling colours apart. Thanks to other adaptations, however, they can see linear, circular and elliptically polarized light – capabilities that must reveal to them things we can only begin to imagine in the movement of water and light reflected from the bodies of their prey.

Weaving and unweaving the rainbow

'Light is something like raindrops.' Richard Feynman

It can be pleasant (and not entirely misleading) to visualize sunshine as raindrops falling to Earth – only, of course, raindrops that are incredibly tiny, incredibly fast and incredibly numerous. Every second, about 10^{45} solar photons strike the Earth – a hundred thousand trillion for each square centimetre in bright sunlight.

Dappled light on a kitchen wall has been enough to arouse a sense of wonder in me. And it hasn't taken much more than this to inspire popular and enduring works of art. The painter Edward Hopper, for example, said his favourite thing to paint was sunlight on the side of a building. But for most of us, most of the time, wonder is more usually associated with something a little more dramatic or unusual. One such phenomenon arises when a shower of photons and a shower of water interact to make a rainbow.

Virtually every culture has stories associated with rainbows. Many Aboriginal Australians revere the Rainbow Serpent, an enormous beast that lives in the deepest waterholes. Descended from an even larger being visible at night as a dark streak in the Milky Way, the Rainbow Serpent reveals itself as it moves through water and rain. It shapes landscapes and sings places into being, and it can be malign or benign. Sometimes it swallows and drowns people, or blights them with weakness, illness and death. At other times it endows them with rainmaking and healing powers.

In Aboriginal belief, every person has a home place in the land from which his or her spirit emerges at the beginning of their life. As one elder puts it: 'We all come from the Rainbow.' And at the end of life, every human spirit must be ritually returned to its home in the dark, invisible underworld. In this cosmology the rainbow is not an arc but a circle, half hidden within the land, connecting the life force in earth and sky. The movement of water through the material world enables the reproduction of organic beings, as well as the spiritual movement of persons from the dark recesses of the invisible world inside the land out into the light and animation of material being and back again – an example of what the geographer Yi-fu Tuan calls a 'hydro-theological cycle'.

Other cultures have diverse stories about the rainbow. In ancient China it was a slit in the sky sealed by the goddess Nüwa using stones of five different colours. In Greenland it was the hem of a god's garment. In Wales it was the chair of a goddess, and in Armenia the belt of the sun-god. The Blackfoot Indians of North America called the rainbow the Rain's Hat or the Old Man's Fishing Line. A Germanic myth describes it as the bowl used at the time of the creation for paint to tint the birds. The root of a Hungarian word for rainbow, *szivárvány,* derives from a verb meaning 'to draw water', and in Hungarian lore the two ends of the rainbow suck water from the world's oceans.

In the Classical Mediterranean world, the rainbow was associated with the relationship between the divine and the earthly realms. The Greeks and Romans said that it was a path made by Iris, a messenger between Heaven and Earth. (Iris is the daughter of Thaumas, a sea-god whose name is linked to the Greek word for wonder, and from which in English we have 'thaumaturgy', or the capability of a magician or saint to work magic or miracles.) Norse myth also describes the rainbow as a bridge that connects Asgard and Midgard, the homes of the gods and men. And across much of Europe and western and southern Asia there is, of course, an association with a bow. For Finns and Lapps, the rainbow was the bow of the thunder god, a skilful archer whose arrow is lightning. In Hindi it is *Indradhanush*, the bow of Indra. In Arabic it is *Qaus Quzaḥ*, the war bow of Quzaḥ, a pre-Islamic god.

Some cultures have regarded the rainbow as malign. The people of Nias, an island in the Indonesian archipelago notable for its megaliths, feared it as a huge net spread by a powerful spirit to catch their souls. And among many indigenous peoples of the Amazon, rainbows were associated with harm. So great was ancient Peruvians' fear of it that they believed it best to stand still in silence until it disappeared.

But in the modern Western imagination – influenced by the biblical account, in which the rainbow

The account of the rainbow in Genesis is in turn shaped by older stories such as the one that appears in the *Epic of Gilgamesh* (*c.*1800 BC), which tells how the goddess Ishtar lifted a jewelled necklace into the sky as a promise that she would never forget the great flood that destroyed her children.

is a covenant from God – its associations are overwhelmingly positive. Political and social movements have adopted it as a symbol of hope and solidarity for hundreds of years. A rainbow featured on a flag in a peasants' revolt in Germany in the sixteenth century (one of the largest and most widespread uprisings in Europe before the French Revolution, it ended with the peasants' defeat and mass slaughter). In the twentieth century a rainbow flag was flown by the cooperative and peace movements in Europe. It was also adopted by Birobidzhan, the Jewish autonomous homeland in Siberia, and by the indigenous peoples' movement in the Andes. Since the late 1970s a version has been associated with LGBT pride.

We'll never know who first suggested that a rainbow is an inanimate phenomenon without inherent significance. But the earliest surviving record of what we would now call a scientific explanation dates from the fourth century BC. In the *Meteorologica,* a treatise on earthquakes and weather phenomena, Aristotle describes a rainbow as a reflection or an echo. Light, he wrote, travels in straight lines, and is reflected back off a raincloud in a circle of angles to create the impression of a bow in the eye. The different colours – for Aristotle just three: red, green and violet – are made by light deflecting at slightly different angles.

Aristotle's explanation is partly right, but it is incomplete. Sunlight *is* reflected in a rainbow in a way broadly analogous to the reflection of sound, but it is also refracted – first when it enters a drop of rainwater, and a second time as it leaves the drop after reflecting off the back of it. Although Aristotle knew that light refracted through a prism or water droplet separates into different colours, he did not apply this to the rainbow and insisted that light was being reflected by an entire cloud. His account does not seem to have been seriously challenged in Europe or the Islamic world for about sixteen hundred years. Only in China (in the *Dream Pool Essays* published in AD 1088 by the statesman and polymath Shen Kuo) is there a record of the idea that rainbows are formed by sunlight refracting through individual drops of rain. New ideas start to appear in Western accounts about two hundred years after Shen Kuo. In 1266 the English monk Roger Bacon wrote that a primary rainbow never appears higher than forty-two degrees above the horizon. Then, around 1300, the Persian scholar Kamāl al-Dīn al-Fārisī and Theodoric of Freiburg independently undertook research that still forms the basis of our modern understanding. Working without any knowledge of each other, they both showed that a rainbow could be made by an ensemble of individual drops and not, as Aristotle had proposed, by a cloud or mist acting like a mirror.

Both men based their conclusions on experiment and measurement. Al-Fārisī put a spherical glass vessel filled with water – his large-scale model of a raindrop – in a camera obscura, projected light into it and deduced that the colours are made by the decomposition of white light. Theodoric used glass spheres but looked closely at natural phenomena. 'In spider's webs,' he wrote, 'which are stretched out and closely covered with many drops of dew in a suitable position with respect to the sun and the eye, the colour . . . yellow . . . appears most plainly between the other colours . . . as in the rainbow.' And then he got right down on the ground and saw the same thing in dew drops on blades of grass – moving his head slightly, his angle of vision changed and each colour of the rainbow appeared in order. Both al-Fārisī and Theodoric also worked out that a secondary rainbow, which sometimes arcs more faintly around and over the main one, is caused by rays of light reflecting one more time within the drops, reversing the order of colours.

Raindrops are usually between 0.1 millimetres and 5 millimetres across. They are shaped like oblate spheroids – that is, spheres with squashed noses. In the foggy conditions that create a 'ghost rainbow' the drops are so small that quantum mechanical effects become important and smear all the colours to white.

Little further progress was made until the 1620s, when René Descartes used the law of refraction to calculate that a rainbow appears at an angle of 41°30' above the viewer. A generation later, Isaac Newton refined these computations. He found that a primary rainbow always occurs between 40°17' and 42°2', and a secondary one between 50°57' and 54°7'. Newton also showed that, in the right conditions, a third rainbow is created behind the viewer in a circle around the Sun. Because sunlight is so bright, however, this will always be invisible to the naked eye.

The law of refraction describes the relation between the angles of incidence and refraction when light passes between different mediums such as air and water. It was discovered by Ibn Sahl around 984, and then rediscovered by Thomas Harriot in 1602 and Willebrord Snel in 1621.

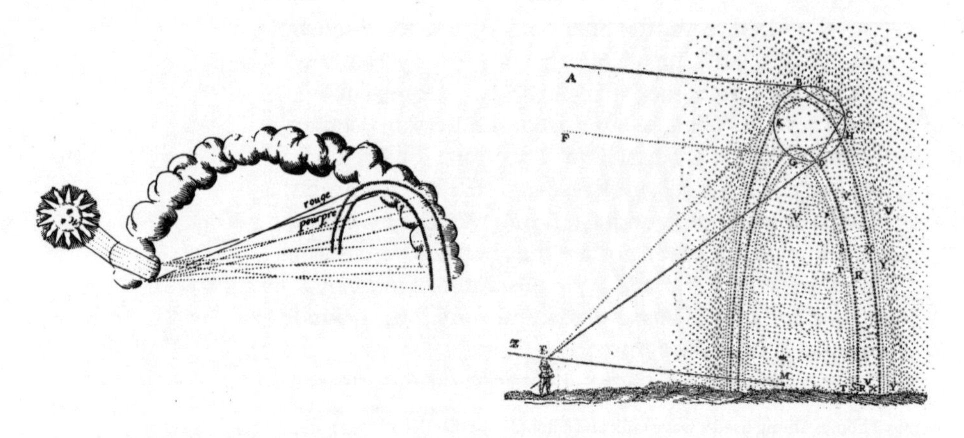

Marin Cureau de la Chambre, a contemporary of Descartes, believed sunlight was refracted and projected to make a rainbow (left). Descartes showed that the light bounced off the back of raindrops and then refracted at an angle towards the viewer (right).

By the nineteenth century rainbows seemed to hold few secrets. An experiment by Thomas Young in 1803 showed that light behaves like a wave, and the way in which these waves can interfere with each other (increasing their joint amplitude when they are in sync, and dampening each other when they are not) explained the additional bands that sometimes form on the inner arc of the primary rainbow. With the growing understanding of the dispersion and polarization of light that followed, a comprehensive explanation of the rainbow appeared to be in sight. But had it been explained away?

As the novelist Philip Pullman explains, single vision for Blake is a literal, rational, dissociated and unemotional view of the world. Twofold vision sees not only with the eye but through it to contexts, associations, emotional meanings, connections. Threefold vision is the place of poetic inspiration and dreams, while fourfold vision is a state of ecstatic or mystical bliss.

After a birdbath of heavenly paint, a sky pathway and a covenant with the Almighty, the rainbow as a trick of the light may seem like a letdown. Does science undo the wonder of the rainbow? Some of Thomas Young's contemporaries certainly thought so. Johann Wolfgang von Goethe attempted a new theory that took more account of human feelings, exploring how we perceive colour in a wide array of conditions, rather than seeking to develop a mathematical model of its behaviour as Newton had. William Blake decried the 'single vision' of natural philosophers since Newton, and John Keats famously wrote that the rainbow, once an object worthy of awe, had been reduced to part of the dull catalogue

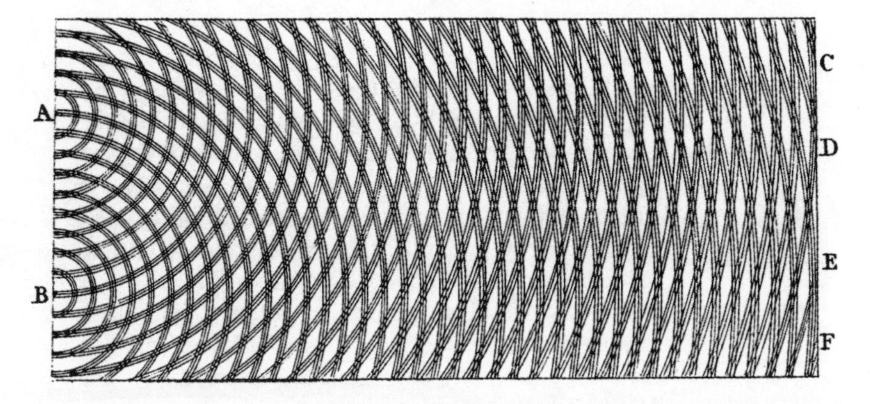

In 1803 Thomas Young used a wave tank and a double-slit interferometer to show that light behaves like a wave. Some physicists see this as the single most important discovery in the study of light.

of common things. But, as the modern scholar Philip Fisher argues, Aristotle's pairing of echo and rainbow is a poetic thought as well as a scientific one. It captures an inner similarity between two things that seem remote from one another, combining them in instantaneous perception by means of language. Nor is the work of al-Fārisī and Theodoric without beauty. When I picture a thirteenth-century monk prostrate in the wet grass and moving his head slowly from side to side in order to catch the changing colours of light refracted through dewdrops, I do not see an irritable reaching after fact or an inability to live with uncertainty, but a delightful, childlike fascination and a willingness to explore the world on its own terms. A scientific approach does not have to be antithetical to a spiritual vision. Descartes' treatise on the rainbow was intended as part of a preamble to his massively ambitious *Discourse on Method* of 1637, in which he strove to determine beyond doubt what could really be known through mathematics, logic and deduction. This, he believed, was a sure route to God, the ground of all being. In that sense it was a profoundly religious project, not a mere exercise in calculation (although he did see a pragmatic and political application of his knowledge and suggested harnessing the rainbow effect in displays such as artificial fountains for the amusement of kings).

'The short night –
on the hairy
caterpillar
beads of dew.'
Buson

Newton was extraordinarily fastidious in his researches into natural philosophy, or what today we would call science, and although he made some of the biggest strides taken by a single human being into our understanding of the laws of nature, he never lost sight of how limited that understanding was. He did argue forcefully that light is made of particles, but he also understood the strength of arguments to the contrary made by his contemporary Christiaan Huygens, and corroborated by Thomas Young a hundred years later, that light was made of waves, and he knew when to stop speculating. 'To determine more absolutely what Light is,' he wrote, 'and by what modes or actions it produceth in our minds the Phantasms of Colours, is not so easie. And I shall not mingle conjectures with certainties.'

'To explain all nature is too difficult a task for any one man or even for any one age . . . 'Tis much better to do a little with certainty & leave the rest for others that come after you than to explain all things by conjecture without making sure of any thing.' Isaac Newton

And yet Newton was also obsessed with biblical prophecy, devoting inordinate amounts of time to tracts with chapter titles such as 'Of the power of the eleventh horn of Daniel's fourth beast, to change times and laws'. He was a practising alchemist too, believing matter to be active and charged with spirit. This side of his character was unknown until John Maynard Keynes discovered his private papers in the early twentieth century, and concluded that 'Newton was not the first of the age of reason. He was the last of the magicians.' And at least one trace of magical thinking seeps into his work on light. Newton divided the colours of the rainbow into seven, a mystical number that he linked to the seven notes of the Western musical scale (and hence to the ancient Pythagorean notion that the world expresses a divine proportion), as well as to the seven classical planets. Today we happily recite these colour names – red, orange, yellow, green, blue, indigo and violet – even though an unprejudiced eye seldom sees them all with this degree of differentiation.

In my experience, such physics as I do understand only adds to my wonder at the all that is. And whatever is on my mind, the great dance of light, in which it passes effortlessly through itself in different directions while travelling through air or water, is occasion for meditation. For in the midst

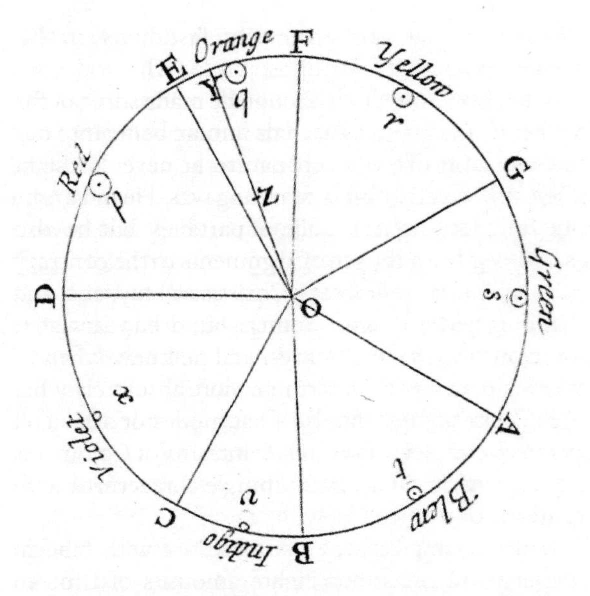

Making an analogy to different tones in music, Newton speculated that rays of light of different magnitude, strength or vigour excited 'vibrations of various bignesses' in the aether, the medium through which light supposedly travelled, and in the retina of the eye. The biggest vibrations, he suggested, corresponded with the 'strongest' colours, reds and yellows; the least with the weakest, blues and violets; the middle with green; and a confusion of all with white. Following this analogy, it made sense to Newton that there should be seven colours like the seven tones of the diatonic musical scale. This beautiful idea was quite wrong.

out of this 'tremendous mess', as Richard Feynman vividly described it, we sometimes find clarity. I live near a river, and kayak on it when I can. This being England, the weather is always changing. When it's sunny, and golden light flashes on the water, I find solace in gliding across its surface. Typically I see a rainbow a few times a year – infrequently enough that I always look forward to another and delight like a child in something so beautiful that is simultaneously near and beyond reach. I also enjoy knowing that I am seeing something as close as I ever will to the spectral colours of what Newton called light's 'differently refrangible rays', and not the electronic composites of artificial images with which we are bombarded for so much of our lives. In a rainbow,

Going on appearances, Philip Larkin would seem to be to lyric poetry is what Eddie the Eagle is to ski-jumping. But Larkin's 'Water', in which he imagines a new religion, shows otherwise:

'My liturgy would employ
Images of sousing,
A furious devout drench,
And I should raise in the east
A glass of water
Where any-angled light
Would congregate endlessly.'

When viewed in ultraviolet, the rings of Saturn, which are made of water ice, appear as a strange rainbow of reds, pinks, turquoises, deep blues and other colours.

rain really is (among other colours) purple, just as Prince said.

A rainbow is not a sign from the supernatural. But neither is it an object that exists independently of an observer. It is like the horizon: an artefact of vision, albeit one that is also a revelation of – or at least a clue towards – what is going on in the world beyond us. Arising from the interplay of sunlight with falling drops of water, it is a seemingly constant but subtly changing pattern, and as such not a bad image of consciousness itself. 'Perception,' wrote the mathematician and astronomer Johannes Kepler, 'belongs not to optics but to the study of the wonderful.' This includes our sensations of colour, which may not really be 'there' in the world but are so precious to us as means of connecting with it.

You can almost hear the drops in György Kurtág's 'Play with Infinity' (1973) or in György Ligeti's étude 'Arc-en-Ciel' (1985–2001).

It may be impossible for humans to entirely escape allegory and metaphor in how we respond to our perceptions of natural phenomena. Recalling that the Rainbow Serpent of Aboriginal Australians is not always a benign presence, for example, and reading that an approaching storm can cause a giant repeating rainbow (the effect is caused by gravity waves in airglow, analogous to when a rock is thrown into calm water), I find myself thinking of the storms of many kinds that are likely in coming years and decades.

The year 2016 was the hottest in human history, and a new high was reached for the third year in a row, bringing the global average temperature to 1.2°C above pre-industrial levels. It is likely that climate change resulting from human emissions of greenhouse gases will significantly increase future storm intensity and frequency.

Light beyond the rainbow

Rainbows are just one of the wonders of all the light we can see. In the optical phenomenon known as heiligenschein, for instance, a whitish glow resembling a halo surrounds the shadow cast by the viewer's head on dew-covered grass. It arises because dewdrops act like lenses, focusing parallel beams of light that pass through the drops closer together so that the leaves beneath them seem brighter. In the case of the phenomenon called a glory, a bright halo appears around the viewer's shadow when that shadow falls on mist or cloud, but it differs from heiligenschein in that it has concentric rings of colour, with blue on the inside, ranging through green, yellow and red to purple on the outside. The major source of the glory's light is reflection from the back or front of the water droplets. Refraction as the light enters and leaves them accounts for the colour. In the mountains, when the Sun is low in the sky, the shadow in the centre of a glory may be greatly magnified and sometimes distorted as it falls over a larger region of rugged ground, giving rise to a Brocken spectre – an apparition of self/not-self that has frightened and fascinated people in equal measure. Today we can include among the resonances of the spectre a sense that our actual bodies are also 'just' projections: geodesics in three-dimensional space of phenomena in four-dimensional space-time.

Turning from shadows towards the source of light itself, a sundog (which is also called a phantom sun or parahelion) projects bright 'echo' Suns on either side of the real thing, frequently with a luminous ring all the way round. It is created by the refraction of sunlight through flat ice crystals that are falling slowly through the air many thousands of metres above the ground. Higher still, and quite different in origin and appearance, an aurora is a rapidly shifting curtain of colour, often mostly green, that appears when charged particles from the Sun hit and ionize atoms in the magnetic field high above the Earth. In footage from the International Space Station you can watch seemingly endless auroras scroll over the curve of the horizon like the puddles of cool flame from brandy burning on a giant Christmas pudding.

Some wonderful effects of light in space are easily visible from the ground. Earthshine – light reflected from the sunlit side of the Earth onto the dark disc within the bright narrow crescent of a new or old Moon – is one of my favourites. Most clearly visible when the Moon is close to the horizon at dawn or dusk in calm and clear weather, it makes it seem especially close – a Janus on the threshold between different realities. At a given moment and place on the ground or at sea, Earthshine does not vary, but its hue and intensity will change according to where you are on Earth because the reflectivity of the ocean, land or cloud around you will be different. Earthshine on the Moon

'In general relativity, bodies always follow straight lines in four-dimensional space-time, but they nevertheless appear to us to move along curved paths in our three-dimensional space. This is rather like watching an aeroplane flying over hilly ground. Although it follows a straight line in three-dimensional space, its shadow follows a curved path on the two-dimensional ground.'
Stephen Hawking

Other planets in the solar system also cast light on their moons. Saturnshine has been photographed with great clarity on its moons Dione and Mimas. A grainy image shows Plutoshine on its moon Charon.

A legend of the Yurok Native Americans says that far out in the Pacific Ocean, but not farther than a canoe can paddle, the rim of the sky makes waves by beating on the surface of the water. On every twelfth upswing, the sky moves a little more slowly, so that a skilled navigator has enough time to slip beneath its rim, reach the outer ocean, and dance all night on the shore of another world.

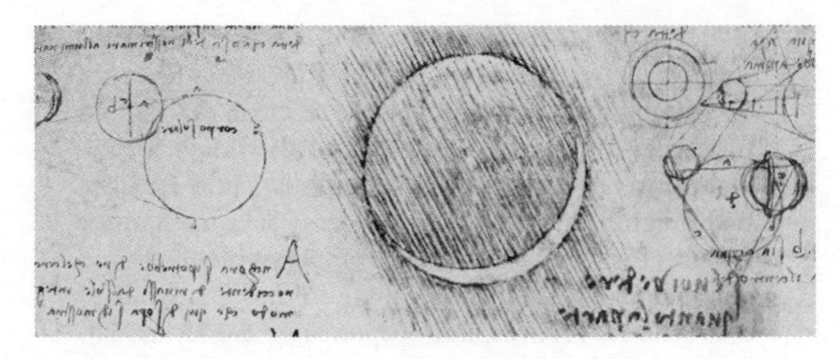

Sketch of Earthshine by Leonardo da Vinci (1506 to 1510).

is a subtle and gentle light to us, but the view from the part of the Moon that is within Earthshine would be breathtaking: a dazzling blue, green and white orb in the sky four times wider than the Sun and fifty times brighter than the full Moon as it appears on Earth.

A catalogue of the wonders of light on Earth could go on for a long time. It could include the appearance of Sirius, the brightest star in the sky, which appears to pass through the entire visible rainbow of colours as its light is refracted through the atmosphere. It could include zodiacal light: the diffuse glow visible in the sky after dusk and before dawn that is a reflection of sunlight from trillions of tiny dust grains – no more than one in every cubic kilometre – that did not make it into planets or asteroids of the solar system. And this before we even got started on the trickster properties of light itself. (One of my favourite examples, highlighted by Richard Feynman, is that you can take a piece of mirror, scratch away part of its surface, and the mirror will reflect light at an angle it didn't before.) But I will pass on to the shared origin of many of them: the Sun.

Zodiacal light is, in the astronomer Caleb Scharf's words, 'a dusty impression of the alignment of the planets in their huge disk of orbital paths, and of all the other objects sharing this same space'.

Staring at the Sun

For most of recorded history (and, we may presume, for long before), people have been in awe of the Sun. Often they have given it a central place in religion. But until less than a hundred years ago almost nobody had a very good idea of what it was or knew what made it shine.

To look directly into the Sun for more than a few moments can cause blindness, but thanks to recent technology we are now beginning to appreciate its true magnificence. A video by NASA's Solar Dynamics Observatory published in 2015 compresses five years of detailed observations into a few minutes. The roiling contortions, vortices and outpourings are so utterly fascinating that for a time you can only stare, forgetting almost everything else.

The NASA film is mesmerizing, but the first movie of the Sun was made in 1613. Over successive days in June and July of that year Galileo projected an image of the Sun though a telescope onto a screen and recorded the position of spots on its surface. The drawings he made rival photographs taken hundreds of years later in their quality and, viewed in sequence like frames from a film or a flick book, show the spots moving smoothly across the star.

People had known about sunspots for a long time. The earliest surviving written record, which was made in China, dates from around 800 BC, and many cultures have stories about them. It is reported, for example, that the ancient people of the Zambezi believed sunspots were mud spattered in the face of the Sun by a jealous Moon. Galileo was not even the first person to observe them through a telescope. The mathematician Thomas Harriot had done so in 1609.

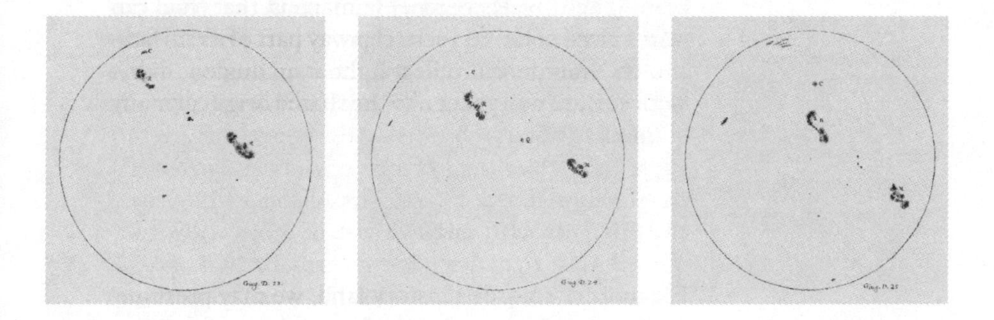

Every day from 2 June to 8 July 1613 Galileo drew the spots he was able to see on the Sun. Here are three in the sequence of thirty-five images

But Galileo and another observer named Johannes Fabricius were the first to surmise that the movement of sunspots across the Sun and their disappearance off one edge followed by their reappearance on the other showed that the bright disc was actually a rotating sphere which carried the spots along as it turned. And it is in Galileo's drawings that we can see most clearly the first step in transforming the Sun from an object of dazzling unknowability into something knowable but no less astonishing.

For a long time after Galileo, astronomers learned very little more about the nature of the Sun. Some of the speculation was, by turns, astute and wildly wrong. Descartes thought that the spots were scum on a primordial ocean. The seventeenth-century polymath Athanasius Kircher depicted a fantastical boiling ocean with spouting lakes of fire. William Herschel, the eighteenth- and early-nineteenth-century astronomer who discovered Uranus, suggested that sunspots were portholes into a darker world where people lived beneath the radiant outer sheath. But almost no one had much by way of actual evidence, and in 1835 the philosopher Auguste Comte concluded that, short of travelling there, it would always be impossible to study the chemical composition of the Sun or other stars. But Comte was wrong. The Sun does tell us what it is made of, if only we know how to look.

In 1802 the chemist William Wollaston found that if he passed sunlight through a narrow slit in a piece of metal before it entered a prism he could see several thin black vertical lines amid the vivid colours of the solar rainbow it projected. The physicist Joseph Fraunhofer repeated the experiment in 1814 with better equipment, and proved that the apparent continuity of a rainbow is an illusion: there are tiny gaps, dim or black arcs of missing colours, too narrow for the human eye to see. Fraunhofer had invented the spectroscope, and with it he eventually catalogued more than five hundred gaps, known today as Fraunhofer lines (or absorption lines), in the spectrum of the Sun. Today, tens of thousands are known. It turns out that the Sun, our universal light, does not emit light across the entirety of the visible spectrum.

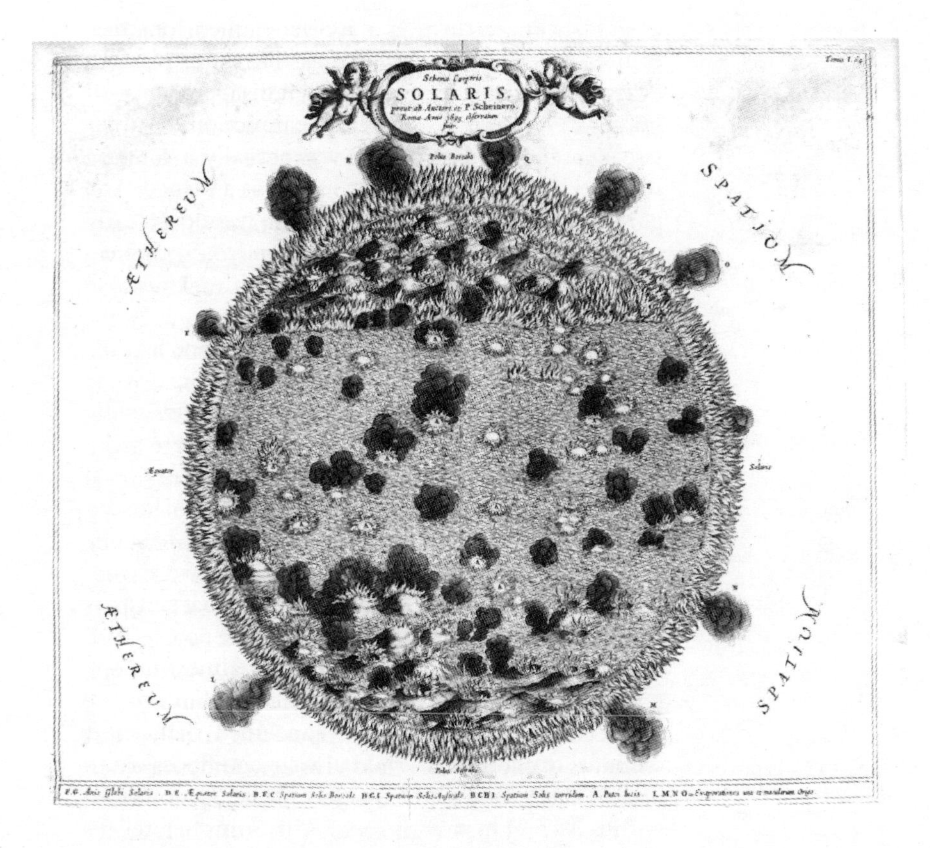

Schema Corporis Solaris, or the Sun, by Athanasius Kircher (1665).

Another revelation followed half a century later when Robert Bunsen and Gustav Kirchhoff heated various chemical elements and observed the light they gave off through a spectroscope. When mercury, for example, is vaporized, the hot gas glows blue to the naked eye. But when this light is passed through a slit and then a prism it appears as violet, green and yellow. Bunsen and Kirchhoff realized that the blue is actually the impression created when the eye blends those three colours. A spectroscope, they now understood, displays the 'true' spectral colours given off by any hot material and, experimenting with different elements, they found that each has its own unique fingerprint. Sodium, for instance, gives off a tightly spaced pair of golden lines. Turning a spectroscope at the Sun, Bunsen and Kirchhoff matched its black-line patterns against the bright-line patterns of dozens of elements, notably hydrogen, oxygen, iron and calcium. By the late 1860s they had matched all of the lines to elements known on Earth except for one series in the yellow part of the spectrum. The mysterious lines were interpreted as the signature of a completely unknown element, and named 'helium' after the Greek sun-god Helios. Many thought that helium existed only in the Sun, but in 1895 it was discovered on Earth when the chemist William Ramsay isolated a gas from deep in a uranium mine and, passing it through a spectroscope, accounted for every remaining unexplained line in the yellow spectrum of the Sun. In this way the chemical composition of the Sun was finally known.

Helium accumulates as a stable end product when uranium decays into lead. A kind of alchemy turns the source material for nuclear bombs into the stuff of the Sun and the element most protective against radioactivity.

The answer to the question of what made the Sun shine came from linking the discoveries of spectroscopy to advances in theoretical physics. Albert Einstein's famous formula $E = mc^2$, published in 1905, showed that matter can be turned into energy and, grasping the significance of this, the astronomer and physicist Arthur Eddington proposed in 1920 that vast amounts of energy are released as light and heat when one of the protons (the nuclei of hydrogen atoms) within it combines with another to form helium. In the 1930s the physicist Hans Bethe and others showed that the process in stars like the Sun is rather more complicated, but this is, in broad outline, what we now call fusion. It turns out that every second the Sun transforms about 4.25 million tonnes of its mass into energy and light in this way. This is the equivalent of more than 90 billion megatonnes of TNT – that's about seventy-five thousand times the yield of a B83, the most powerful hydrogen bomb in service with the US military today, or 6 trillion Hiroshima atomic bombs. That said, there are huge differences between the Sun and a nuclear device. For a start, the Sun releases its energy relatively slowly and constantly, and the fury within is tamed by hundreds of thousands of kilometres of gas between its core – where the reactions take place – and its surface.

In reality, fusion is a complex, multistep process. The closer the two protons get, the more strongly their positive charges push them apart. Only an effect known as quantum tunnelling enables them to bond. It is as if they don't have enough energy to open the door and walk through, but will occasionally teleport straight through a wall. Further, the helium-2 produced when two protons do bond is extremely unstable and usually splits back into two separate protons. One time in ten thousand, however, one of the protons will spontaneously transform into a neutron, and the atom then becomes deuterium, a stable isotope of hydrogen. Deuterium and hydrogen can fuse to make a stable form of helium, and it is this that releases the energy that powers stars.

Nearly all of the fusion in the Sun takes place in its

core, which occupies about the same proportion of the whole as the stone does of a peach. Here, under enormous pressure and gravity, the temperature exceeds 15 million degrees Celsius, and atoms exist as plasma – the fourth state of matter, in which atoms are too hot to be solid, liquid or gas. The surrounding sphere heats up as energy is transferred outwards through successive layers before it reaches the surface of the Sun and escapes into space. But this does not happen quickly. Hydrogen and helium are so densely packed in the core that the photons produced by fusion move tiny distances before bumping into other atoms, where they are absorbed and then emitted again. As a result, photons take a random walk for as much as two hundred thousand years before they reach the surface of the Sun about 500,000 kilometres above the core. Once in space, they can cover that same distance unimpeded in less than two seconds, and those that fly towards the Earth cross the intervening 150 million kilometres in eight minutes and twenty seconds. This means that the light from the Sun that you see began its existence in a reaction that took place as long ago as the time that anatomically modern humans evolved.

You may think that 'plasma' is a hard word to rhyme, but the band They Might Be Giants manage it in 'Why Does the Sun Really Shine?': their solution was 'miasma'.

The temperature at the visible surface of the Sun, which is known as the photosphere, is less than four ten-thousandths of that in the core. But, at an average of 5,500°C, it is still more than three times the temperature of molten lava or the melting point of iron. Like some superheated bowl (or ball) of soup, however, this is a turbulent place, and conditions at the surface vary. Large bubbles of hot plasma move upwards to the photosphere in some places but in others strong magnetic fields slow or block their rise. Slow-moving matter is cooler – in this case, as low as 4,000°C – and cooler means darker. These areas of stronger magnetism on the surface are the sunspots that Galileo drew with such care.

The weather on the Sun

The idea that the Sun affects the weather on Earth is familiar. Less so is the fact that the Sun itself has incredibly vast and violent weather of its own. Above the Sun's bright visible surface, magnetic field lines flow outward and curve back towards the surface in arcs that dwarf the Earth in scale. Highly charged plasma follows these arcs and when, sometimes, the magnetic lines connect with others or disconnect, the plasma snaps together or breaks apart, like a rope breaking under tension, at great speed (and fast means hot: temperatures can rise to millions of degrees). These whips of highly charged, fast-moving plasma are the Sun's corona – so called because when seen during a total eclipse they look like the points on a crown. As the solar wind cools the corona, some matter escapes into space. But much of it falls back down to the surface. In other words, it rains on the Sun – with droplets of slightly cooler (but still very hot) plasma the size of countries falling from altitudes of over 60,000 kilometres at speeds of more than 200,000 kilometres per hour.

There are also tornadoes on the Sun. Swirling plasma creates vortices, which cause magnetic fields to twist and spiral into super-tornadoes that reach from the surface into the Sun's upper atmosphere. Even more tremendous are solar flares and coronal mass ejections. In flares, plasma heated to millions of degrees along magnetic arcs ejects electrons, ions and atoms (as well as electromagnetic radiation mostly outside the visible range) into space at near the speed of light. Coronal mass ejections squirt particles out into space at about only two per cent of the speed of light, but this is still millions of kilometres an hour, and they eject much more mass than flares. 'Limpid jets of love hot and enormous, quivering jelly of love, white-blow and delirious juice,' wrote the poet Walt Whitman in a very different context. When, some days later, the outer edges of these ejections lick the Earth's magnetosphere they cause geomagnetic storms and unusually strong auroras at the Earth's poles.

The Sun is not the only cause of auroras in the solar system. In the case of Jupiter's moon Io, a green, red and blue aurora is caused by interaction with the magnetosphere of the giant planet rather than with our star.

All this activity is ultimately linked to processes deep within the Sun and in particular to what goes on at the boundary between its inner and outer spheres. The inner sphere of the Sun – its core and radiative zone – rotates almost as uniformly as a solid ball. But about two-thirds of the way from the centre to the surface – above a thin layer called the tachocline, and in what is known as the convective zone – movement is more turbulent. Here, plasma takes about twenty-five days to rotate about the equator but longer to do so at the poles. This differential sets up eddies and meridional flows – huge conveyor belts of magnetized material that flow towards the higher latitudes, where they sink and then return towards the equator. Normally, these currents dive down and start to head back at a latitude of about 60 degrees, but sometimes they travel all the way to the poles before doing so. When this happens the return is slower and the Sun becomes less active, with the result that there are fewer sunspots. Over a cycle of eleven Earth-years, the Sun 'breathes' or pulses – varying its output of solar wind, X-rays, ultraviolet and visible light.

The birth and death of the Sun

It is sometimes said that the Sun had a mother and father. Mum, in this simplified account, was a giant molecular cloud made of mostly hydrogen, and Dad was a shockwave from the explosion of a giant blue star – much bigger and denser than the Sun but tiny compared to the cloud. The maternal cloud contained almost all the other elements besides hydrogen, mixed in with the debris from previous generations of stars in a galaxy that had already been swirling and fizzing with supernova explosions for billions of years. There was nothing inevitable or unique about the particular circumstances in which the cloud met the star, but about 4.6 billion years ago the two happened to be close enough to be drawn together by gravity and they precipitated from the cloud a new body: a dense ball of hydrogen known as the solar nebula. This proto-Sun was a ten-millionth of the size of the original cloud: comparatively, an apple seed to a football stadium. Myriad similar events are visible today in images such as those of the Pillars of Creation in the Eagle Nebula.

'Stars, like thoughts, are not inevitable. Out of the diffuse disorder something may or may not coalesce, and floating specks in space find each other very escapable.'
Amy Leach

As for the end, well, many of us have a general idea

David Bedford tried it in *Star's End* (1974), but it's not an easy listen. For the Sun's main sequence, Brian Eno's *Lux* (2012) could be a starting point. Or, encompassing all, *Sun Star* by John Coltrane (1967).

that, one day, the Sun will become a red giant which will incinerate and swallow the Earth. No more cloud capp'd towers and gorgeous palaces for you, pal. The full story, though, is even more awesome and beautiful. It deserves, at the very least, a great musical score. I'd like to imagine something beyond the final chord of 'Der Abscheid' in Gustav Mahler's *Das Lied von der Erde* (which, Benjamin Britten suggested, was imprinted on the atmosphere) and surpassing recent works by John Luther Adams such as *Become Ocean* (described by music critic Alex Ross as 'the loveliest apocalypse in musical history') not to mention his *Sky with Four Suns*. All I can offer, however, is a précis of the physics – a sketch for the programme notes.

Barring intervention by intelligent beings with stupendous powers (which seems like a stretch) or collision with another star (which is highly unlikely), the Sun's future trajectory is predictable in almost every respect. As the astronomer Martin Rees says, 'even the smallest insect is far more complex than a star'. But there can still be something magnificent in what is all but inevitable. Human events are unpredictable in their details but the fate of our planet is not, and maybe there is some truth in the old Norse idea that the future is determined in fibres that have already been selected and are being woven.

Fusion, which powers the Sun, will continue for as long as a fuel supply remains. The Sun is about three hundred and thirty thousand times the mass of the Earth and a million times its volume – a basketball compared to a pea. At present it is about 4.6 billion years old, and almost halfway through what is known as the main sequence in which it generates energy by fusing hydrogen into helium. At two octillion (two followed by 27 zeroes, or 2,000,000,000,000,000,000,000,000,000) tonnes, it has enough fuel to burn for billions of years yet. All through this time it will get hotter and brighter. In little more than a billion years from now the heat will be enough to evaporate away all the Earth's oceans, and the average temperature on our planet will reach over 370°C – more than hot enough to bake pizza. 'I think we are inexterminable, like flies or bedbugs,' the poet Robert Frost once said, but these conditions will trump the thermonuclear armageddon that he and his contemporaries feared. Macroscopic life on Earth, including human life, will have long since ceased.

Even at this point, however, the Sun's life as a main sequence star will still be hardly more than half over, and for another four billion years it will remain just about the same size, and radiate white light just as it does today. Only about five billion years from now will hydrogen in the core finally be exhausted, and the Sun start to expand. To begin with the expansion will be slow, and the Sun will take about half a billion years to double in size. Then it will grow more quickly until it becomes a red giant over two hundred times larger and two thousand times brighter than it is today. By this time it will have swallowed and incinerated Mercury and Venus, but the Earth will probably have been pushed outwards by the expansion and continue its orbit unconsumed (though much too hot for life). It is possible that Saturn's moon Titan, which is so cold at present that liquid methane on its surface flows through its deep canyons, will have warmed to temperatures within the range comfortable for life as we know it, making it, conceivably, a refuge for our distant descendants, assuming they somehow escape the heat on Earth in time.

The last complex multicellular life on Earth less than a billion years from now may be, or resemble, tardigrades – the 'water bears' that subsist on bacteria and on smaller tardigrades – and/or something like the tubeworms found at hydrothermal vents on the ocean floor. Microbes may persist deep within the Earth for another 2 billion to 3 billion years.

The Sun will be a red giant for about a billion years. During this time it will gradually burn away a third of its mass and then, suddenly, helium in the core will ignite violently in what is known as the helium flash and more than a third of what remains will turn into carbon in a few minutes. After that it will shrink from more than two hundred to around ten times its current size, and burn helium for about a hundred million years. When all the helium is finally exhausted the Sun will repeat the expansion it followed when its core was hydrogen, except that this time the expansion will be much faster, and the new giant will only last about twenty million years. The Sun will then become increasingly unstable. Over the following few hundred thousand years it will pulse about four times, like a lightbulb on the blink, only a little brighter each time, before it finally blows.

The Sun is too small to turn into a supernova

One of the most beautiful nebulas, the Ring (M57), is a circular rainbow with a sky-blue centre surrounded by green, orange and then red. Planetary nebulas play a significant role in galactic evolution, expelling heavy elements such as carbon and nitrogen forged from hydrogen and helium by their parent stars into the interstellar medium where they become part of the next generation. The carbon in every living thing on Earth probably comes from nebulas like these.

when it explodes, and only a small fraction of its mass will blast away into space. The rest will shrink to a superdense core about the size of the Earth – an ultra-crushed ball made mostly of carbon and oxygen. This core will emit intense ultraviolet light which will make the expanding bubble of gas from the explosion glow mostly green and red. For a few tens of thousands of years this remnant will be surrounded by a planetary nebula as beautiful as any of the marvels visible today. It may be a moment comparable to the penultimate bar in Handel's 1713 *Eternal Source of Light Divine* when, after the conclusion of the duet with voice, the trumpet rises to a top D, the tonic, at a pitch not previously reached in the piece.

After that, the gas bubble will disperse and, with no fusion taking place in the core, the Sun will remain a white dwarf about the size of the Earth today. Tiny contractions under its own gravity will be enough to generate light for many trillions of years until it becomes a black dwarf – a remnant that emits no heat or light at all. (The feel of this, as far as human imagination can extend, may be something like the adagio of György Kurtág's 1994 piece *Stele*.) The most likely fate of the black dwarf that was once our Sun will be that, after around 10^{19} years, it will be ejected (together with the remaining bound planets) from the galaxy into intergalactic space. If this doesn't happen then it may collide with another black dwarf in about 10^{21} years and produce a Type Ia supernova explosion that will destroy whatever remains of the solar system. If neither of these things happens, the black Sun will continue to orbit the galaxy, slowly falling towards the black hole at its centre. But before it can get there – in about 10^{100} years – the black hole will have evaporated. In this eventuality, the Earth will finally spiral into what was once the Sun – unless some unpredictable gravitational interaction knocks it out of our Sun's orbit into the depths of a cold, empty universe. Here, at no extra charge, is your cut-out-and-keep guide to the past, present and future of the Sun:

```
MS MS MS MS MS MS*MS MS MS MS
MS MS MS MS MS MS MS MS MS MS
MS MS MS MS MS MS MS MS MS MS
MS MS MS MS MS MS MS MS MS MS
MS MS MS MS MS MS^MS MS MS MS
MS MS MS MS MS MS MSxMS MS MS
MS MS MS MS MS MS MS MS MS MS
MS MS MS MS MS MS MS MS MS MS
MS MS MS MS MS MS MS MS MS MS
MS MS MS MS MS MS MS MS MS MS
RG RG RG RG RG RG RG RG RG RG
h CHB o
WD WD WD WD WD WD WD WD WD WD
WD WD WD WD WD WD WD WD WD WD
WD WD WD WD WD WD WD WD WD WD
```

(repeat thousands of times)
BD *ad finem*

Each pair of capital letters represents 100 million years in the life of the Sun. 'MS' stands for 'main sequence'. The beginning of life on Earth is marked by * at between 600 million and 700 million years after the formation of the solar system (and about 4 billion years before the present). The ^ is for 'you are here', and the x marks the likely end point of Earth as a viable home for life as we know it. 'RG' stands for 'red giant' and 'CHB' for 'core helium burning' (although this will last less than 100 million years). If the typeface for RG was proportional to MS in the same way that a red giant is proportional to a main sequence star each letter would be the best part of a metre high. The 'h' is the helium flash and the 'o' represents the tens of thousands of years in which the Sun is a planetary nebula. 'WD' is for 'white dwarf', 'BD' for 'black dwarf'.

Dark wonder: neutrino

'The light tells us much,' said the nineteenth-century nature writer Richard Jefferies, 'but I think in the course of time still more delicate and subtle mediums will be found to exist, and through these we shall see into the shadows of the sky.' Beyond the light there are many kinds of darkness at the edge of knowledge.

Invisible light was discovered by William Herschel in 1800. He noticed that a thermometer placed in darkness just beyond the edge of the red light of a rainbow pattern projected by a prism heated up, and he concluded, correctly, that this was caused by 'calorific rays' – or what we now call infrared light. And it turns out that only a little over forty per cent of the photons hitting the Earth's surface are in the part of the spectrum visible to humans. Infrared and ultraviolet light make up almost all of the rest, with more than fifty and less than five per cent respectively. All three are important to life on Earth. Most significantly, perhaps, infrared light helps keeps the planet warm enough for life as we know it. Some snakes can detect infrared light – heat emitted by their prey – while many birds and insects can see into the ultraviolet part of the spectrum. This gives a richness to their perception of colour that is hard for us to imagine – enabling them, for example, to see things such as patterns in the petals of flowers which are invisible to us. But there is something else even stranger than light pouring from the Sun, and it is entirely hidden from us.

In certain circumstances, humans can see some infrared light. It happens when pairs of infrared photons 'double up', and hit the same pigment protein in the eye at the same time. Subjects report seeing infrared light from a low-energy laser.

Neutrinos, like photons, are elementary particles generated during fusion in the core of the Sun, as well as by other events in the universe. But, unlike photons, neutrinos pass straight through us with no discernible effect. Escaping instantaneously from the Sun without any of the delay experienced by photons, and travelling at fractionally less than the speed of light, trillions of them are flying through you every second. Even at night neutrinos from the Sun are whizzing through you, but this time from below, having first passed through the Earth. To neutrinos we may as well be ghosts.

Neutrinos were dreamed of before they were detected. In 1930, seeking to explain the conservation of energy and momentum when a proton is transformed into a neutron (a phenomenon called beta decay), the physicist Wolfgang Pauli found that he needed to posit the existence of an entirely new, invisible and hitherto unimagined particle. It was a wild idea at the time, and neutrinos themselves are no less strange. For one thing they are amazingly small – a tiny fraction of the mass of the next least massive elementary particle, the electron. For another, once created they interact with very little else. It would take, for instance, an average of a thousand light years (9,500 trillion kilometres) of lead to stop one. That's just the average, however, and very occasionally a neutrino will strike an atom in a much less massive and dense object. It is by detecting these rare events that we know for sure that they exist.

The first neutrinos were observed in the 1950s. They came not from the Sun but from the explosions of supernovas – massive stars at the end of their lives – in deep space. When a typical supernova explodes it unleashes an octodecillion, or 10^{57}, neutrinos. The Sun's neutrinos were not detected until the 1960s, when a physicist who shares his name with *Kinks* frontman Ray Davis oversaw the construction of a 'telescope' consisting of a hundred thousand gallons of cleaning fluid deep in an old mine in South Dakota. On average, one neutrino each day would interact with an atom of chlorine in the fluid, turning it into an atom of argon. Amazingly, Davis worked out a way to find the argon.

There is, on average, one supernova explosion per galaxy per century. In the observable universe about a billion explode every year. That is, thirty per second. The universe bubbles like champagne.

But there are even stranger things about neutrinos than their tiny size and their elusiveness. One is their changeable nature. The type, or 'flavour', of a given neutrino is never fixed. Instead, it oscillates between three different states as it flies through space. If a neutrino does interact with ordinary matter, it converts into one of three different types of charged particles with different properties depending on which part of its oscillation it happens to be in. Another mystery is the question of how neutrinos (unlike photons) have mass. According to the Standard Model of particle physics, particles must exist in both 'left-handed' and 'right-handed' versions if they are to have mass. But only left-handed neutrinos have been observed – a riddle to match the koan about the sound of one hand clapping. A solution to this, if there is one, may help reveal why is there more matter than antimatter in the universe.

Dark wonder: black holes

A simplistic description of Hawking radiation goes something like this: every cubic millimetre of space in the universe, no matter how empty it seems, is actually a chaotic arena of fluctuating fields, with pairs of particles and anti-particles such as positrons and electrons flickering in and out of existence. (To adapt Heidegger, *das Nichts etwast*, or 'the nothing somethings'.) Normally, the particle–antiparticle pair annihilate each other within about a billionth of a billionth of a second. But near the horizon of a black hole it's possible for one of the pair to fall in before the annihilation can happen, in which case the other escapes as Hawking radiation.

Another darkness at the edge of understanding concerns the nature of black holes. One way these celebrated anti-objects come into being (or non-being) is when a star of sufficient mass – typically more than about twenty-five times that of the Sun – burns up all its fuel and, with fusion no longer pushing energy outwards, collapses in on itself. It then explodes as a supernova, flinging electromagnetic radiation and neutrinos into space in huge quantities; but at the same time, the core collapses inwards until it becomes a singularity: a region where matter is infinitely dense and space-time is infinitely curved. At this stage the laws of physics as we know them run into trouble. General relativity predicts black holes, but quantum mechanics predicts something called Hawking radiation at their event horizons (the boundary between the black hole and the rest of the universe) – a phenomenon that appears to be incompatible with general relativity. At any rate it is unclear how to reconcile the two. The puzzles go further, and have led physicists to astonishing hypotheses. Some have suggested that black holes may end their lives by transforming into their exact opposite – 'white holes' that explosively pour all the material they ever swallowed back into space. Others have argued, variously, that our universe could look like a black hole to people in another universe; that new universes are continually being created within black holes; and that a hyper-black hole spawned our universe – meaning that the Big Bang was a mirage created by a collapsing higher-dimensional star. Yet others have computed the internal energy of a black hole, the position of its event horizon and other properties to indicate that gravity arises from infinitely thin vibrating strings which exist in ten dimensions, with our universe merely a 'hologram'.

Some black holes also create the brightest known objects in the universe. They spin, and in doing so they twist the encircling space-time around themselves, creating a maelstrom around an infinitely thin ring instead of a point. This pulls the mass of nearby gas, dust, stars and planets from the surrounding

In the giant galaxy Centaurus A (NGC 5128), jets ejected perpendicular to its plane are signatures of a supermassive black hole at its centre.

galaxy towards them, setting them spinning, accelerating them and tearing them apart, and in the process, they create vast magnetic fields and enormous heat. The magnetic fields shoot jets of particles out into intergalactic space at right angles to the plane of rotation, and at close to the speed of light, for thousands and sometimes millions of light years. If the black hole is massive enough, the gravitational shearing and friction in its accretion disc can produce more heat and light than anything else in the universe. This is called a quasar and its radiation covers the entire electromagnetic spectrum, from radio waves and microwaves at low frequencies, through infrared, ultraviolet and X-rays, to high-frequency gamma rays.

The brightest known quasar, memorably named S5 0014+81, is three hundred trillion times brighter than the Sun, or more than twenty thousand times brighter than all the stars in the Milky Way galaxy combined. At the quasar's centre is a black hole 40 billion times the mass of the Sun and about ten thousand times more massive than the black hole at the centre of our galaxy. Actually we should say 'was', because S5 0014+81 is over twelve billion light years away, so telescopes show it to us as it was more than twelve billion years ago.

At the time of writing, every image of a black hole is a product of human ingenuity and imagination rather than a picture of the real thing. The rendition in the 2014 film *Interstellar*, created with the help of the physicist Kip Thorne, shows a funnel or sphere of absence that bends the light from the stars behind it and to its sides (a phenomenon called gravitational lensing), and is surrounded by a glowing accretion disc of gas around its equator which (by the same lensing effect) appears to bend into a 'rainbow of fire' across its top. This is probably a fairly accurate representation of at least part of what you would see, except that the light around a spinning black hole would appear much brighter on the side turning towards you than on the side turning away, creating an effect more like a smooshed crescent moon than a halo. Thorne deliberately avoided this asymmetry, fearing it would confuse cinemagoers, but it can be seen clearly in an image created back in 1978 by the astrophysicist, writer and poet Jean-Pierre Luminet. Employing what was already a long-obsolete 1960s IBM 7040 computer which used punch cards, Luminet had no way to visualize the results on a screen so he used the data to draw an image by hand, putting individual dots of India ink onto a photographic negative.

It may be that astronomers will have captured the first actual pictures of a black hole by the time you read these words. The technical challenge is enormous: the nearest black hole is thought to be hidden in a bright and compact astronomical radio source called Sagittarius A* about 26,000 light years away in the centre of our galaxy. At that distance it is about as big as a bagel on the surface of the Moon, and it will require a telescope with a resolution more than a thousand times better than the Hubble Space Telescope to produce an image. As this book went to

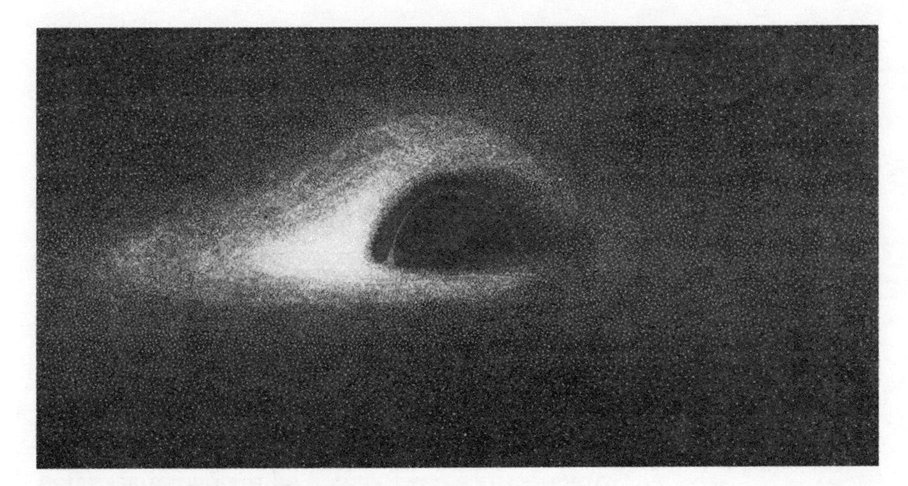

Black hole by Jean-Pierre Luminet (1978).

press, a global collaboration called the Event Horizon Telescope had created just that. With radio-telescopes distributed from Hawaii to Spain, and from Arizona and Chile to the South Pole, it is hoping to create from their pooled data what would in effect be a single telescope with an aperture as wide as the diameter of the Earth.

Other techniques are likely to greatly increase understanding of black holes and in turn the capability to visualize them. In 2016 a team of nearly a thousand scientists astonished and delighted almost everybody by recording a 'bleep' thought to be caused by the merger of two black holes 1.3 billion light years away. The Laser Interferometer Gravitational-Wave Observatory, or LIGO – an apparatus consisting of two sets of two four-kilometre-long arms set at right angles to each other and located nearly five thousand kilometres apart – recorded a change in the position of one array with respect to the other by a ten-thousandth the diameter of a proton. It was equivalent to measuring a change in distance to the nearest star by the width of a hair. With their findings, the team confirmed the existence of gravity waves – ripples in the curvature of space-time that propagate at the speed of light and were first predicted by Einstein a

Electromagnetic waves can also be observed by artificially transforming them into sound waves. The astrophysicist Wanda Diaz Merced, who became blind in her twenties, investigates the energy and light released by gamma-ray bursts, the most violent events in the universe, by transforming light curves and data sets into sound. By listening to variations in pitch, duration and other sound qualities, she decodes patterns in burst-like interstellar light.

hundred years ago. They also created a simulation of the supposed cause of the bleep – an animation in which two black holes circle each other ever closer and faster until suddenly they merge into one, like bubbles meeting in the vortex above a plughole. In future, LIGO and systems that exploit the same principles are likely to make it possible to investigate black holes, neutron stars and other 'dark' phenomena with even greater precision and detail. Astronomers will increasingly listen as well as look at the universe.

Black holes are not the only kinds of darkness at the edge of understanding. If some theories are right, dark matter and dark energy are also on that horizon. Together, these two are thought to account for more than 95 per cent of all the mass and energy in the universe, and yet both currently escape direct detection. Dark matter does not interact with any part of the electromagnetic spectrum and is therefore invisible. Its existence is inferred from gravitational effects – notably, from the fact that galaxies move more quickly and create greater distortion through gravitational lensing than the ordinary matter in them can account for. Similarly, the existence of dark energy is inferred only from its effect, which is to accelerate the expansion of the universe. A leading hypothesis on the nature of dark matter, at least until recently, has been that it is composed of weakly interacting massive particles, or WIMPs, that exert influence only through gravity and the weak nuclear force. This idea has proved to be extremely successful in accounting for the observed history of the cosmos. But the failure, so far, to find any trace of these particles has raised doubts. Perhaps, it has been suggested, dark matter is a superfluid: a Bose-Einstein condensate that could account for quantum entanglement (or what Einstein called 'spooky action at a distance'), which, being instantaneous, seems to be faster than light (though it may, in fact, be more meaningfully characterized as existing outside time). Some physicists now doubt that dark matter and dark energy actually exist, suggesting that theories such as Modified Newtonian Dynamics, in which gravity no longer weakens with distance, will account for observed effects. Future research may tell. For now, we may be almost as much in the dark as those who, thousands of years ago, wondered about the nature of light.

Many other questions are yet to be answered. What, for example, lies in darkness beyond the edge of the visible universe? Is there an infinite extension of the same – ever more galaxies arranged into a cosmic web of stupendous beauty? Or is the universe as we know it ultimately limited in extent – although, like the surface of a sphere, unbounded? If the latter is the case then could our universe, of which we see only a small part, be one bubble among many?

Henry Thoreau describes walking on a November day just before dark when the Sun breaks through grey cloud, and the land, touched with 'such a light as we could not have imagined a moment before', becomes a paradise. For Thoreau, the fact that these same conditions will recur on an infinite number of evenings in the future makes it more glorious still. His sentiment is easy to share: the Sun may be finite within a cosmological time frame, but measured against the scale of human history it is unending.

'late sunlight enters
the deep wood,
shining over the
green moss again.'
Wang Wei

When you descend more than about two hundred metres below the surface of the sea, the water is said to turn the deepest blue imaginable, described by the deep-sea pioneer William Beebe as 'luminous black'.

On the morning in early October when I write this, unusually bright sunshine pours down. Almost a month has passed since a day on which I last stepped into light of comparable brilliance, wheeling my father, who had been close to death, out of his hospital ward into the open air. Rising briefly from the oblivion into which he had almost disappeared, he said how beautiful it was. Today, again, it feels like the dearest, most wonderful gift. 'I cannot understand time,' wrote Richard Jefferies. 'It is eternity now. I am in the midst of it. It is about me in the sunshine.' The deep sky above me must be what the nineteenth-century art critic John Ruskin called a visible heaven, and I stand here, hoping to store some of its strength for the dark months ahead.

For a particle of light, or photon, this is literally true. Time stands still such that past, present and future all collapse down to one eternal moment.

'Half our days we pass in the shadow of the earth,' wrote the seventeenth-century physician Thomas Browne, 'and the brother of death exacteth a third part of our lives.' For we the living, sleep is the death of each day's life, but on the morrow, governed by the circadian rhythms that our ancestors have followed

'The unfathomable
deep
Forest where all
must lose
Their way.'
Edward Thomas

since the Proterozoic, we wake and the morning light is a daily grace. Light from the kind old Sun cannot restore the dead to life (although the near-infrared part of its spectrum can help heal wounds and relieve pain). One day the Sun itself will die. For now, it shines in glory and allows us to see the light in the eyes of other waking souls.

2

THE GATHERING OF THE UNIVERSAL LIGHT INTO LUMINOUS BODIES

Life

The world, though made, is yet being made . . .This is still the morning of creation.

John Muir

Life is a self-sustaining chemical system capable of incorporating novelty and undergoing Darwinian evolution.

Gerald Joyce

It rains the same old rain, the same old rain that it rained on the dinosaurs.

Nick Cope

As the lights go down the final members of the audience take their seats around a circular stage. The theatre is full of the liquid notes of tropical birds and, distantly, the growls of strange animals. Edging the stage and facing inwards are large electric fans: propellers inside iron cages the size of car wheels. Inside the circle, a woman in a long black coat puts a thin plastic bag on the ground and, with scissors and tape, reconfigures it into a two-dimensional figure whose arms are flattened tubes and whose head is a small square sack. The bag's handles become its legs and feet, which she weighs down with two small coins. Then the woman places the plastic figure in the centre of the circle, and the fans begin to turn. Over the sound system a flute slides up and down through a tritone twice before stretching upwards and falling through an octave. It is the opening of Debussy's *L'après-midi d'un faune*. Hesitantly and slowly – as harps, woodwind and strings join the flute and the music slips through A sharp, E, G sharp, C sharp, B flat, D and back to E, where an oboe takes up and develops the theme before passing it back to the flute – the bag starts to pivot around its weighted feet in the breeze rising from the fans. By the time the flute slides onto C sharp to restate the opening theme for a fourth time and is echoed by a second flute, the little bag has stood up and begun to move around. Soon, it starts to leap and dance for all the world like a sprightly homunculus. Over the next few minutes, as the music drives forward and becomes ever more animated, with a new melody carrying the entire orchestra higher and higher before gently falling away in swirling phrases, the woman whips one plastic-bag man after another from her pockets, like a magician whipping out kerchiefs, and each one inflates and becomes part of a great column of figures rising, dancing and playing around and above her head. The spectacle is comic – a few old plastic bags in a breeze are somehow matching Debussy's gorgeous, playful sound world – but it is also entrancing, and a thing of wonder.

In the beginning

Many creation myths suppose a supernatural agent – like a thaumaturge or theatrical director, but with vastly greater powers – who fashions life out of base materials. In China, for example, the goddess Nüwa made the first humans from yellow earth and mud. The English word for human derives from the Indo-European word for soil, from which we also get 'humus'. In Hebrew *adam*, meaning 'man', is cognate with *adamah*, 'ground': biblical man is made from clay – undifferentiated material into which the deity breathes life.

'I have scarcely touched the clay, and I am made of it.' Antonio Porchia

These days we can be more specific with regard to the materials from which life on Earth is made. We can list our contents like a tin of soup, making us mostly (about 57 per cent) water, plus proteins, carbohydrates (or sugars), lipids and nucleic acids. Or we can itemize our elemental ingredients according to the quantity of each kind of atom. Just three – hydrogen, oxygen and carbon – account for 93 per cent of our bodies, and a further three – nitrogen, calcium and phosphorus – for a further 6 per cent. The last of these was among the first elements to be identified. In 1685 the diarist John Evelyn described a demonstration with phosphorus that had been extracted from human blood and urine:

Most of the atoms in our bodies – about 62 per cent – are hydrogen, but because they are so much smaller than other atoms they are only about 10 per cent of our mass. About 24 per cent of our atoms are oxygen but they are 65 per cent of our mass. Carbon atoms are 12 per cent of all the atoms in our bodies and 18 per cent of our mass.

> Dr Slayer showed us an experiment of a wonderful nature, pouring first a very cold liquor into a glass, and superfusing on it another, to appearance cold and clear liquor also; it first produced a white cloud, then boiling, divers coruscations and actual flames of fire mingled with the liquor,

which being a little shaken together, fixed divers suns and stars of real fire, perfectly globular, on the sides of the glass, and which there stuck like so many constellations, burning most vehemently, and resembling stars and heavenly bodies, and that for a long space. It seemed to exhibit a theory of the eduction of light out of the chaos, and the fixing or gathering of the universal light into luminous bodies.

The smallness of the proportions of some elements does not mean they are insignificant. We are only 0.0042 per cent iron, for example, but without it we would die immediately.

Six elements – hydrogen, oxygen, carbon, nitrogen, calcium and phosphorus – make up 99 per cent of the human body. Most of the final one per cent is potassium, sulphur, sodium, chlorine and magnesium, while more than half a dozen other elements – including iron, copper, zinc, selenium, nickel, manganese, molybdenum and cobalt – are present in trace amounts. But where did these elements come from?

The story of stuff

The Big Bang might actually have been a Big Bounce. 'Our universe,' says the physicist Carlo Rovelli, 'could be the result of a previous contracting universe passing across a quantum phase, where space and time dissolved into probabilities.'

If you want to make an apple pie, said the astronomer Carl Sagan, you must first invent the universe. To this you can add: if you want to understand the origin of life on Earth you must first know the story of stuff. At present the furthest we can go back in that regard is to the Big Bang, about 13.8 billion years ago. At that time the universe was almost inconceivably hot and dense. Quarks and other elementary particles such as electrons started to form within a billionth of a second, and after a few millionths of a second, quarks aggregated in threes to make protons and neutrons. About three minutes later, in a process called primordial nucleosynthesis, the nuclei of what would become hydrogen and helium began to coalesce from these protons and neutrons. As the universe continued to expand and cool, things began to happen more slowly. By 380,000 years after the Big Bang most electrons became bound in orbits around these nuclei, forming hydrogen and helium atoms. And, with electrons now bound into atoms, photons were able to travel freely. This was the time of first light, and its trace is still visible as the cosmic microwave background.

Millions of years later, as the universe cooled further, gravity began to pull together stars and galaxies out of clouds of molecular gas. Over time and ever since, heavier elements such as carbon, oxygen, nitrogen and iron have been forming, transmuting continuously in the alembics of successive generations of stars. In the triple-alpha process, which takes place inside main sequence stars, three helium atoms (which have two protons and two neutrons each) are transformed into a carbon atom (six protons and six neutrons), which may then fuse with an additional helium atom to produce oxygen (eight protons and eight neutrons). In the explosive stages towards the end of a star's life, it forges elements of increasing atomic mass all the way up to iron (twenty-six protons and, typically, thirty neutrons).

If a star is massive enough, the collapse leads to a rebound and to the explosion called a supernova, which briefly outshines an entire galaxy, radiating as much energy in weeks or months as an ordinary star such as our Sun does in billions of years. The pressures and temperatures in the short period before a supernova fades vastly exceed anything during the star's previous existence, and produce lots of iron, as well as more massive elements, including at least one – iodine (fifty-three protons) – which is also essential to life as we know it. Phosphorus, another element essential to life, is made in especially large supernovas called hypernovas. Boron – an element that plants (and possibly animals) need – is created when cosmic rays, which are the highest-known energy particles in the universe and which originate in massive explosions of this kind, strike a heavier element and blow it apart. Gold, which is in the nice-to-have rather than the essential-to-life category, is probably made in the ultraviolent collision of neutron stars, the densest-known things in the universe short of black holes.

As supernovas and hypernovas explode, they hurl the elements they have created across space into the interstellar medium – a 'mist' that is mostly hydrogen and helium but also one part in a hundred heavier

The cosmic microwave background is detectable as very slight differences in temperature across the entirety of deep space. It is the oldest light in the universe: an echo, and a map, of the distribution of matter and energy about 380,000 years after the Big Bang. It presents a cosmological analogue for something the psychologist William James said about thunder: 'The feeling of the thunder is also a feeling of the silence as just gone.'

The oldest known stars, such as HD 140283 (which is nicknamed the Methuselah Star) and SM0313, are about 13.6 billion years old.

The Origin of the Solar System Elements

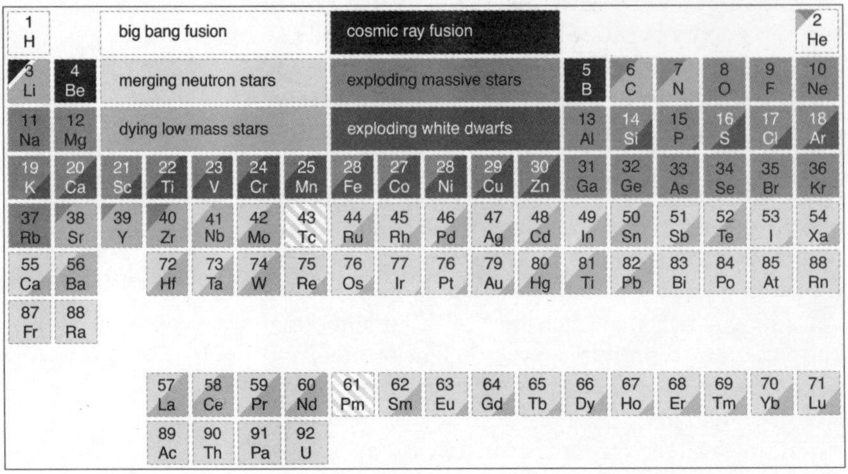

1 H		big bang fusion					cosmic ray fusion										2 He
3 Li	4 Be	merging neutron stars					exploding massive stars					5 B	6 C	7 N	8 O	9 F	10 Ne
11 Na	12 Mg	dying low mass stars					exploding white dwarfs					13 Al	14 Si	15 P	16 S	17 Cl	18 Ar
19 K	20 Ca	21 Sc	22 Ti	23 V	24 Cr	25 Mn	28 Fe	27 Co	28 Ni	29 Cu	30 Zn	31 Ga	32 Ge	33 As	34 Se	35 Br	36 Kr
37 Rb	38 Sr	39 Y	40 Zr	41 Nb	42 Mo	43 Tc	44 Ru	45 Rh	46 Pd	47 Ag	48 Cd	49 In	50 Sn	51 Sb	52 Te	53 I	54 Xa
55 Ca	56 Ba		72 Hf	73 Ta	74 W	75 Re	76 Os	77 Ir	76 Pt	79 Au	80 Hg	81 Ti	82 Pb	83 Bi	84 Po	85 At	88 Rn
87 Fr	88 Ra																
			57 La	58 Ce	59 Pr	60 Nd	61 Pm	62 Sm	63 Eu	64 Gd	65 Tb	66 Dy	67 Ho	68 Er	69 Tm	70 Yb	71 Lu
			89 Ac	90 Th	91 Pa	92 U											

The elements in our solar system were made in collapsing, merging and burping stars as well as exploding ones.

atoms. The mist is thin – at about one atom per cubic centimetre, a more complete vacuum than has ever been achieved on Earth – but it is a hundred thousand times denser than the space between galaxies. And where it is relatively concentrated in molecular clouds, the constituents begin to exert mutual gravitational attraction and sometimes draw together enough material to collapse into a new generation of stars and planets.

The story of the Earth begins around 4.6 billion years ago when the 0.04 per cent of the mass of solar system that was not a part of the Sun formed a disc of dust around it. Amazingly, astronomers have recently photographed a protoplanetary disc that probably looks like ours did at that time. HL Tauri, which is about 450 million light years away from Earth, is only about a million years old (as we see it), but already its disc appears to be full of forming planets. And it was from such a disc that the proto-Earth, which some call Tellus, first formed out of the debris into

A protoplanetary disc around the star HL Tauri..

a sphere about the size of Venus, or a little smaller than the Earth today. Then, some tens of millions of years later – according to the giant impact hypothesis – another planet about the size of Mars, which astronomers call Theia after the Greek goddess who was mother of the Moon, struck the Earth in what is now thought to have been a head-on collision rather than a glancing blow. The impact by an object about a tenth of the Earth's present mass released about a hundred million times more energy than the Chicxulub impact that wiped out the dinosaurs, and it was enough to melt both planets and mix them together. Think of the punches to the face taken by Robert De Niro as Jake LaMotta in *Raging Bull* and then some. A great chunk sheared off to become the Moon, while the remaining mass quickly regained its spherical shape because of gravity. The blow had, however, tilted the axis of the planet to a 23.5° angle – giving rise, ultimately, to what we know in high latitudes as the seasons.

All this took place at the beginning of Earth's first eon, the Hadean, which lasted from 4.54 billion to 4 billion years ago. The name, familiar to many as Hades for the Greek underworld and its god, actually derives from a word for 'the unseen', and this is particularly appropriate because the enormous stretch of Hadean time has left next to no visible traces. Unseen does not, however, mean completely unknowable, and researchers can construct plausible scenarios for the momentous changes that must have taken place – making reasonable inferences to suggest subdivisions to the eon with names such as the Procrustean and the Promethean. A detailed animation or a virtual-reality production based on these deductions would be at least as compelling as anything in Terrence Malick's *Voyage of Time* or Werner Herzog's *Into the Inferno*.

'The Earth rocks: thunder, echoing from the depth, roars in answer; fiery lightnings twist and flash . . . Sky and sea rage indistinguishably.' Aeschylus, *Prometheus Bound*

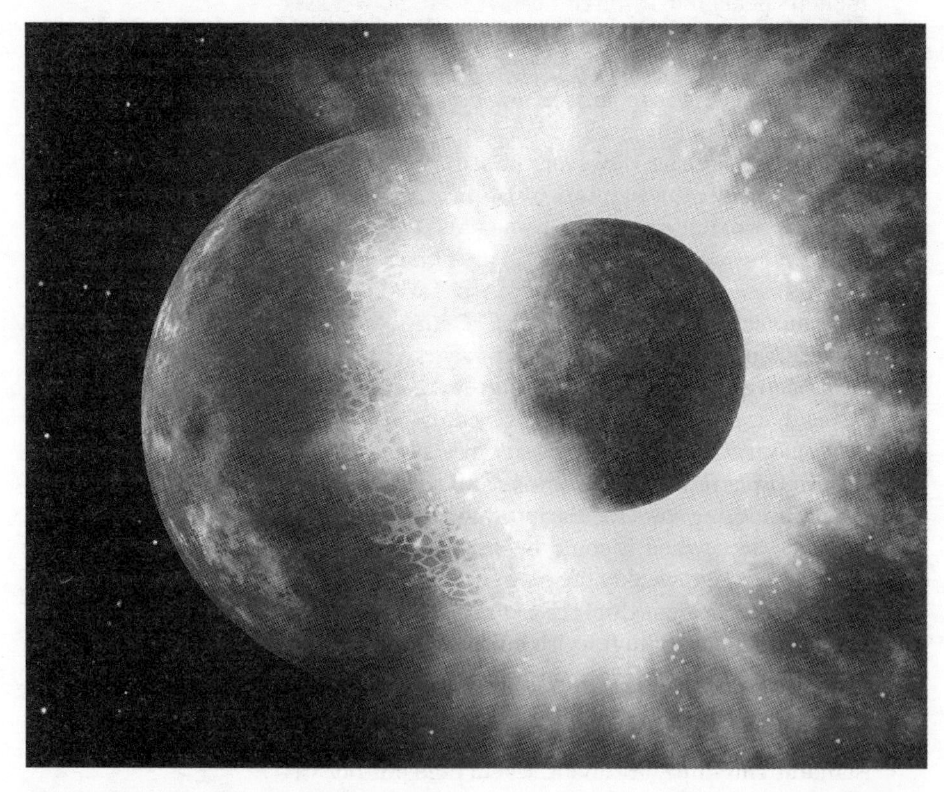

Hypothesized impact of Theia on Earth

Initially, a thick cloud of intensely hot vaporized rock surrounded the reformed Earth, but after a few thousand years this cooled and condensed, and as little as 10 million years later the planet itself had cooled enough for rock to form a solid surface crust mostly covered by liquid water and surrounded by an atmosphere composed largely of nitrogen and carbon dioxide. But the Earth was not out of the wars yet. In a period known as the Late Heavy Bombardment, from about 4.1 billion to 3.8 billion years ago, it was probably hit repeatedly by planetesimals of various sizes. Thousands of them were around twenty kilometres across – tiny compared to Theia, but much bigger than Chicxulub. A few may have been five hundred kilometres or more across: big enough to vaporize huge regions of the Earth's rocky surface and to evaporate much or even all of the ocean, leaving only molten salt behind. If you could have seen through the opaque atmosphere, the sight might have resembled the desolation that Moomintroll and his friends find on the seabed in Tove Jansson's *Comet in Moominland* when the ocean burns away. Within a few thousand years of each impact, however, the atmosphere would cool again and rains of unimaginable intensity would refill the oceans.

Our view of the origin of life is as cloudy as a sky in the late Hadean or the early Archaean (the eon that succeeded it). But one thing is sure. Stepping onto this planet about four billion years ago would have been quite an experience. The Earth spun much faster than today, and night followed day every five or six hours. The stars were seldom visible because the atmosphere was full of smoke and dust, but spectacular shooting stars regularly whizzed through the sky. The Sun, when it could be seen at all through the smog, shone weakly. The Moon was only about a third of its present distance from the Earth, and it would have looked huge, with an apparent diameter almost three times larger and an area eight times larger than today.

Ever since it was sheared off from the Earth, the Moon has been gradually getting farther away as it endlessly circles us. At present it is about 384,400km away, and retreating at 3.8 centimetres a year, meaning it has retreated nearly 2 metres since the Apollo landings. If the Moon suddenly disappeared, a lot of the water it currently attracts towards the Earth's equator would be redistributed to the polar regions and the Earth's rotation would become much more erratic, with drastic impacts on regional climates.

Imagine standing on the rocky shore of a volcanic island at this time. At about seventy degrees centigrade, the air is hotter than Death Valley but cooler

The starting temperature at the World Sauna Championships in Finland in 2010, the year before it was abolished, was 110°C. Half a litre of water was poured on the stove every thirty seconds. The winner was the last person to stay in the sauna and walk out without outside help. Alcohol consumption was prohibited prior to and during the competition.

than a sauna. The atmosphere is mostly nitrogen and carbon dioxide. You need both a cooling and a breathing apparatus. In the distance, you can see other islands rising from the sea, some of them active volcanoes. The rocks beneath your feet are made of dark larva, and volcanic ash fills the crevices. Hot springs boil nearby. The sea water has a greenish tint from all the unoxidized iron it contains. White deposits of dried salt on the lava rocks show where small tide pools have evaporated. Freshwater ponds a few metres above the beach are constantly being filled by small streams of rainwater cascading down the hillside, then drying out in the heat. Suddenly the landscape is brilliantly illuminated as a blinding white streak silently crosses the sky and falls into the sea just over the horizon. An asteroid about a hundred metres across has penetrated the atmosphere at twenty kilometres per second and crashed into the ocean several miles away – one of many such impacts to occur every day. A thin dark line on the ocean advancing towards you is the resulting tsunami. If you move to higher ground in time and escape the flood you may live to witness the colossal tides, ebbing and flowing a hundred vertical metres or more, under the pull of the colossal Moon.

Water

'Like all profound mysteries,' wrote Nan Shepherd in *The Living Mountain*, her 1940s meditation on the Cairngorms, 'water is so simple that it frightens me. It wells from the rock, and flows away. For unnumbered years it has welled from the rock and flowed away. It does nothing, absolutely nothing, but be itself.' Shepherd's vision is compelling but – in this passage at least – she overlooks a key property of water. For although water seems to flows endlessly, indifferent to our presence, it is also involved in everything.

'notice . . . how the little animal wins its way up against the stream, by alternate pulses of active and passive motion, now resisting the current and now yielding to it in order to gather further strength and a momentary fulcrum for a further propulsion.'

Recalling childhood days by the river Otter in Devon, Samuel Taylor Coleridge (1772–1834) compared the mind's self-experience in the act of remembering to the movement of water boatmen as they skate across the surface of a pool. The image is apt in more ways than he knew. A typical human brain floats within a bubble of cerebrospinal fluid, which is 99 per cent water, and thus exists at neutral buoyancy. Our thoughts, like water boatmen, are suspended on water.

The properties of water that make both the insects and humans possible arise from the configuration of the atoms in each molecule. The much larger single oxygen atom sports the two smaller hydrogen atoms like Mickey Mouse wears his ears and, having borrowed an electron from each of them, assumes a negative electrical charge while the hydrogens assume a positive one. The result is a molecule with negative and positive poles (Mickey's chin and ears respectively) and hence a readiness to bind, negative to positive, to other water molecules. This intermolecular bonding is what creates phenomena such as the surface tension which supports pond skaters and capillary action, which plants and trees use to lift sap up their stems. It also facilitates the rapid atmospheric water cycle on Earth because it holds raindrops together, and is what makes ice so strong.

Nonpolar volatile molecules, like methane on Titan and carbon dioxide on Venus, can't form droplets, so 'rain' does not exist there, just a constant, dreary mist . . . like England in February but with more charm.

Water molecules like to bind to each other, but when mixed with molecules of many other substances in a liquid state they exhibit another remarkable property: they are an excellent solvent. The negative and positive ends of each molecule attach to charged atoms (ions) in other molecules. This happens quickly with familiar substances such as sugar or salt, but it also happens in the long run with most types of rock, carrying them away in tiny increments over what humans perceive as vast periods of time. This unrivalled ability to dissolve and transport other chemicals makes water an ideal medium for life to come into being and to evolve.

'Wind and sea. Everything else is provisional. A wing's beat and it's gone.'
Kathleen Jamie

The origin of water was a mystery at the time when Shepherd was writing, and until the 1980s astronomers had little idea how it was created or how it was distributed in the universe. Now we know it to be an abundant by-product of star formation. This process is accompanied by a strong outrush of gas and dust, which creates shock waves that compress and heat the surrounding interstellar medium, producing molecular water (two parts hydrogen, one part oxygen) from atomic constituents of the medium. This newly created water can persist in space almost indefinitely or be swept up in the formation of new stars and their surrounding planets. Some research suggests that up to half the water on Earth is older than the solar system. Half the molecules in every raindrop may be older than the Sun.

In the last few decades astronomers have found the signature of water almost everywhere they have looked in space. A cloud of water surrounding a quasar in the ultra-luminous infrared galaxy APM 08279+5255 contains about 140 trillion times more water (surrounding a massive black hole) than all of Earth's oceans combined. The Orion nebula in our galaxy creates enough water every day to fill Earth's ocean sixty times over. Water also exists in giant molecular clouds between stars, in proto-planetary discs of dust around stars, and (it is thought) on many of the billions of planets orbiting other stars, known as exoplanets. Though many of these billions are much larger than the Earth, a fair number are likely to be Earth-sized and largely or wholly covered in water, either frozen solid or rolling in global oceans tens or even hundreds of kilometres deep.

Water is abundant in the outer planets of our solar system and their moons. Jupiter, Saturn, Uranus and Neptune all have significant amounts of water vapour in their dense atmospheres. Jupiter's moons Europa and Callisto have shells of ice many kilometres thick over deep oceans of water encircling a rocky core. The rings of Saturn are almost entirely made from shards of water ice, forever turning and catching the light. Saturn's moon Enceladus is also sheathed in ice. But beneath this ice is an ocean, heated by the force of gravity from the giant planet that grinds the moon's interior, and Enceladus sprays liquid water into space through geysers that crack its surface. The water either falls back to the surface or escapes to form Saturn's E Ring. Miranda, the smallest and innermost of the five moons of Uranus, is mostly made of water. Its five-kilometre-high ice cliffs are the tallest known in the solar system.

In 2015 the Cassini spacecraft flew through the water plumes of Enceladus and found them to be rich in molecular hydrogen – evidence that the Moon's underlying sea could support microbial life.

Pluto and its largest moon Charon, turning around a common centre like a couple jitterbugging on a dance floor, are both clad in rugged landscapes that astonished astronomers when they first saw them in

2015. On Pluto, rock-hard mountains of water ice float in relatively soft, mushy nitrogen glaciers, like blancmange. (Far beneath the surface there may be a liquid ocean of salty water some one hundred kilometres deep.) Unlike the mythical beings after which they are named, Pluto and Charon are a vision not of death (which would imply they had once supported life), but rather of a world where life has never dawned. This is not the locked-in un-death which the spirit of Achilles in Hades describes to still-living Odysseus, or the terrible underworld of Philip Pullman's *Amber Spyglass*. It is pure process: serene and simple.

Surface water on most of the inner rocky planets of the solar system tends to be scarce – although there is, amazingly, ice at Mercury's southern and northern poles, permanently shaded in craters from daytime temperatures of over 400°C. Venus may once have had abundant water but virtually all of it boiled away in a runaway greenhouse effect long ago. And the northern plane of Mars was once covered by a great ocean. When meteorites slammed into it, tsunamis more than 120 metres high swept over its shores. But that ocean evaporated into space billions of years ago and what water remains on Mars is, as far as we know, almost entirely ice. The ice caps at the poles of Mars, which were first spotted by Giovanni Cassini in 1666, have a combined volume a bit greater than that of the ice cap on Greenland. Significant amounts of ice also sit just below the surface in high latitudes and there are scattered patches of surface ice in mid-regions as well as a small frozen sea, equivalent in volume to the North Sea, near the equator. Mars has some glaciers too, but the flowing liquid water that was discovered in 2015 seems limited to a few trickles of thick brine in crater walls and gullies. For the most part the landscape is one of dry channels, canyons and craters, rocks and dust. The only seas are sand.

Earth is, of course, unique in the solar system in that it is covered – seven-tenths of the surface to an average depth of four kilometres – in liquid water. For conditions favourable to life, our planet is the Goldilocks, while Venus is too hot and Mars too small. But the water on Earth's surface is only part of the

The only known body in the solar system besides Earth with large amounts of liquid on its surface is Titan, Saturn's largest moon. But the liquid of its lakes and seas is methane, ethane and propane, oozing at a balmy −179°C and possibly erupting every now and then with dramatic patches of bubbles.

story. Its presence in the planet's inward and nether parts plays a key role in the origin of life.

Though some arrived in comets and asteroids, water was probably present in and on the material from which the Earth formed, gradually escaping to the surface through volcanoes on the new planet. But a part of the water around us may also be a matter of continuous creation. Deep within the mantle, at 20,000 times atmospheric pressure and temperatures of around 1,400°C, silicon dioxide is thought to react with liquid hydrogen to form silicon hydride and liquid water that is then released in earthquakes.

On an extended visit to Malta in 1635 a Jesuit priest and scholar named Athanasius Kircher was entranced by the island's inland sea and, in particular, by its caves and long natural passageways winding deep into the rock. Stopping over in Sicily on his way back to Rome in 1638, he witnessed phenomena which, together with his experience on Malta, were to shape his outlook for life. Sailing through the Strait of Messina, his party encountered a whirlpool that Kircher described as a 'vast hollow' and then watched, astonished, as Etna issued huge clouds of thick smoke that entirely hid the island. Putting ashore, they heard a noise resembling 'an infinite number of chariots driven fiercely forward' and soon after were thrown onto the ground by 'a most dreadful earthquake'. With 'universal ruin all around', his party continued along the coast by sea, finding 'nothing but scenes of horror', and saw the volcano island of Stromboli belching flames with a rumble that was clearly audible even though it was a hundred kilometres away. Terrible as these events were – thousands of people died – they only increased Kircher's fascination with the 'miracles of subterraneous nature' and how the phenomena he had seen in Malta and Sicily might be linked together.

Kircher went on to became a polymath – 'the master of a hundred arts', including a proposed cat piano that would play a different note by stretching the tail of a different cat for each key stroke. But among his greatest achievements is the *Mundus Subterraneus* (*Underground World*) a strange and beautiful work published in 1665 that laid out, along with much else, a theory of the deep workings of the Earth. Volcanism, Kircher proposed, was caused by the circulation of great fires in the 'hollowed rooms and hidden burrows' all the way down to the Earth's core. Volcanos were the 'fire-vomiting vent-holes, or breath-pipes of Nature'. In addition, he said, the tides pushed immense quantities of water through various 'hidden and occult passages at the bottom of the Ocean' and thrust them 'forcibly into the intimate bowels of the Earth'. Somewhere off the coast of Norway, he claimed, the sea drained down a huge maelstrom,

The Moskenstraumen, a system of eddies and whirlpools in the Lofoten archipelago off the Norwegian coast, appears on Olaus Magnus's Carta Marina of 1539 and in Edgar Allan Poe's 1841 story 'A Descent into the Maelström'. It is one of the strongest such systems in the world, but nothing like as strong as Poe imagined. Nor does it drain the sea as Kircher supposed.

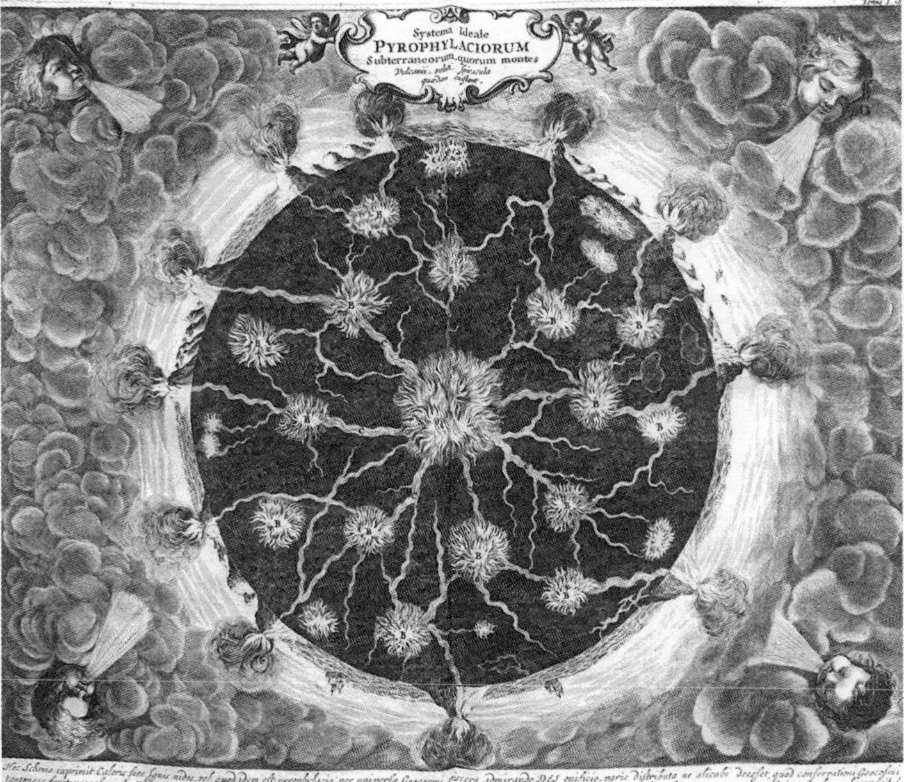

'Vulcan hath his elaboratories, shops and forges in the profoundest Bowels of Nature' *Mundus Subterraneus* (1665).

from where it ran through the Earth, which cooled it and provided it with nutrients, before being expelled through a nether opening at the South Pole. More than once, Kircher compares the movement of the Earth's water to the circulation of the blood in the body as described by his near-contemporary the physician William Harvey.

Though it was wildly wrong in many particulars (water does not, for example, rush in immense liquid flows deep inside the Earth), Kircher's vision was a step in the right direction. Volcanoes along the same plate boundary, such as Etna and Stromboli, *are*

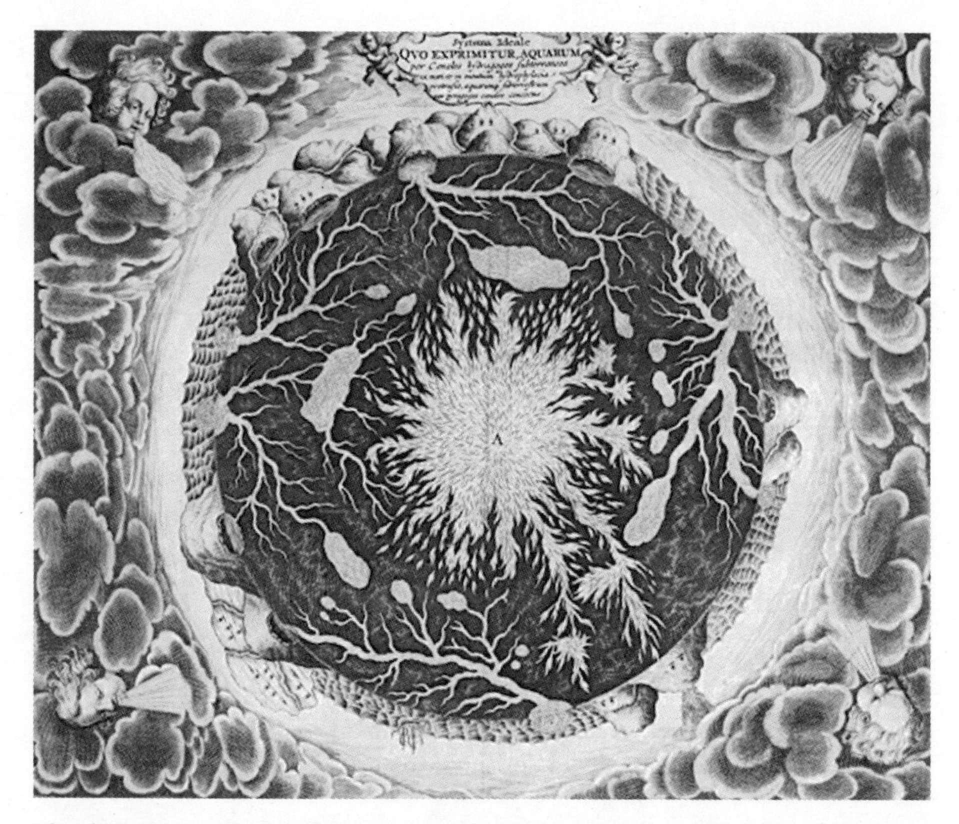

'The tides push an immense bulk of water through hidden and occult passages at the bottom of the Ocean, and thrust it forcibly into the intimate bowels of the Earth.' *Mundus Subterraneus* (1665).

connected. And in the last few decades scientists have determined that, on a geological timescale, water *is* carried down beneath the Earth's crust into the mantle. Something like one to three oceans' worth is packed into rocks in the transition zone some 400 to 650 kilometres (and perhaps deeper) below the surface, where the pressure exceeds a hundred thousand atmospheres and the temperature a thousand degrees centigrade. Carried there by subduction as one continental plate slides under another, the water makes rock more viscous, rather as adding liquid to a cake mixture can do. As we will see later in this chapter, this process is linked to the origin and continued existence of life on Earth.

A diamond that was spat out of a volcano in Brazil 90 million years ago appears to have been created in the presence of water as far as 1,000 kilometres down.

You have to go very deep indeed to get to a place with no water. And even the Moon, which at first sight is as dry as a bone (and is in fact much drier than a bone, which is 20 per cent water), has enough water near its surface for NASA and others to be studying how it can be mined in order to extract oxygen and hydrogen fuel for future missions. Sampling undertaken by Wallace and Gromit during the Grand Day Out expedition of 1989 that indicated the presence of Wensleydale and other cheeses in the lunar regolith remains unexplained.

Three thousand kilometres beneath the Earth's surface is the core: a sea of liquid metal surrounding a super dense ball of solid iron and nickel a little smaller than the Moon, and bristling with a forest of iron crystals up to one hundred kilometres long.

Carbon

If water is the ideal medium for life then, to a first approximation, carbon *is* life as we know it. A fundamental reason for this is that carbon is exceptionally versatile: uniquely, its atoms stick both to each other and to other elements – notably, hydrogen, oxygen, nitrogen and sulphur with up to four bonds at once. Water ice always configures into the same crystal structure, but at a wide range of temperatures carbon atoms can make long chains, or interlocked rings, or complex branching arrangements, or almost any other shape, and these can be foundations for molecules with very different properties. Around ten million configurations are known – from the crystals that make diamond, one of the hardest substances there is, to those of graphite, which is soft and almost greasy to the touch – and this is a small fraction of the carbon compounds that are theoretically possible. In living forms, carbon is the backbone of amino acids, proteins, carbohydrates and lipids. The versatility and transitions Debussy achieved in his music pale by comparison.

All the carbon on Earth was forged from lighter elements in stars billions of years ago. This has been known for several decades. But one of the most extraordinary discoveries of the last few years has been just how widespread in the cosmos are carbon-rich molecules that could act as building blocks for life. For example, more than twenty per cent of the carbon in space is thought to be associated with polycyclic aromatic hydrocarbons, or PAHs, which feature interlocking rings of six carbon atoms. PAHs are widespread in interstellar dust, and probably started forming shortly after the Big Bang. Areas of galaxies such as the Milky Way, meanwhile, are rich with ethyl formate, a carbon-based molecule that gives raspberries and rum their distinctive scents. In 2012 researchers showed that PAHs can be transformed in the conditions associated with new stars and exoplanets into even more complex precursors to amino acids and nucleotides – the raw materials of molecules essential to life such as proteins and DNA. Also in 2012 astronomers identified a sugar called glycoaldehyde in a binary star system about four hundred light years from Earth. Glycoaldehyde, a sugar molecule, is needed to form RNA, the likely precursor to DNA. In 2014 researchers reported the first discovery in the interstellar medium of a carbon-rich molecule with a branched structure. This finding, the researchers wrote, boded well for the presence in interstellar space of amino acids, for which a branched structure is a defining feature. And in 2015 NASA scientists announced that samples of pyrimidine, a ring-shaped molecule of carbon and nitrogen that is found in meteorites, are transformed by high-energy ultraviolet light into three of the key components of DNA.

The emergence of order

'The self-assembly process seems to defy our intuitive expectation from the laws of physics that everything on average becomes more disordered.'
David Deamer

Life is more than a set of chemicals, however remarkable they may be. As the graphic on pages 90 and 91 shows, even the simplest living forms such as bacterial cells are mind-bogglingly complex, running intricate interactions among millions of molecules in myriad different, interdependent chemical reactions. How could such things possibly come into being without the help of some external organizing power? A further puzzle is that, at first sight, life seems to contradict the second law of thermodynamics – a fundamental tenet of physics that says that everything becomes more disordered over time: coffee gets cold, flowers wilt, time pulverizes our dreams, and socks vanish even though you definitely had them right *here*. How, then, does life endlessly renew itself through the generation, growth and evolution of new beings?

There are good answers to these questions, with the exception of the socks. With regard to complexity and order, the fact is that in the right circumstances relatively simple parts do interact and self-organize to produce patterns, processes and behaviours that are novel, complex and not necessarily predictable from the parts. Further, the apparent contradiction of the second law of thermodynamics actually turns out to be a way of facilitating its action. New, more complex and ordered patterns, processes and behaviours sustain themselves by creating a little local order within themselves at the expense of greater disorder in the wider environment. Like tax havens, but in a good way. It's called emergence.

Some researchers say it may one day be possible to characterize emergence as a physical law. Robert Hazen suggests it may be something along the lines of $C \geq \Sigma [n, i, \Delta E (t)]$. At any rate, emergence happens, and it has what the physicist Frank Wilczek calls the beautiful exuberance and productivity of a physical law.

Put like this it sounds simple, but the nature of emergence is actually poorly understood in many instances, and even when or if it is better explained, I am ready to bet that, like great music, it will still be amazing and wonderful for many of us.

Pattern and complexity emerge in non-living systems at almost every conceivable scale in the universe. Over billions of years galaxies tend to become more ordered, as their speed of rotation increases and their constituent material settles into a spinning disc. On Earth, in the right conditions, snowflakes form spontaneously by the million within seconds and minutes, their tiny ice crystals with sixfold symmetry precipitating out of water vapour in cold air, typically around tiny particles such as dust. Counterintuitively, the newly forming ice is actually hotter than its immediate environment because freezing releases heat. Following the second law of thermodynamics, the heat 'wants' to dissipate as fast as it can. New crystals accrete fastest at those places where heat is removed the fastest, and these tend to be at the corners of the hexagons. This creates positive feedback: the locations on a crystal that grow the fastest become spikier, which allows better heat transfer away from these locations, allowing them to

Ice crystals are hexagonal because individual molecules are shaped like tetrahedrons. As water freezes, these tetrahedrons come closer together and crystallize into a hexagonal structure.

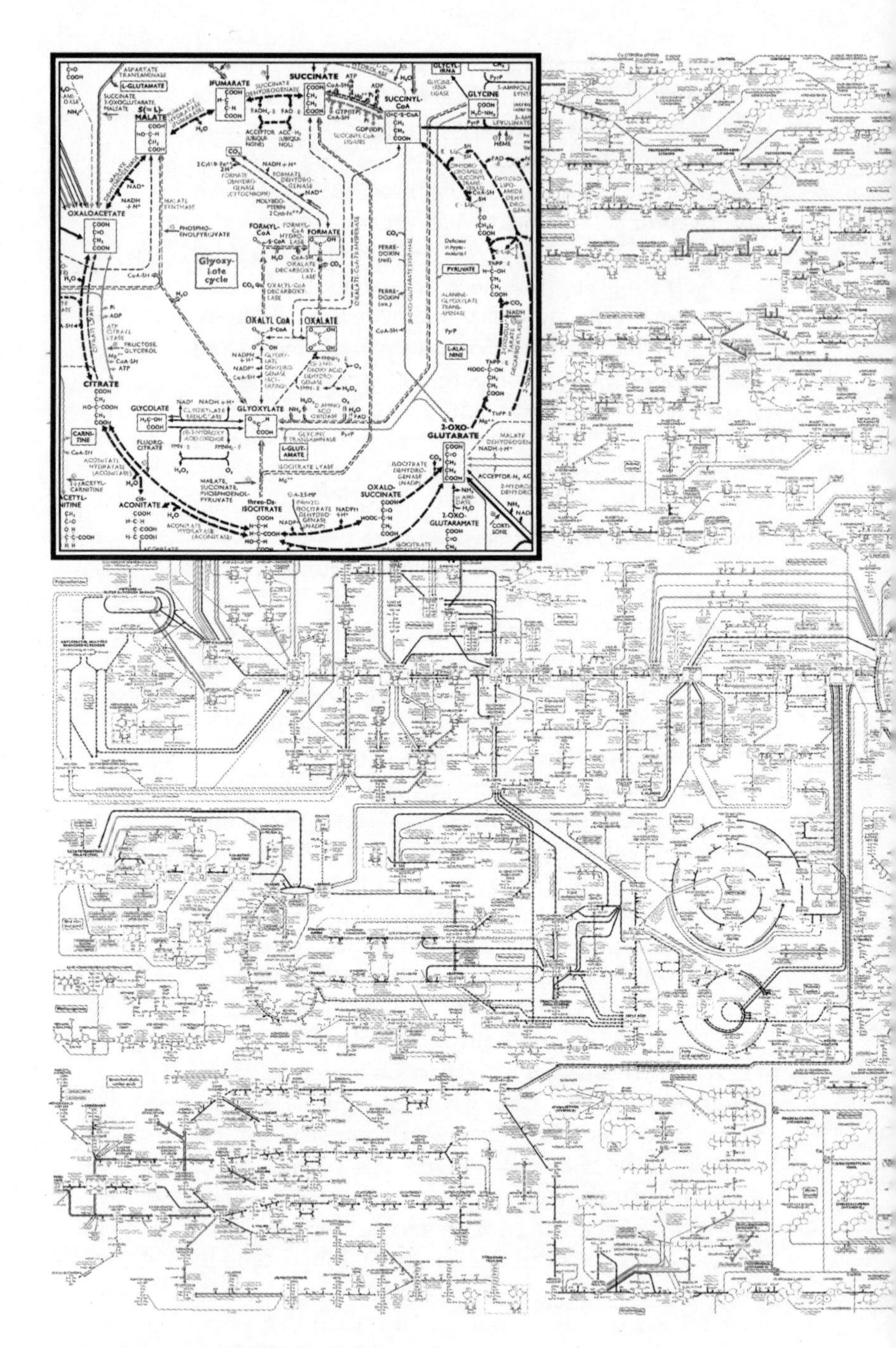

Major pathways in a cell, with detail of citric acid cycle.

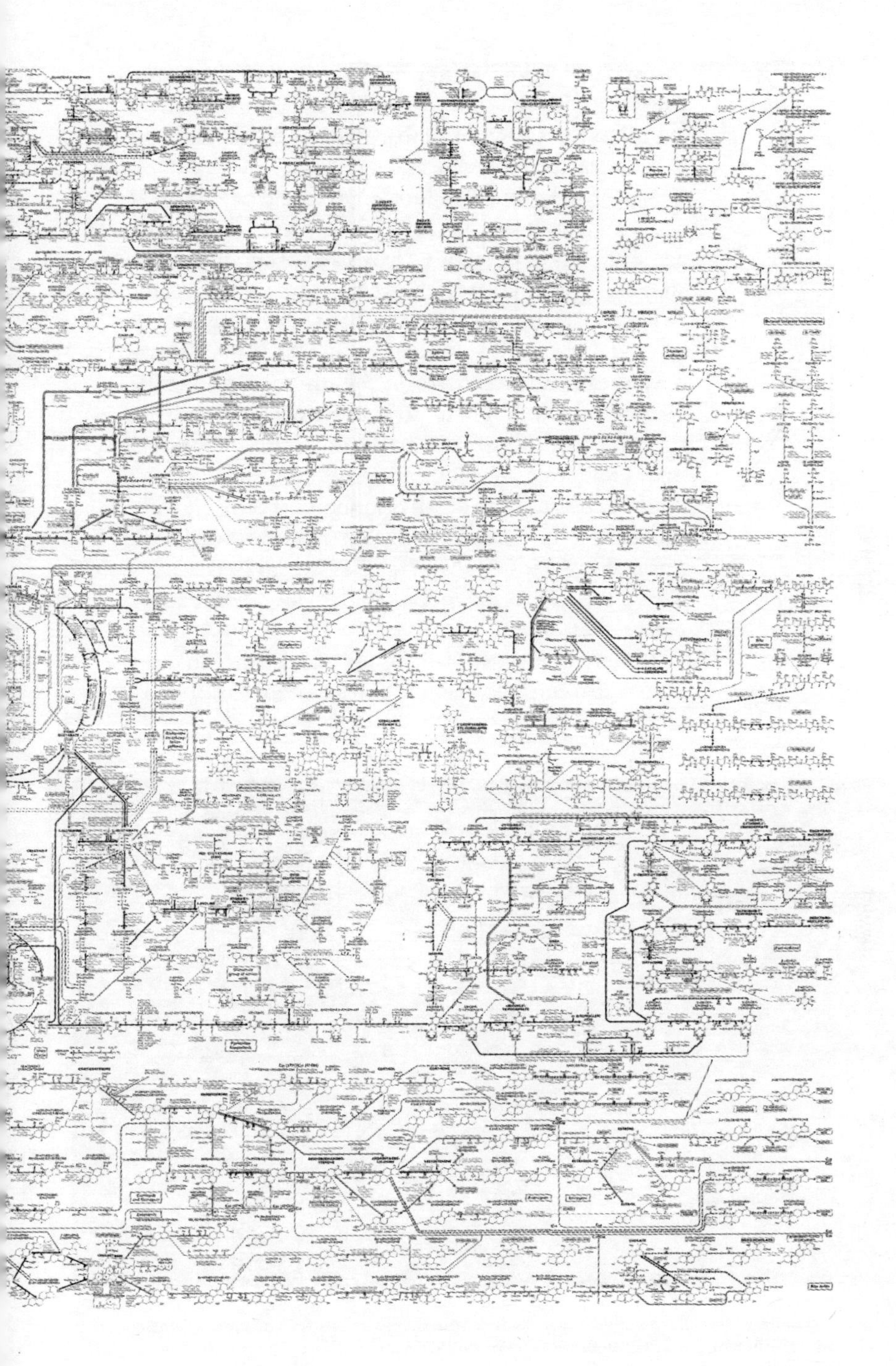

grow faster again. Minor irregularities in structure, together with changing humidity and temperature in the air as the snowflake falls, give rise to their almost infinite variety of shapes.

Patterned ground is another example of emergent order in non-living systems. Soil, gravel and stones are shaped into polygons, circles and other intriguing shapes, typically through cycles of freeze and thaw. No less striking are the varied but repeating shapes of sand dunes which emerge on Earth, Mars and, we may presume, elsewhere in the universe under the agency of wind. Their diverse names only hint at their beauty: dome and barchan, transverse ridge, linear and longitudinal, reversing, star, sheet, streak and shadow, lunette and parabolic, climbing and falling, echo and reflection. On Earth, dunes can reach astonishing sizes. Many in the Baidain Jaran Desert in China are 200 to 300 metres high. The tallest exceed 460 metres.

Robert Hazen, a mineralogist and astrobiologist who researches life origins, notes that four things are

The hexagonal crystals of snowflakes are just one example of a self-created pattern in nature. Others include anfractuous, branch, brachia, cellular, concentra, contornare, crackle, filices, labyrinthine, lichen form, nebulous, phyllotaxy, polygonal, retiform, rivas, trigons, variegates, vascular and vermiculate.

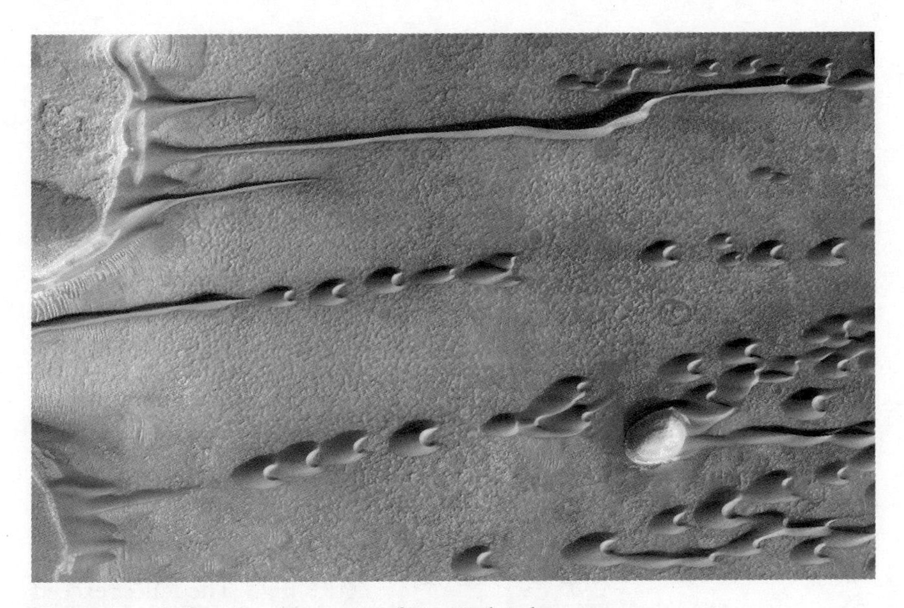

Barchan dunes in the Hellespontus region of Mars. Each is about sixty metres across.

necessary for simple and regular but often surprising patterns to emerge in non-living materials such as sand. First, there must be concentration: the individual particles must be present in sufficient numbers. Below a critical threshold, no patterns are seen. As particle concentrations increase, so too does complexity, but only up to a point. Second, there must be a flow of energy through the system: sand grains that may form lines on a beach or bar, for example, will not start moving without a certain minimum water-wave speed, though every complex patterned system also has a maximum limit to the energy flow it can tolerate. Third, complex patterns often tend to emerge when there is a cycling of energy flow such as freeze/thaw, wet/dry or day/night. And fourth, the particles need to be able to interact. Sand grains do this in very simple ways, such as by sticking or not sticking to each other, while the parts of more complex emergent systems have more ways of connecting and interacting with each other.

But the number of ways in which the parts in an emergent system connect does not have to be very

Sound waves create beautiful patterns in sand scattered on an even surface known as a Chladni plate.

Cells on the grid are either alive (on) or dead (off). A live cell with zero neighbours or one live neighbour dies; a live cell with two or three live neighbours remains alive; a live cell with four or more live neighbours dies; a dead cell with three live neighbours comes alive; and in all other cases a dead cell stays dead.

large, and the complexity with which they interact does not have to be very great, for remarkable things to happen. This is clearly shown in an artificial example: a computer program called the Game of Life. With five simple rules governing the status of squares on a grid, the Game of Life generates dynamic patterns that behave in extraordinary ways – a magic trick that performs itself. 'Gliders' travel steadily across the screen.

'Eaters' consume any 'gliders' that pass. 'Breeders' grow bigger and bigger, replicating faster and faster. Patterns can even embody a universal Turing machine and a universal constructor, meaning that they can process information as well as any computer and build copies of themselves or any other pattern.

The Game of Life depends on an external agent to build the computer on which it runs (or on the board and stones of the game of Go, which is where it actually started). The agent must also write and run the rules or program. But if the material world around us is not governed by any external agent or programmer, the pieces have to come together and interact of their own accord.

For people educated in traditions which posit an external creator, the idea that nature can self-organize even to the extent of becoming living things can be troubling or implausible or both. By contrast, those educated in traditions in which pattern and process are seen as inherent in things have less trouble with the idea. ('The great Tao flows everywhere,' wrote Lao Tzu. 'It loves and nourishes all things, but does not lord it over them.') Scientists, typically preferring the simplest possible explanation of any given phenomenon until proven wrong, favour a self-organized origin to life. Here are parts of some of the stories and hypotheses that scientists have developed over the last few decades to explore and test how life may have emerged on Earth.

'The study of non-equilibrium thermodynamics seems to be telling us that . . . the appearance of life on a planet like the early Earth, imbued with energy sources such as sunlight and volcanic activity that keep things churning out of equilibrium, starts to seem not an extremely unlikely event . . . but virtually inevitable.'
Philip Ball

First signs of life

Let's begin with the When? Until quite recently, it was widely thought that life on Earth began no earlier than about 3.5 billion years ago, some 500 million years into the Archaean eon, which succeeded the Hadean. In the last decade or so, however, evidence based on chemical signatures associated with life has suggested an earlier date – between about 3.8 billion and 4.2 billion years ago. If this earliest estimate proves to be correct then it would mean that life emerged – and perhaps re-emerged – quite quickly after events such as the Late Heavy Bombardment may have sterilized the planet's surface. And this raises the intriguing possibility that in the right circumstances the emergence of (simple) life is almost inevitable.

At the reductive end we have the Nobel Prize-winning physiologist Albert Szent-Györgyi: 'Life is nothing but an electron looking for a place to rest', and the geologist and chemist Michael Russell: 'The "purpose" of life is to hydrogenate carbon dioxide.' The physicist Sean Carroll observes: 'Every organism . . . acts to increase the entropy of the universe.' *Kurzgesagt*, a popular video series, suggests that life is 'an openness to creating new patterns'.

Every living thing on Earth shares the same chemistry, and can be traced back to a single Last Universal Common Ancestor, or LUCA, which is estimated to have lived between about 3.8 billion and 3.5 billion years ago. But LUCA was already quite a complex organism – something like a modern bacterium – and must therefore have evolved from simpler beginnings. Moreover, LUCA (or whatever it evolved from) may not have been alone. Rather than coming into existence just once, first life may have had many origins and many forms, emerging and evolving over and over again in different forms during millions of years, giving rise to the common ancestor of all we see today only when everything else was wiped out in the first mass extinction event.

Second, the What? The answer to this question depends on what you mean by life. Among many definitions I have read, the one credited to the researcher

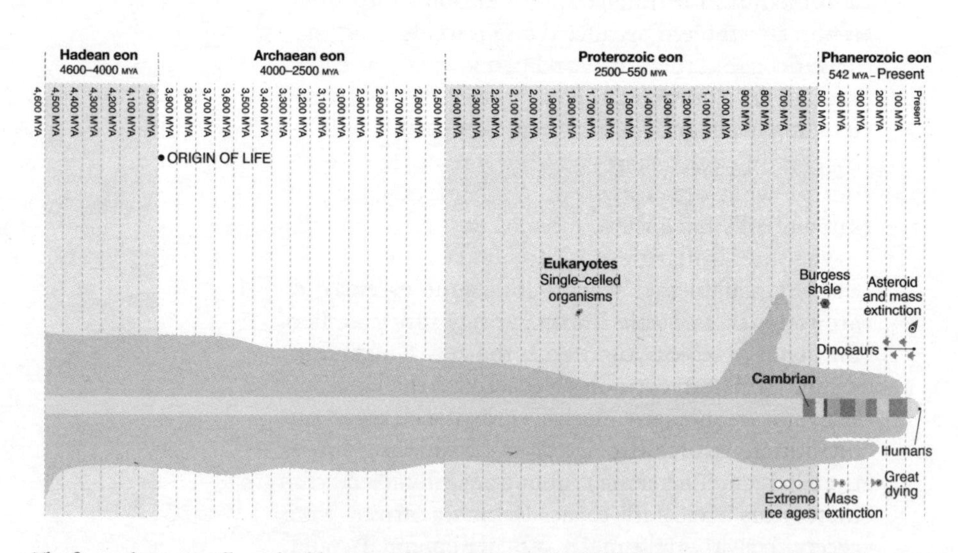

The first eukaryotic cells evolved between about 2 billion and 1.6 billion years ago, but the first large and complex multicellular forms evolved less than 600 million years ago.

Gerald Joyce is as good as any: 'Life is a self-sustaining chemical system capable of incorporating novelty and undergoing Darwinian evolution.' But taking or leaving that definition, scientists agree that all the living things we know – and perhaps all those we can imagine describing as alive – share at least three properties: a physical boundary between inside and outside; the capacity to store (and vary) information; and metabolism, or the ability to extract and use energy to maintain and grow.

Third, the How? The good news is that the scenarios currently under investigation are becoming increasingly comprehensive. Each contains much beauty and wonder. In addition, it is quite possible that experiments run over the next few years will lead to a robust and enduring explanation of the origin of life on Earth. To this it should be added that life can only exist as part of, and interacting with, larger systems.

Take the riddle of how living systems first acquired boundaries. Cells are mostly water on the inside, and most are surrounded by water on the outside. This makes sense because water is an excellent medium for transporting other molecules, but it poses a problem because water is also a very good solvent, liable to absorb and carry away the pieces needed for a complex system. It really wouldn't do to dissolve every time it rained. Life found a solution by creating cells surrounded by membranes made out of what are known as lipid bilayers, and all known cells have these membranes. Lipid bilayers have some similarities to soap bubbles – not least in that, like bubbles, they can self-assemble – but they are tougher, and able to incorporate tiny gateways for material to enter and leave the cell. Researchers have found that the molecules needed to make lipid bilayers are present in the kind of carbonaceous meteorites that fell to Earth in abundance during the Hadean. The same compounds extracted from meteorites that fall today (which, despite their recent arrival, are equally old) will spontaneously self-assemble into bilayers in conditions like those on the early Earth. They may be the source of the

At the risk of sounding like the Spanish Inquisition, some accounts list six essential properties shared by all life on Earth:

(1) compartmentalization – a cell-like structure that separates the inside from the outside;
(2) hereditary material – RNA, DNA or equivalent to specify form and function;
(3) catalysts to speed up and channel these
(4) metabolic reactions; a supply of free energy to drive metabolic biochemistry – the formation of new proteins, DNA, etc.;
(5) a continuous supply of reactive carbon for synthesizing new organics;
(6) excretion of waste.

The biochemist Pier Luigi Luisi and his colleagues have shown that lipid vesicles can grow, gradually incorporating new lipid molecules from solution, and that they are autocatalytic – that is, they can act as templates that trigger the formation of more vesicles and, in a kind of self-replication, divide. Luisi has proposed a 'Lipid World' scenario for life's origin, in which prebiotic lipids formed abundantly on Earth and self-organized into cell-like vesicles that captured an early information-bearing molecule.

first cell boundaries, having formed the vesicles, or tiny containers, for a common ancestor of all living beings. And if that is so then every one of the tens of trillions of cells in your body is sheathed in a membrane that owes its structure to extracts from rocks that fall out of the sky.

A friend of mine tells how one time on a country road in Ireland she was getting close to exhaustion after cycling into a strong wind. A kindly passing motorist noticed her distress and stopped to asked if anything was wrong. My friend thanked him and explained that it was simply a question of the wind. The motorist paused for a moment, looked in both directions and, with a twinkle in his eye, said, 'Why don't you go the other way then?' I'm afraid that some readers may find the next few paragraphs a little hard going, but I'd encourage them to keep peddling all the same. You'll be glad you didn't turn round . . .

The question of how life developed the ability to store and deploy information may be more of a challenge than the question of how it first acquired boundaries. In everything alive today DNA, or deoxyribonucleic acid, does the job. But DNA – the famous double helix – is a huge and hugely complex molecule, and it must surely have emerged from something simpler.

Most researchers believe that DNA evolved within a system that used RNA, or ribonucleic acid, to store genetic information and as a catalyst. RNA is smaller than DNA, with a single helix, and has a much shorter chain of the nucleotides – the molecules that are the building blocks of both RNA and DNA. It is also less stable, but it can store and transport information: in all cells alive today it is involved in the coding, decoding, regulation and expression of genes. Before 'DNA world' – a world in which all life used DNA to store genetic information – was 'RNA world', in which small chains of RNA catalysed all the chemical groups and information transfers required for proto-life.

Some viruses today still use RNA for heredity. They may provide clues to what an 'RNA world' was like.

The precursor of the Last Universal Common Ancestor of life today may well have run on RNA, but how RNA itself emerged has puzzled scientists for decades. In 1999 Gerald Joyce and Leslie Orgel, two leading researchers in the field, wrote that the spontaneous appearance of nucleotides would have been 'a near miracle'. But in 2009 the chemist John Sutherland showed how it could have happened without one. A nucleotide is made of two parts – a sugar and a base – and, rather like two separate metal rings, there seems to be no way short of a magician's legerdemain that the two can be made to interlock. Thinking laterally, Sutherland mixed the precursors to the sugar and the base in a different order so that they formed a 'half-sugar' and a 'half-base', rather like two half-rings. When he added another 'half-sugar' and 'half-base', an RNA nucleotide called ribocytidine phosphate emerged. A second nucleotide was created when ultraviolet light was shone on the mixture. When I visited Sutherland in 2015, he and his colleagues had recently shown that RNA could emerge from even simpler materials, starting with hydrogen cyanide, hydrogen sulphide and ultraviolet light.

But even if the question of how of RNA emerged is answered, the question of where remains open. The biophysicist David Deamer has suggested volcanic

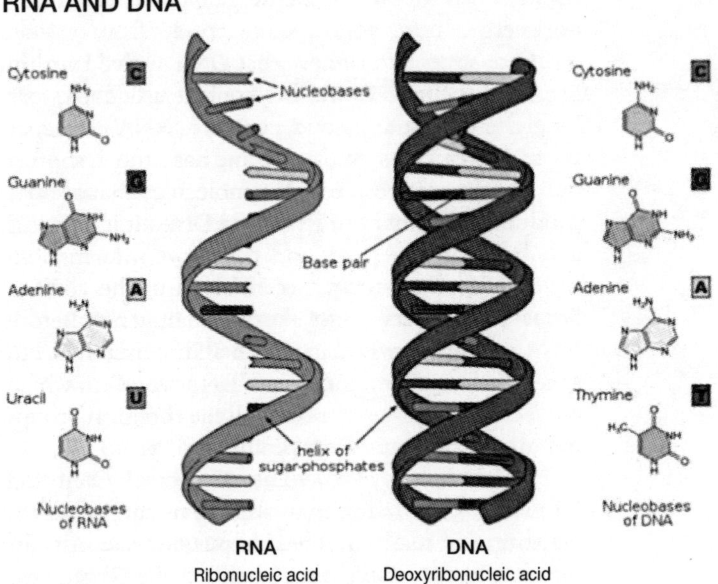

rocks exposed to liquid water and air. This environment would be a little like some places in Hawaii or Iceland today, where hot pools constantly go through cycles of wetting and drying, and very like the ancient beach we imagined earlier in the chapter. Its onshore pools, exposed to an atmosphere very different from ours, would contain complex mixtures of dilute organic compounds from a variety of sources, including carbonaceous meteorites, and other compounds produced by chemical reactions associated with volcanoes and atmospheric reactions on Earth. Because of the fluctuating environment, these compounds would be undergoing continuous cycles in which they were dried and concentrated and then diluted upon rewetting. During the drying cycle, the dilute mixtures would form thin films on mineral surfaces. In these conditions compounds would react with one another and the products would be encapsulated in self-assembled membranes. In this way, vast numbers of what researchers call protocells – precursors of the first living cells – would have appeared all over early Earth.

Most of the protocells in this scenario would remain inert, but a few would contain a mixture that could be driven towards greater complexity by capturing energy, amino acids and nucleotides from outside. As these small molecules were transported into the protocell, energy would be used to link them into long chains (proteins and nucleic acids). Life began when one or a few of the protocells found a way not only to grow but also to incorporate a cycle involving catalytic functions and genetic information, presumably from RNA.

It sounds convincing, but there may be a catch. Some researchers argue that the energy accessible from sunlight and volcanic heat in conditions like this is insufficient to drive the emergence of, first, protocells, and then more complex forms. In a rival hypothesis, proposed by the chemist Michael Russell and championed by the evolutionary biochemist Nick Lane and others, life began at alkaline hydrothermal vents on the seafloor. These strange formations are quite different from the piping-hot 'black smokers' where yeti crabs and giant tubeworms live today. Rather than having a volcanic origin, alkaline vents are formed by a chemical reaction between seawater and rock newly exposed by the movement (at the speed of a growing toenail) of continental plates. The reaction produces methane and hydrogen-rich water, and expands and heats the rock, causing it to crack and fracture. This in turn permits more seawater to penetrate into the rock. At alkaline vent sites on Earth today, such as the 'Lost City' in the mid-Atlantic, the reaction extrudes twisted and precipitous limestone towers. The towers are filled with tiny cavities that happen to be about the size of bacterial cells, and methane and hydrogen bubble through and out of them into the water column above at between 40°C and 90°C. And, according to this scenario, it was in cavities like these that life began.

Russell, Lane and others claim that an alkaline vent origin is the only hypothesis that solves the question of metabolism: the puzzle of how proto-life was able to capture enough energy to assemble and evolve. Raw hydrogen bubbling from the ground as gas, as it does at an alkaline vent, is 'a free lunch you are paid to eat'. The temperature and acidity differences across the tower walls create a weak but vast battery. The vents are therefore an environment that favours the emergence of greater complexity because they provide a steady stream of free energy that is essential for the energy-hungry reactions needed to make complex polymers such as proteins, lipids, RNA and DNA. And, exploiting the electron and proton gradients across these vent walls, the Last Universal Common Ancestor of everything on Earth alive today set the pattern for all future organisms, which recapitulate across their cell walls the chemistry of oxidation and reduction found in warm alkaline vents today. If this is right, then all of us carry within a memory of an ancient 'Lost City' on the ocean floor.

Recent research supports the hypothesis that chemical reactions occurring spontaneously in Earth's early chemical environments provided the foundations upon which life evolved – in other words, that metabolism is older than life itself. Markus Ralser and others have discovered that a version of the cirtic acid cycle (a series of chemical reactions used by all aerobic organisms to generate energy) can proceed in the absence of cellular proteins called enzymes.

Back in the seventeenth century Athanasius Kircher envisaged water and fire coursing through the inward parts of the Earth, driving turbulence and change. We now know this to be a distorted but not entirely misguided intuition of reality. We have known since the

'The fire and water sweetly conspire together in mutual service.'

mid-twentieth century that heat from the planet's core drives tectonics: the movement of continental plates across its surface, cycling and recycling rock and water over billions of years. Thermal activity at alkaline vents and hot smokers are small parts of these much larger loops. But since then an even more astonishing reality has become apparent.

Life – Earth scientists now generally agree – moderates the planetary system to its advantage. Its influence extends as far as the upper atmosphere, where it has created the ozone layer that blocks high-energy ultraviolet rays and so allowed the spread of plants and everything that depends on them across the continents. Over billions of years, life has altered not just the skin and sky but also the Earth's deep subterranean realms, pulling carbon from the mantle and piling it on the surface in the form of sedimentary rock, and sequestering huge amounts of nitrogen from the air into ammonia stored inside the crystals of mantle rocks. By controlling the chemical state of the atmosphere, life has also altered the rocks it comes into contact with, and so oxygenated the crust and mantle of Earth. This changes the material properties of the rocks – how they bend and break, squish, fold and melt under various forces and conditions. The clay minerals produced by Earth's biosphere soften Earth's crust and have helped to lubricate the plate tectonic system that would otherwise have ground to a halt. Life is not a minor afterthought on an already functioning Earth, but has become a central part of its nature and process.

Stupendous contrivances: wonders of the cell

After the best part of 2 billion years of only microbial life on Earth, an archaeon engulfed a bacterium without digesting it and gave rise to larger, more complex cells called eukaryotes, which are the ancestors of plants, animals and fungi.

For the initial 1.7 billion or so years of its existence, life consisted solely of microbes – the precursors of today's bacteria and archaea. Innovations made early on by these relatively simple creatures still power every cell in our bodies and those of everything else that lives. But it is only very recently that we have begun to see and fully appreciate their reality.

The first glimpses of a tiny invisible world came in the seventeenth century. In the 1670s the microscopist Antonie van Leeuwenhoek used tiny lenses – drops or globules of glass – that magnified by anything from about 275 to 500 times to observe 'animacules' present in countless numbers in even the most innocuous body of water (as well as in places less innocuous, such as saliva from the mouths of old men who had never brushed their teeth). These animacules we now know as protists – broadly, pond life – and bacteria. Van Leeuwenhoek's contemporary Robert Hooke had inferior microscopes but superior powers of communication, at least in English, and in his *Micrographia* of 1665 he published images of marvels never before imagined: tiny 'cells' in cork (named for their supposed resemblance to the cells in which monks lived in a monastery); the eyes of a fly revealed by magnification as monstrous compounds; and the alien body-armour and mouth parts of a flea. 'Nature', wrote Hooke, 'not only work[s] mechanically, but by such excellent and . . . stupendous contrivances, that it were impossible for all the reason in the world to find out any [more ingenious].'

In recent decades researchers have found that the mechanical workings of nature operate at a much, much smaller scale than anything contemplated by van Leeuwenhoek or Hooke. For at the nanoscale (where measurements are made in nanometres – billionths of a metre), life is made of molecular machines. And 'machines' here is not a figure of speech: these entities are assemblages of moveable parts that rearrange other molecules in ways as regular as clockwork. These machines differ from those we are more familiar with, however, in being capable of continuous self-assembly, repair and disassembly, and in being much more reliable than anything humans have made, having endured essentially unchanged for billions of years. They are, I think, truly stupendous: as great a wonder for our generation as anything discovered by van Leeuwenhoek and Hooke was for theirs. Here are two examples.

The first is the ribosome: a tiny 'factory' that makes all the proteins essential for life. (There are about twenty-five thousand different kinds of protein in a human body. Most individual protein molecules last only a few days so they need to be steadily replaced.) Ribosomes are so fundamental that it is hard to see how cells as we know them could ever have come to be without them. They are found in every cell of everything alive, and have the same essential structure in all of them. Their active sites and central cores are built of entirely of RNA, so ribosomes could be evolved relics or adaptations from an RNA world. At the base of everything that is your material presence in the world are their numberless goings-on, inaudible as dreams.

Compared with molecules such as water and simple sugars, ribosomes are enormous, consisting of about a million atoms; but compared with cells they are tiny, and a typical human cell can contain many millions of them. Each ribosome, which consists of large and small subunits like parts of a robotic press, reads information conveyed to it from DNA in the cell nucleus by messenger RNA rather as if it were reading brail, and uses the information to select and then stamp together amino acids so that they form new proteins. It does this at the rate of about forty per second, and with an error rate of less than one in ten thousand – far better than humans achieve in high-quality manufacturing. All in a space just twenty to thirty nanometres across.

The physicist Neil Gershenfeld calls ribosomes the original digital fabricators, 4 billion years ahead of 3D printers, and vastly more capable and reliable. With a 3D printer, the design is determined by a computer program, which is digital, but the material with which it works, such as a resin, has no self-organizing properties: it is just kind of smooshy. In the case of a ribosome, however, the twenty amino acids it 'prints' with come in regular and repeating shapes – a fair analogy is Lego bricks – and this makes fabrication repeatable and precise even as a very large number of configurations is possible. To an extent, the 'code' is also in the material, because the shape of the parts directs them to configure in a limited number of ways.

There's an old joke about two men taking a break

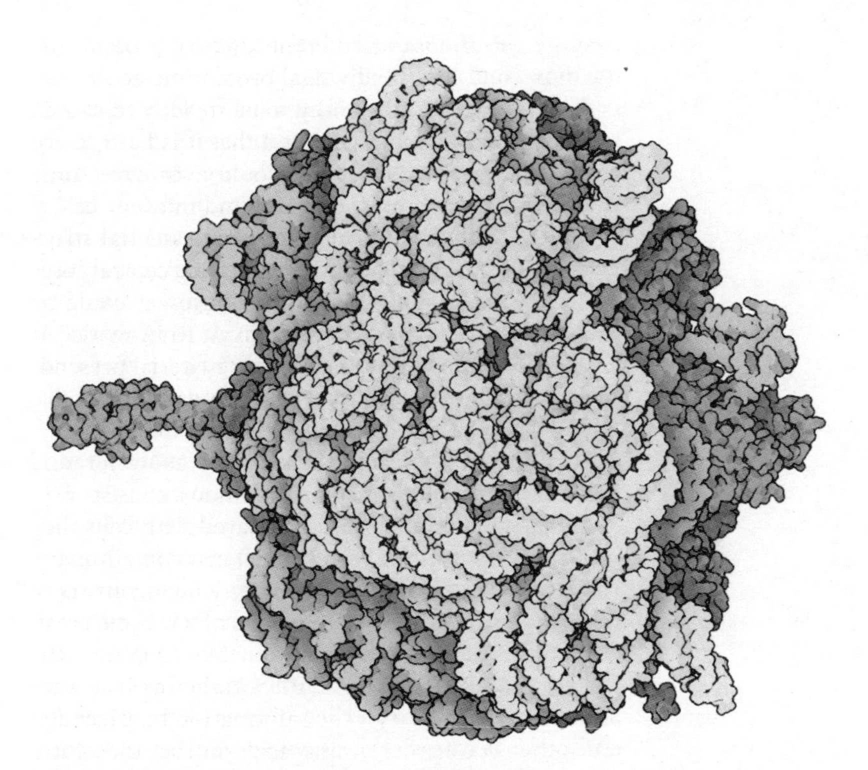

The ribosome manufactures all the proteins essential to life in every living thing, and likely predates life as we know it.

outside on a cold day. One pours hot tea from a thermos into two cups. 'You know,' he says, 'this flask is incredible. In winter it keeps the tea hot, but in summer, when I fill it with iced tea, it keeps it cold. The question I ask myself is, how does it know?'

The answer to the question of how many molecules inside a cell 'know' where to go is a little like this: they don't. The fluid within a cell, which is called the cytoplasm, is mostly water, and in a simple cell bacterium many molecules and large assemblies of molecules float freely within it. At any temperature above absolute zero, all molecules vibrate. That is what heat is. At room temperature, a medium-sized protein floating in the watery medium of a cell and jiggled by random motion moves at about five metres per second – the speed of a fastish runner. If placed alone

Molecules in the air around us jiggle much faster than they do in water at the same temperature. If we shrank to the size of molecule, says the physicist Peter Hoffman, we would be bombarded by a molecular storm so fierce it would make a hurricane look like a breeze. But despite these high speeds, individual molecules in air in the room around you do not go very far because they collide frequently with other molecules travelling in different directions.

in space, the protein will travel its own length in about a nanosecond (a billionth of a second), but inside the cell, battered from all sides by water molecules, it will take a thousand times longer (a millionth of a second) to move that same distance. Cells are, however, tiny. A bacterium such as *E. coli*, for example, is about seven-thousandths of a millimetre long and less than two-thousandths of a millimetre wide. And because of this, random motion is fast enough to transport many amino acids and proteins inside to where they are needed simply by bumping around until they find the right place. Any molecule in a typical bacterial cell, which has few if any internal barriers, will, during its chaotic journey through the cell, encounter almost every other molecule in a matter of seconds.

Eukaryotic cells – those with a single nucleus in which all their DNA is stored – are more complex than bacteria and archaea (prokaryotic cells, which do not contain such a nucleus), and usually much larger. Sometimes these eukaryotes cooperate to make and sustain the likes of you and me. Unlike bacteria and archaea, every eukaryote has internal compartments and other features such as microtubules, which act like train tracks for motor proteins called kinesins to transport important loads around the cell. If you search for animations of this online you will see the motor proteins high-stepping along on a pair of outsize feet like goofy cartoon characters. No less amazing is that they harness the energy from the random bombardment of water molecules to power their walk, ratcheting useful work out of chaos. But kinesin is not the most wondrous molecular machine I have come across besides the ribosome. Within every fibre of our being is a contraption even more stupendous, and my second example among molecular machines.

Molecular machines are smaller than a wavelength of visible light, so cannot be viewed directly. Their structure and function can be determined in other ways, including X-ray crystallography. Researchers create animations based on these findings.

'Life is a pure flame,' wrote Thomas Browne in the mid-seventeenth century, 'and we live by an invisible Sun within us.' This sounds like poetry but it is actually an understatement of the truth. Being human, and staying that way, requires about 2 milliwatts of energy per gram of body mass – about 130 watts for a typical adult. All that you think, do and are has a power rating of little more than an incandescent light bulb. It doesn't sound like a lot, but per unit of mass it is about ten thousand times more than that of the Sun. How is this possible?

The Sun is 10^{21} as big as a human but it gives off just 2.8 ergs per second per cubic centimetre, while a human gives off more than 100,000. (An erg is a unit of energy equal to 10^{-7} joules.)

The short answer is cellular respiration: a metabolic process in which organisms combine glucose with oxygen to release energy (with carbon dioxide and water as waste products). But the short answer leaves a lot out. For one thing, cells do not exploit all of this energy directly, but rather use a molecule called adenosine triphosphate, or ATP, to transport the energy to where it is needed – by, for example, ribosomes.

To power a single cell requires a lot of ATP. A complex multicellular form of life such as a human being – who consists of around 37 trillion cells even without her bacterial coterie – needs a lot more. And to make it all, each cell contains millions of tiny bean-shaped organelles, or organs, called mitochondria. In total you have more than a quadrillion (a thousand trillion) mitochondria inside you, with a combined surface area of about 14,000 square metres, or about two football fields. Together they pump more than a billion quadrillion, or 10^{21}, protons across their membranes every second, using energy extracted from sugars to do so. Each proton (which is the nucleus of a hydrogen atom like those in the heart of the Sun) carries a positive electrical charge. Pumping them across the membrane of the mitochondria generates a difference between their concentrations on the two sides and hence a difference in electrical charge of 150 to 200 millivolts. That difference sounds small but because the membrane is so thin – about 6 nanometres – it scales to 30 million volts per metre – equivalent to a bolt of lightning. This gradient has been carried within you and every one of your ancestors since the earliest days of life.

Mitochondria originate in bacteria from a group called the alphaproteobacteria, which were engulfed about 2.5 billion years ago by archaea and, providing them with an extra source of energy, gave rise to the eukaryotic cells found in animals, plants and fungi.

And here we get to the second wondrous molecular machine. The membrane of each mitochondrion is studded with thousands of units called ATP synthase which – as their name indicates – make ATP. Each one works like a tiny hydroelectric turbine. Protons that have built up behind the external membrane pour through like water cascading downhill and turn a rotary assembly, which acts exactly like a motor. This is connected by an axle to a second rotary assembly, which acts as the catalytic head, stamping phosphate (P) and adenosine diphosphate (ADP) groups together to make ATP. The second rotary assembly consists of nine or more subunits, six of them arranged like segments of an orange which clasp the phosphate molecules as they turn. In normal operation, the entire assembly spins at several hundred revolutions per second. Typically, it produces three ATP for every complete revolution.

Almost incredibly, the rotations can be observed by attaching a tiny gold nanoparticle to the rotating parts of ATP synthase. Indeed it was such observations that confirmed its existence.

Tucked away in the offices of the Mitochondrial Biology Unit in Cambridge is a scale model of ATP synthase made, as it happens, out of Lego bricks.

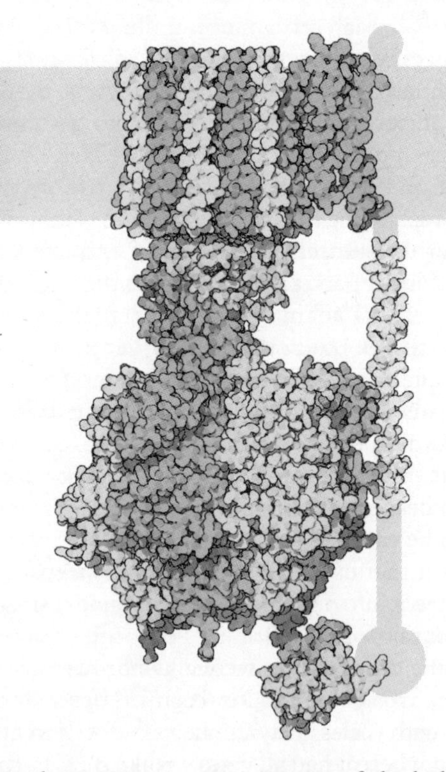

ATP synthase: precision nano-engineering of the highest order. 'The more we learn about it,' says the biologist Nick Lane, 'the more marvellous it becomes.'

Over a metre high and enormously detailed, it is, in a sense, just a gimmick – something to help children and ignorant adults like me get a superficial sense of one of life's most astonishing machines. All the same, it blew me away when I first saw it, and I stood almost speechless as the man who, perhaps more than any other, helped to elucidate the structure of this extraordinary molecular machine showed it to me. John Walker, who won a Nobel Prize for his work, was generous to a fault with his time and, despite the Nobel accolade, direct and humble. He remains someone for whom the work and the wonder matter above all. 'Working this thing out,' he told me quietly, 'was quite an effort.'

Life itself

In the introduction to this book I described a patch of sunlight on a kitchen ceiling. The light had passed between the branches of a tree which was moving in the breeze and casting ripples on the patch of light. The tree and the human perceiving the dappling shadows cast by the tree have a common origin. They have their differences. The tree, for example, 'eats' light, performing amazing tricks such as exploiting quantum effects to maximize the efficiency with which it does so. But both tree and human share the same biochemistry and many of the same essential mechanisms working away without cease, including the ribosome and ATP synthase.

Two or three years ago a friend invited me to join him on a descent of a small river in mid-Wales. The Llyfnant, he said, was just about the only truly wild river left in that excellent country, and we would be going offtrack into one of the last remaining fragments of Atlantic rainforest in Britain. Few people ventured there. Some days later, as we walked beside the river on a tarmac road behind a local council rubbish truck with a Keep Wales Tidy sticker on the bumper, I wondered where it had all gone wrong. But gradually the shape and feel of the valley began to change. The lorry pulled ahead, leaving only the sound of the river and the wind in the trees. As we walked, the river fell away from the road into a gorge, and the understory of the woods between us and the river became increasingly dense and luxuriant. Liverworts, lichens and epiphytic ferns spread across rock faces and the nooks and branches of trees. My friend observed with joy that the trees soaring above us were not oaks but small-leaved limes, *Tilia cordata*. These, he explained, were trees of the ancient woodland – evidence that the place had flourished undisturbed for hundreds of years and even since prehistoric times.

We cut down the slope, slipping and falling but landing on soft moss and thick soil until, with a bit of clambering, we reached the river at a point where it poured between large rocks over a stone lip and down a steep channel into a pool four or five metres below. Here we sat down, ate our sandwiches, and stared into the water. Then, without warning, a salmon jumped from the pool at the base of the fall, thrusting upwards as if it were trying to get some purchase on the air itself before it fell back in the downward rush of the thick pipe of falling water. It tried and failed repeatedly, and when it did finally succeed in jumping to the upper pool I found myself cheering. I was, I realized, completely happy and at home in this place. That night, as on many nights since, I dreamed of the wildwood. It is a real place but also an inscape, a vibrancy in the soul.

The sun came out and I looked at the ring on my finger. A familiar fact came to mind but more as a feeling than as an idea: all the matter in and around me, including the gold in that ring, *really was* formed billions of years ago in stars and supernovas – and, if we are wise and generous enough, will continue to play at being endless forms most beautiful long after I am gone.

3

THREE BILLION BEATS

Heart

Wonder is like a systole of the heart.

Albertus Magnus

Thanks to the human heart by which we live.

William Wordsworth

For the heart, life is simple: it beats for as long as it can. Then it stops.

Karl Ove Knausgaard

We hear first. Five months after conception, when a human foetus is typically about half the size it will be at full term, its eardrums and the inner bones of the ear are already near adult size. Its acoustic nerves are mature and can conduct signals, and the temporal lobes of the brain, which process sound, are also functioning. The foetus can hear low sounds, and one of the first it hears is the swell of its mother's blood as it is pumped through her aorta with every beat of her heart. By six to seven months a foetus can hear pretty much the full range of its mother's voice, though in a muffled sort of way. Certain tones and patterns, spoken or sung, may prompt it to move or stay still, and sometimes the mother and baby start 'talking' to each other – the foetus moving to certain sounds, especially song, and the mother sensing this and repeating those sounds.

Humans are not the only animals to hear their mothers' voices before they are born. Australian red-backed fairywrens, for example, learn their mothers' calls before they hatch.

Thanks to ultrasound, a tiny heartbeat from within the womb is the first sound made by their baby that many parents hear. The first time I heard it as a father-to-be I found its speed, at well over a hundred beats per minute, both thrilling and terrifying. Seeing my anxiety, the nurse gently explained that the rate was perfectly normal for the baby's stage of development. I calmed down a little but continued to listen in awe.

Seen in the context of the spectrum of heart rates across the entire animal kingdom, a human baby's heartbeat is neither very fast nor very slow. Hummingbird hearts can beat well over a thousand times per minute. The heart of a clam beats twice a minute when it is a calm clam, rising to twenty times per minute at party time. Nevertheless, the rapid beat of the heart of a foetus, a baby or a young child – the rhythm of a life just beginning – remains one of the most sublime things I know: beautiful but also disturbing in its relentlessness.

In situations of stress, joy or arousal, we are quite often aware – or imagine that we are aware – of the heart beating within us. We don't perceive any other organ in this way: the wrenching or turning of the guts is quite different. This contributes to a sense that the heart is at the centre of some of the things that matter most in our being. Clearly, this goes a long way back. In ancient Egypt the heart, *Ib*, was believed to be the most important manifestation of the soul, and surpassing happiness was *Awt-ib*: 'wideness of heart'. The heart was the only major organ left in a mummy after death so that it might be weighed against a feather of Maat, the goddess of truth, harmony and justice, to judge how well a person had lived. Still today in various traditions the heart is seen as central to what is most precious in us. In Laghunyasa, a meditation in the Hindu tradition, Shiva – the Supreme Being who creates, protects and transforms the universe – is visualized as residing in the heart. In the whirling meditation of Sufis, dervishes revolve from right to left around the beat of the heart in order to express an embrace of all humanity. And in the European Romantic tradition, many feel strongly the truth of John Keats's declaration, 'I am certain of nothing but the holiness of the heart's affections and the truth of the imagination.' Recent scientific research suggests that people who are aware of their own heartbeat are better at perceiving the emotions of others.

Attempts to retain a sense of decency in dark times have been mindful of these resonances. In the 1960s, fearing that distance and abstract language could blind the US President to the enormity of a nuclear strike, the lawyer Roger Fisher suggested that, instead of having the launch codes in an attaché case carried by a young officer constantly at the President's side, the codes be surgically implanted in a capsule beneath the officer's heart. Then, if the President decided that the murder of tens of millions of people was necessary, he would himself have to access the codes by using a butcher's knife to gouge out the young man's heart. Fisher reported that when he put this proposal to friends in the Pentagon high command, they said, 'My God, that's terrible ... [The President] might never push the button.' But the heart can also be recruited and enchained in systems of insidious control and oppression. In Dave Eggers's 2013 satirical novel *The Circle*, the protagonist Mae Holland wears a 'SeeChange' camera – part of the new universal surveillance system – in a lovely pendant that sits on her breastbone directly in front of her heart. In real life, sociometric badges, which hang around employees' necks in this fashion and monitor their every move and interaction, are already being piloted by several companies.

The discovery of the heart

For most of human history people have had little idea what the heart is for or how it works. This may seem odd to those of us educated in modern industrial societies, but there is nothing intuitive in the idea that the main purpose of the heart is to pump blood around the body. The circulation of the blood is no more obvious to the naked eye than is the fact that the Earth orbits the Sun. Arteries and veins appear to peter out in the tissues of the body, and the capillaries that we now know join them to complete the circuit through which our blood travels are so fine that they are invisible without a microscope.

The Roman physician Galen, who flourished in the second century AD and whose ideas dominated European medicine for more than fifteen hundred years, was impressed by the size and central position of the liver in the human frame, and he placed this organ rather than the heart at the centre of the life of the body. He taught that blood is one of four humours, the other three being yellow bile, black bile and phlegm. Supposedly, food digested in the gut passed through the portal vein into the liver, where it was transformed into blood and imbued with what Galen called 'natural spirits'. The great veins then carried this brew sluggishly to the tissues of the body, which consumed the spirits before returning it for fresh nourishment along those same veins. Meanwhile, according to Galen, some blood from the liver travelled into the right side of the heart, where it met air from the lungs. The encounter produced a kind of fire, and hence the warmth so characteristic of a living body. But the blood was not consumed by this fire; rather it was refined and, passing somehow through the septum (the dividing wall in the middle of the heart) to the chambers of the left side, produced 'vital sprits' that then flowed through the arteries to quicken movement in the body and thought in the brain.

Galen's understanding of the heart was wildly wrong and has long since been rejected by medical science. But his doctrine of the four humours has – in its corollary of the four temperaments – had a remarkable afterlife. It lurks behind typologies such as the Myers–Briggs Type Indicator, which supposedly distinguishes different personality types and was quite widely used until at least the late twentieth century. It even endures in the minds of scientists in an imagined future. In Kim Stanley Robinson's 1993 science fiction novel *Red Mars*, a chronicle of twenty-first-century planetary settlement, the psychologist Michel Deval finds to his surprise that it offers a good lens for analysing the different personalities of the first colonists. Perhaps the enduring appeal of Galen's doctrine is that it seems to readily unlock the mystery of corporeal being, and do away with doubts and uncertainties – something that is especially welcome in the face of illness or anxiety. By contrast, modern medicine at its best accepts complexity and uncertainty. The human body, says the physician Atul Gawande, can be 'scarily intricate, unfathomable, hard to read', and this recognition is sometimes less comforting than false hope.

The four temperaments are: sanguine (outgoing and active); choleric (irritable and quick to anger); melancholic (analytical and wise), and phlegmatic (calm).

Some of the biggest steps towards a modern understanding the heart were taken by Leonardo da Vinci in the early years of the sixteenth century. Indeed, he began to understand the heart in ways that were unsurpassed in some respects until the late twentieth century. From around 1508 to 1513, six years before he died, Leonardo undertook detailed study of the inner anatomy of the human body – the skeleton, muscles,

In a job application to Ludovico Sforza, Duke of Milan, back in the 1480s, Leonardo outlined his skills, including the following: 'I will make cannon, mortar and light ordnance of very beautiful and functional design that are quite out of the ordinary. Where the operation of bombardment might fail, I would contrive catapults, mangonels, trebuchets, and other machines of marvellous efficacy not in common use.'

tendons and nerves, the reproductive system and the major organs, notably the heart. A military engineer as well as a supremely skilled artist, he applied his understanding of levers and fluid flow and of the subtle movements of living beings to his investigations. He produced anatomical drawings that have seldom if ever been surpassed for detail or accuracy, let alone beauty. Perhaps, after decades attempting to capture the sublime in outward appearance in his paintings, he was now searching for beauty that, in the writer Ursula Le Guin's phrase, is not just skin-deep but life-deep.

Working for the most part with the hearts of oxen and pigs, and only later with those of humans, Leonardo realized that the heart is first and foremost a muscle. He saw that it had four chambers where Galen had said there were two, and that the upper two – the atria – contract at the same time, followed by a simultaneous contraction of the lower two, the ventricles. He saw that the pulse in the wrist keeps time with the beat of the heart, and he attempted to calculate cardiac output (the amount of blood that leaves the heart each minute). He appreciated that the valves were one-way structures – something that was incompatible with the Galenic belief in the continuous flux and reflux of the blood. He also worked out that turbulent movement in the blood helps the heart valves to open and close – a fact that was again fully understood only in the late twentieth century. He discovered and drew bronchial arteries and also described the moderator band, rightly identifying it as a muscular bridge between the walls of the right ventricle that prevents overdistention. His insights were so many and so deep that it is hard to believe he did not realize that the heart pumps blood around the body. There is, however, no clear statement to this effect in his surviving notes. In any case, Leonardo never published his work on the heart, and it was unknown to his contemporaries and all but forgotten until nearly five hundred years later, when his sketches and notes were finally examined by expert eyes. As a result, his successors had to grope their way without the benefit of his discoveries.

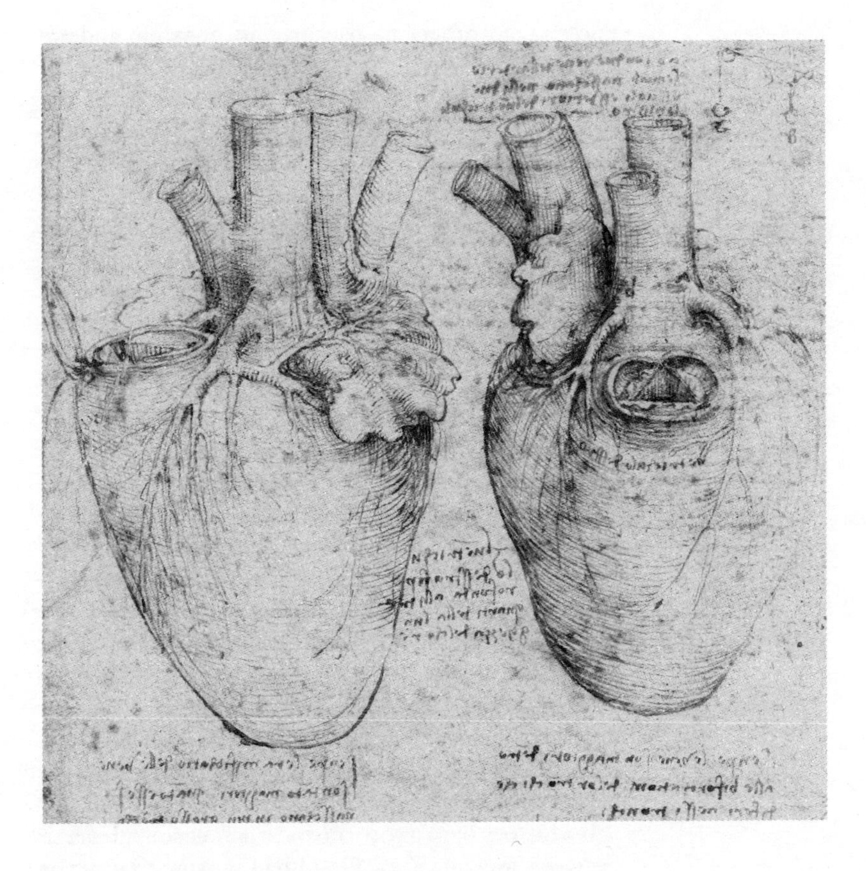

In *On the Fabric of the Human Body*, published in 1543, the anatomist Andreas Vesalius brought the standards of fine art and the rapidly advancing field of evidence-based cartography to his new maps of the inner world. Vesalius, who became professor of anatomy at Padua University, took great pains in his observations of the corpses he dissected and he was able to correct many of Galen's errors, such as the idea that the great blood vessels originate in the liver. Vesalius also questioned the doctrine that blood passed through unseen pores in the septum of the heart. But he couldn't entirely shake off the weight of tradition, and held to the Galenic idea that different types of blood flow through veins and arteries. Nevertheless, Vesalius's scepticism and his confidence in first-hand observations encouraged others to continue to question Galen's authority.

The decisive break with Galen's teachings on the heart was made by William Harvey, who studied medicine at Padua in the early 1600s with teachers in a direct line from Vesalius. Harvey performed some bold if gruesome experiments. He severed the aortas of dogs and measured the blood squirting out from their still-beating hearts. It was clear that the amount pumped out in a few minutes, let alone in an hour or day, was far more than the liver could possibly produce. Over a day it would have to be tonnes. The only explanation, Harvey concluded in 1628, was that the blood must be flowing in a complete loop.

A typical adult human heart pumps around 70 millilitres (that is, 70 grams, or about 2.5 ounces) of blood each time it beats. At 70 beats a minute, that is nearly 5 litres of blood per minute, or more than 7,000 litres – 7 metric tonnes – per day.

Harvey has been acclaimed as the Galileo of medicine for this insight. But his theory was still vague in certain respects, adhering to some aspects of Galen's teaching, and retaining mystical elements. Having no knowledge of the existence or nature of oxygen, for instance, Harvey wrote that the blood was 'vaporous, full of spirit' that was heated and cooled according to the Galenic principles. He also shared a notion, championed by his friend, the physician and hermetic philosopher Robert Fludd, that circular motion was an expression of divine will. In *Utruisque Cosmi*, a labyrinthine history of macrocosm and microcosm published between 1617 and 1621, Fludd had written that when God first breathed divine spirit into the air to create the world it moved in a circle, and it was in imitation of this that the Sun moved in a circle across the sky. And as above, so below: the Sun distributed divine wind to the Earth in circulatory air currents, where – as 'aerial nitre', or 'quintessence' – it was taken up into the human body via the lungs and from there to the heart, which distributed it in yet another circle.

René Descartes was an early and influential champion of Harvey's theory of the circulation of the blood. He highlighted the mechanical aspects of the theory, comparing the heart to a pump and a clock. This may sound too simple today, but there is much truth in both comparisons. Indeed, the regularity of the beat of a healthy heart at rest is so reliable that it is said to have led Galileo to one of his most ingenious ideas. Decades before Harvey discovered the circulation of the blood, the young Galileo had, supposedly, watched chandeliers swinging from a church ceiling and, relying on the measurement of time taken with his pulse, found it to be the same whether the chandelier swung in a small or a wide arc. This gave him the idea of regulating the mechanism of a clock with a pendulum, and in 1637 – late in life, when he was under house arrest for saying that the Earth went around the Sun – he drew up a design for one. He did not live to build it, but in 1656, fourteen years after his death, Christiaan Huygens did. The new device reduced clock error from about fifteen minutes to fifteen seconds per day.

For Descartes, the body was a machine, a material object that received instruction from the immaterial mind and soul by means of the pineal gland, a structure at the centre of the brain that acted like some kind of seventeenth-century Wi-Fi receiver. This model is not accepted today even by those who draw a sharp distinction between body and soul, but this doesn't mean that the body and the organs within it, including the heart, are not mechanisms. Rather, they are mechanisms of a vastly subtler and more sophisticated nature than Descartes ever dreamed, and, whatever else is true about them, we can extend the realm of wonder by looking at them as such.

The pineal gland, which in humans produces melatonin to modulate circadian cycles, is actually an atrophied photoreceptor. In our distant evolutionary ancestors, it was linked to a light-sensing organ called the parietal eye. The leatherback turtle, with whom we shared an ancestor hundreds of millions of years ago, has a 'skylight' in its skull – an unusually thin area of bone – that allows light to impinge directly upon its pineal gland. This enables the turtle's brain to compute day length, which helps it to navigate.

A short tour of the heart

Form a loose fist with your right hand and place it against your chest about a thumb's length below the point at which your collarbones meet. That gives you a fair idea of your heart's size and its location on the other side of your ribcage. Squeeze your fingers together, using about as much force as you would need to squeeze a tennis ball, and then relax them. This produces a movement similar to the one your heart makes every time it beats (which, at the time of writing, mine is doing about fifty times a minute). A healthy heart beats without tiring about a hundred thousand times a day, and will beat some three billion times in a life of eighty years. Heart muscle does not tire because

'Teach us to number our days that we may apply our hearts unto wisdom.' *Psalm 90:12*

it is exceptionally rich in myoglobin (a short-term source of stored oxygen for aerobic respiration) and glycogen (stored energy). It also has especially large mitochondria which, as we saw in Chapter 2, are the powerhouses in each cell, and about ten times as many of them in each cell as in other muscles.

Now listen to your heart – or to someone else's – through a stethoscope. You'll hear the double sound sometimes described as 'lub dub'. The 'lub' is made by two large valves called the mitral and tricuspid closing at the same time during the active part of the beat, known as systole, when blood is squeezed out of the two larger chambers of the heart called the ventricles. The 'dub' is made by the other two valves, called the pulmonary and aortic, closing to prevent backflow while the ventricles refill. These soft percussive sounds have been compared to a gloved finger tapping on a leather-topped desk. That's a good description but it goes only so far. For these are such intimate, primal sounds unique to every individual in their subtle details. 'Lub dub' is neither trochee nor iamb but 'I am' repeated again and again.

Though we cannot tell it by ear, each human heart has a unique beat. An electrocardiogram can be a unique biometric marker.

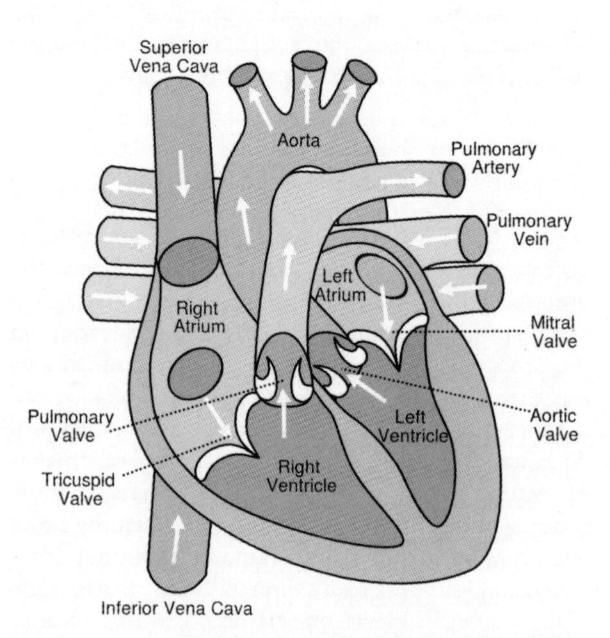

The heart pumps blood around two circuits of unequal size at the same time. The circuits cross over in the heart in one endless loop, like a lopsided infinity sign or Möbius strip. The heart's larger side, which is situated on the left in all but the one in ten thousand of us, drives the systemic circuit, pumping newly oxygenated blood through the arteries to tissues all around the body. Once in those tissues, the blood unloads oxygen to mobilize cellular respiration, and absorbs carbon dioxide and water, before it returns through the veins to the smaller side of the heart. This smaller side powers the pulmonary circuit, which drives the blood through tiny vessels lining the walls of the lungs where the carbon dioxide and water are released (to be breathed out) even as fresh oxygen is absorbed, returning the blood to the larger side of the heart, where it will be pumped around the body once more. At any one time about 12 per cent of the blood is in the chambers of the heart itself, while 70 per cent is in the systemic circuit and 18 per cent in the pulmonary circuit.

The details are complex, but in outline, respiration is simple: glucose (sugar) plus oxygen produces energy (stored as adenosine triphosphate), with carbon dioxide and water as waste products.

One of the wonders of the heart, says the cardiologist Vivek Muthurangu, is simply that it contracts as strongly and efficiently as it does. An isolated heart muscle cell or cardiomyocyte contracts by around only fifteen per cent each time it twitches, but many in combination give rise to a much greater total contraction in the heart's volume. The general principles behind this are fairly clear: bundles of muscle coil around the heart resulting in a twisting motion that results in an extra squeeze when the chambers contract. But recently scientists have realized that this was not nearly enough to account for the extent of the contraction. It appears that both the micro- and macrostructures in the muscle behave in more complex and subtle ways than was realized. Sheets of muscle may turn on edge as they contract, for example, amplifying the twist of the larger structures of which they are part. The detail of how this all works has yet to be worked out.

Isolated individual heart muscle cells twitch spontaneously and, without a strong external signal to coordinate them, the two billion or so in the walls of the heart are liable to fall out of phase. When they do so the heart goes into fibrillation – chaotic spasms

Fibrillation kills hundreds of thousands of people prematurely every year, often suddenly and without warning, including many who have no history of heart disease. With good organization and training many lives can be saved. You are two to three times more likely to survive out-of-hospital cardiac events in the Netherlands or Norway than in England.

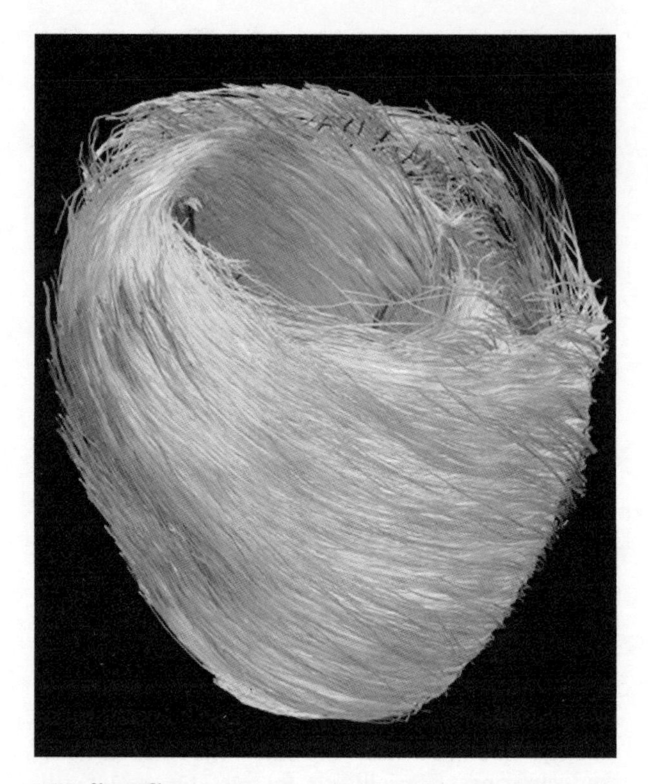

MRI of heart fibres

If left to its own devices the sinoatrial node will normally fire about a hundred times per minute. The parasympathetic nervous system slows it down; the sympathetic nervous system speeds it up.

without rhythm – and ceases to pump. The beat that keeps us alive is generated by a natural pacemaker called the sinoatrial node, a group of a few thousand specialized cells in the wall of the heart over the right atrium. To initiate a beat the node contracts, generating an electrical impulse: a signal that it then discharges through the tissues of the heart. The signal travels first to the atria of the heart, causing them to contract. It also travels to a second node, called the atrioventricular, which delays it until the atria have fully contracted. Once they have done so and pushed blood into the ventricles, this second node forwards the signal to the ventricles, causing them to contract in turn.

The signal from the sinoatrial node is strong enough to initiate this because all its cells fire at exactly the same time. This is an example of synchrony, a phenomenon that occurs across a wide range of living and non-living systems. In all such cases, individual oscillators are coupled through physical, chemical or – as in the heart – electrochemical processes that allow them to influence one another in a sufficiently short period of time. In non-living systems, synchrony can be brought about by forces ranging from tiny vibrations (which explains the tendency of pendulum clocks or mechanical metronomes placed on the same table top to fall into line) to gravity across empty space (which accounts for certain planetary orbits). In the living world, one of the most remarkable examples besides the heart is said to be fireflies flashing in unison for miles along riverbanks in tropical Southeast Asia at night. The mathematician Steven Strogatz writes that he finds this 'beautiful and strange in a way that can only be described as religious . . . a wonderful and terrifying thing, [touching us] at a primal level'. Such spontaneous emergent order, he suggests, can seem like the secret of the universe.

I am content to gaze at jellyfish for longer than many people would consider sane. I especially like the weird ones. Box jellyfish, for example, which are largely found in the shallow waters of the tropical Indo-Pacific, have eyes complete with retinas, corneas and lenses that look humanoid but for the trifurcation that, as if in some radiation-spawned mutation, enables them to look in three directions at once. Perhaps, brainless as they are, they even gaze back at me. Meanwhile, far away in marine lakes that pit the mushroom-shaped islands in Palau, golden jellyfish, dense as soup, circuit endlessly in a pilgrimage of unknowing. Out in the open ocean, atolla jellyfish glow as they pulse through the deep water beneath the reach of sunlight: red-whiskered flying saucers.

I find beauty in all this strangeness. But even the most ordinary jellyfish fascinate me simply by the way they move – pulsing and beating just as their ancestors did more than five hundred million years ago in the Cambrian period, if not before. Looking at these creatures, we glimpse one of the earliest kinds of animal movement, now continuous across more than 500 million years. The pulse of a jellyfish likely even predates fire: the combustion of carbon and other materials in the presence of oxygen probably began on Earth no earlier than about 470 million years ago in the Ordovician. In 2012 scientists at Harvard attached mammalian heart muscle cells to a silicone substrate to create a cyborg medusoid. Its pulse was almost indistinguishable from that of an actual jellyfish. One point is clear: the last common ancestor of jellyfish and humans may have lived 600 million years ago, but the striated muscle fibres in jellyfish and humans are minutely comparable. The beat or pulse predates the heart itself.

The signalling systems in human neurons and in the neural net of jellyfish (which have no brains) are also the same. And, like humans, jellyfish sleep.

If we are to fully appreciate the heart in air-breathing creatures such as ourselves, we also need to know the lungs. Because the heart – the most muscular organ in the body – exists to serve as a conduit between the lungs (the most aethereal) and all the cells in our bodies. Every time we breathe in, the air passes through 2,400 kilometres of branching tubes into 500 million tiny sacs in the lungs before we breathe it out again. The surfaces of each of these sacs, which are called alveoli (not to be confused with aioli, which is garlic mayonnaise), are lined with blood vessels that absorb oxygen from the air and release carbon dioxide and water into it. Their total surface area is some seventy square metres in an adult – supposedly about the same as the leaf area of a fifteen- to twenty-year-old oak tree. And, in a sense, the lungs are a negative of a tree: the alveoli, empty where the leaves are solid, absorb oxygen and leak carbon dioxide, while the leaves, when photosynthesizing, do the reverse. Our lungs embrace the sky and bring it within our galumphing frames.

The biologist David George Haskell compares the daily flow and cease of water in a tree to the beat of a heart: 'All summer long the forest throbs with the water-blood heartbeat of twigs . . . systole and diastole surge and draw back, the forest's subsonic hum.'

Most of the living beings on Earth that need oxygen are tiny, and actually get along fine without a heart or lungs. In respiring bacteria and single-celled organisms such as amoebas, oxygen simply diffuses through the cell wall, and is carried by random motion to where it is needed. Some multicellular organisms such as flatworms, which have no lungs, heart or gut, also allow oxygen and nutrients to pass into and through their bodies by diffusion. Even quite big animals such as frogs absorb some of the oxygen they need through their skin. Once you get beyond a certain size, though, oxygen cannot get all the way to the innermost cells fast enough simply by random motion or diffusion. There are simply too many other cells in the way. Sponges, which can grow to be bigger than a man, solve this problem by assembling themselves around a network of tubes that allow water currents to transport oxygen, nutrients and waste through them. Virtually all insects absorb much of the oxygen they need directly through little tunnels that allow air through their bodies. They breathe by gently expanding and contracting their whole bodies, sucking air in and expelling it again through these holes. Some, such as grasshoppers, also have tube-like hearts running from the head to the base of their bodies, contracting in peristalsis, a tiny Mexican wave, along their length. But these organs pump something called haemolymph, a fluid that contains ions, carbohydrates, lipids, glycerol, amino acids and hormones but not oxygen, around the body.

Blood is a very special substance

Oxygen is a highly reactive gas (as in: fire), and it can do lot of damage to body tissue. So it helps to have something that keeps it from reacting with any passing body part until it gets to where it is needed. For this reason animals evolved what Mephistopheles in Goethe's *Faust* calls 'a very special substance': a fluid containing special cells that lock away inhaled oxygen and release it only in response to certain cues, such as the more acidic environment created by a high concentration of carbon dioxide. In vertebrates like us, these cells are, of course, red blood cells. At any one time, more than twenty trillion of them (which, leaving aside the bacteria that live within us and on us, is well over half of all the cells in our bodies) are coursing around inside each of us, taking about a minute to complete a circuit.

Picturing this ensemble of 20 trillion blood cells is almost beyond human powers of imagination, but zooming in to look at a tiny section has its rewards. 'The direct observation of phenomena has an indescribably disturbing and leavening effect on our mental inertia,' wrote the nineteenth-century anatomist Santiago Ramón y Cajal; 'a certain exciting and revitalizing quality altogether absent, or barely perceptible, in even the most faithful copies and descriptions of reality.' Recalling his first year of medical studies in the early 1870s, when he observed the circulation of the blood through a microscope for the first time, Ramón y Cajal is transported:

> During the sublime spectacle, I felt as though I were witnessing a revelation. Enraptured on seeing the red and white blood cells move about like pebbles caught up in the force of a torrent; on seeing how the elastic properties of red corpuscles allow them to suddenly regain their shape like a spring after laboriously passing through the finest capillaries . . . it seemed to me as though a veil had been lifted from my soul.

At just 7.5 millionths of a metre across, red blood cells, which are also called erythrocytes, are among our smallest cells. Amazingly, the malaria parasite – a complex protozoan – spends parts of its life cycle within this tiny space as it hitches a ride before taking up residence in the liver, where it will reproduce. And, as the young Ramón y Cajal observed, the capillaries through which the cells pass are narrower than the cells themselves, so that they have to squeeze through. This helps the release of oxygen and the absorption of carbon dioxide and water before they bounce back into shape like pop-up frisbees that rebound after being scrunched. Recently, researchers have discovered that, powered by the energy-bearing molecule ATP, red blood cells are perpetually vibrating like tightly stretched drums. This helps the cells to maintain their characteristic flattened oval or disc shape. The frequency of the vibrations, which is in the millions per second, indicates the health of the cells.

Each healthy red blood cell contains about 280 million molecules of haemoglobin, and each molecule

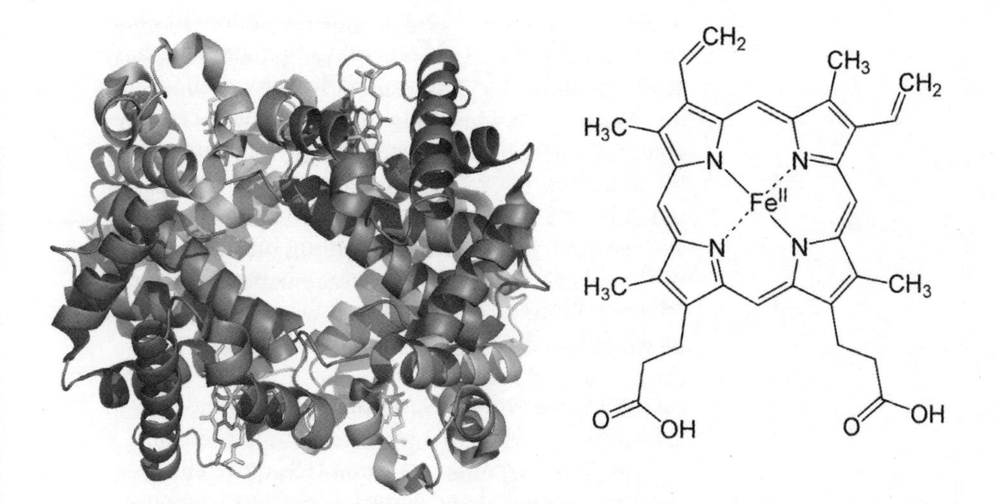

Schematic representation of haemoglobin and one of its four haem groups.

carries four oxygen molecules in special 'cages'. In some simplified two-dimensional graphical representations, haemoglobin, which contains about ten thousand atoms, looks like a bird's nest made of fusilli or a basket fashioned from bedsprings. It is in fact composed of four identical subunits, each woven out of helical amino acids and non-helical strips, grasping each other in a tetrahedral configuration. Each subunit holds a haem group: a lattice as regular as a snowflake but with square and pentagonal configurations to its atoms rather than hexagonal. And at the core of each haem group is a quincunx: four nitrogen atoms surrounding a single atom of iron. This is the binding site for oxygen, and the entire huge molecule shifts its alignment, like some monstrously complex Rubik's Cube, when the tiny oxygen molecule – just two atoms – sticks there, returning to its former alignment when it releases the oxygen again. (Red blood cells also carry away carbon dioxide but this attaches to a different part of haemoglobin.) The bond between oxygen and iron in haemoglobin is what makes blood red. We run on rust.

Invertebrates know no such restrictions. Horseshoe crabs and octopuses have blue blood thanks to haemocyanin, which binds oxygen to copper. The 'blood' of ringed and segmented worms is green, while that of brachiopods and penis worms is violet-pink.

The beat of the heart not only delivers oxygen to body tissues while removing carbon dioxide and water from them, it also brings nutrients absorbed in the digestive tract to those tissues, and transports wastes to the kidneys for removal. It moves hormones and stem cells to where they are needed, takes clotting factors to sites of injury, and carries many different proteins that have roles that are only partly understood. In addition, it regulates body alkalinity/acidity and temperature. But of all the things riding around in the bloodstream nothing is more remarkable than the adaptive immune system. Made up of vast numbers of white blood cells, it has no nerves and no central processor but it acts like a giant distributed brain, searching out threats, probing and learning, taking action and storing memories that can last for our entire lifetimes. Some of its white blood cells are constantly shuffling their genetic material like random number generators and creating new cells with the potential to recognize an almost endless diversity of potential pathogens. And you don't have to be a mathematical genius to have this capability, which is known as somatic hypermutation. It first evolved more than 400 million years ago in jawed fish.

Making a heart

It takes just thirty-six weeks for a human egg cell fertilized by a sperm cell to give rise to trillions of cells, of hundreds of different kinds, all cooperating to make a human being. How do a few thousand genes and proteins within those cells, none of which holds a concept of the structure and function of the organs or the human body, manage to do this? The processes astonishes most those who understand it best. 'As we learn about embryological development,' says the experimental anatomist Jamie A. Davies, 'our collective sense of awe has only increased.' Yoshiki Sasai, a pioneer in the use of stem cells to grow eyes and other organs comparable to the heart in complexity, said the process of self-organization makes him 'completely in awe of life'.

Still, some basic principles are readily discernible. One of the first things to understand, says Davies, is that no one set of components – such as the DNA, the RNA or the proteins – is in overall charge of embryological development. Proteins in a particular cell are made because active genes specify that they should be made. But those genes are active because proteins already present in a particular cell tell them to be. Thus 'control' is located nowhere and everywhere. Another important point is that the assembly of many large and complex proteins – and haemoglobin is a good example – follows the same basic principles as the assembly of inorganic crystals, where 'information' and structure are two sides of the same thing. As with crystals, the results are reliable, reproducible and inflexible, though vastly more intricate. And when it comes to larger scales, such as the shape and arrangement of cells, other factors come into play. Researchers are beginning to appreciate the significance of, for example, bioelectricity, whereby existing cells send signals to guide the deposition and structure of new ones. In addition, the overall shape of a new cell must also adapt to the space it must fill in a tissue. None of this is to say that genes are unimportant, however, and when it comes to the development of the heart there is some astounding genetic conservation across evolutionary time. For example, one gene vital to heart formation (mischievously named tinman after the character in *The Wizard of Oz* who has no heart) is common to humans, fish and fruit flies.

Other genes with important roles in embryological development have names like *sunday driver* (after a mutation that leads molecules to get lost as they journey through nerve cells); *ken and barbie* (after a mutation in which the external genitalia are missing); and *british rail* (after a gene that suppresses the expression of another gene called *always early*). A mutation to a gene called *cheap date* renders the animal especially sensitive to alcohol.

Eighteen days after conception a human embryo resembles a tiny jam sandwich: a disc with three distinct cell layers. Soon it will soon roll up to make three nested cylinders, but before this happens some of the cells in the middle layer begin to assemble into a crescent-shaped tube around the edge of the disc. Then, as the disc starts to curl, the crescent is brought together in the front of the forming body. This is the precursor to the heart. By twenty-two days the two ends of the tube have fused and the wall between disappears, rather as the wall between two adjacent soap bubbles can disappear. And by twenty-eight days the tube starts to beat – our first moving part, and our last – pumping newly formed blood cells through newly formed blood vessels. At this stage the

When it first starts to beat, the embryonic heart rate is about 75–80 beats per minute (bpm) – a similar rate to the mother's. It then accelerates, peaking at 165–185bpm during the seventh week, and then decelerates to about 150bpm by the fifteenth week. At term it is typically around 145bpm. The average adult heart rate is 60–100bpm.

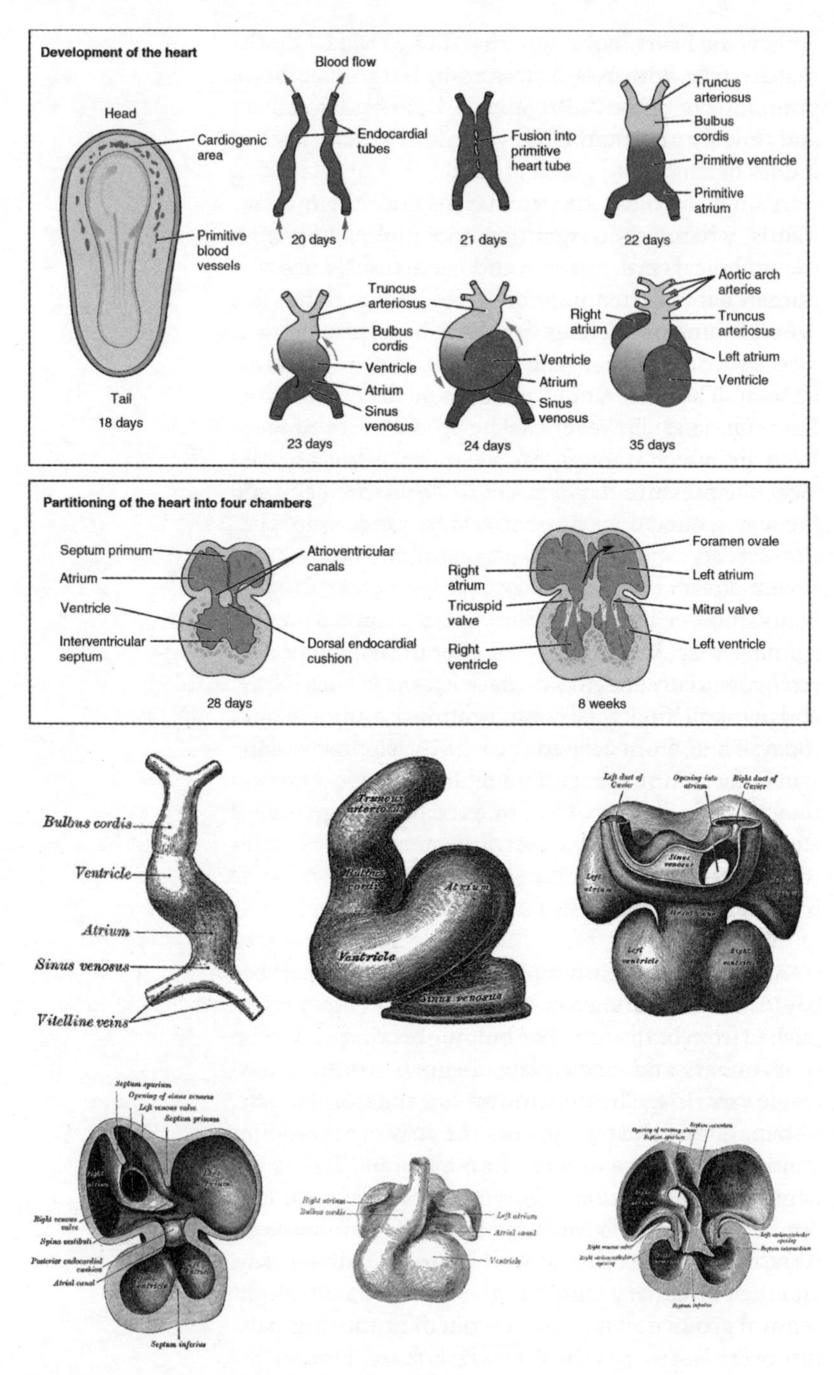

Two depictions of stages in the development of the heart.

embryonic heart looks very much like the heart of a mature fish, with blood coming in at the 'tail' end, pumping out of the 'head' end as the heart contracts, and flowing around the body in a single loop, just as it does in a fish.

A single-loop blood circulation is fine for a human foetus, which gets oxygen from its mother through the umbilical cord, but it would mean death once the human baby started breathing air. The reason is to do with pressure. The gills of a fish extract oxygen directly from water, which is as dense as the blood in its body. The same thing applies to a foetus bathed in fluid. On land, however, we have to extract oxygen from air, which is much less dense. If we had a single loop, the pressure that is necessary to drive blood all the way around our bodies would be too high for the thin-walled capillaries in the air spaces of our lungs. It would push blood out into those air spaces and we would drown. This is why land animals had to evolve a double loop, with lower pressure in the pulmonary circuit. Initially they made do with three chambers, and we still find this configuration in amphibians today. Then, from around the time of the first archosaurs (the common ancestors of dinosaurs and birds), they evolved four. Four-chambered hearts have been enough to power every kind of beast, from the nimble wren to the mighty blue whale, and since then there have been few changes to the essential architecture of the heart.

This is what happens when a pathological condition called pulmonary venous hypertension is untreated.

At four weeks the simple tube of a human heart begins to bulge and twist a little like an empty sock pushed from both ends. The bulging produces a series of chambers and, principally, a single atrium and a single ventricle. The twisting pushes the tube into an S-shape at first and then pushes the atrium behind the ventricle. Then the division into a left and right side begins, with partitions growing across like sliding doors. The ventricle is completely divided in two, as is the outflow tract that will form the ascending aorta and the pulmonary trunk. In the atrium the dividing septum grows down from the roof of the atrium, but just before it reaches the floor it starts to perforate at the top, leaving an oval opening called the foramen between the right and left atria. A second septum grows down next to the first. This one stops short of the floor, leaving a hole at its base. Because of the hole, blood can flow from one side of the heart to the other and still function like the two-chambered heart of a fish.

When the baby is born and takes its first breath of air, its lungs expand and the pressure inside them falls. This draws blood into the network of capillaries surrounding the expanding spaces in the lungs, and a tide of oxygenated blood then rushes through the pulmonary veins into the left atrium. The rise in pressure in this atrium pushes the second septum against the first, permanently shutting the valve between the two. Over the coming months, the septa fuse together, although traces of the original openings remain, and in some unlucky individuals the seal is not tight – a condition known as a ventricular septal defect, or a hole in the heart. But if all goes well, there we have it: a fully functioning four-chambered heart, set going like a fat gold watch the first time we smell the air.

Our hearts and the blood they pump follow rhythms that span all timescales of our existence. The red blood cells vibrating millions of times a second are also constituents in a continuous cycle of birth and death. Formed in the bone marrow, they jettison their nucleus before entering the bloodstream and, because of this, typically last only about 120 days before degrading and being flushed away to be replaced by new ones. With trillions of them in the body at any one time, the churn of creation and dissolution is about 2.5 million every second, or hundreds of billions per day. In addition, our bodies also produce many hundreds of billions of new platelets (which stop bleeding by clumping and clotting blood vessel injuries) every day and about ten billion white blood cells (immune system cells), and destroy an equivalent number.

At the other end of the timescale of our lives, a few of the immune cells in the blood endure most or all of a lifetime. And most of the cells in the heart itself endure from the time we are born to when we die. Indeed, it used to be thought that cell division in the heart came to a standstill shortly after we are born, and that we lived and died almost entirely with the heart cells we were born with. Within the last decade, however, it has transpired that this is not quite true. Even as the heart pumps without ceasing, a few of its cells are being replaced. The turnover is very slow – about one per cent of the heart per year up to the age of twenty, and declining gradually thereafter – but it is enough for a transformation. This discovery may be important in the development of therapies to increase the rate of regeneration. Still, the truth remains that for the most part our hearts are solid, enduring and unbelievably tough even as the body around them whirls through change.

The rhythms of daily life occur between these extremes of millions of times per second and less than once in a lifetime. And, for all our analytical and technological capabilities, not to mention our capacity for alienation, we never escape these beats.

The seat of the emotions

In the 1990s researchers discovered 'mirror neurons' in the brains of macaque monkeys that fired when they saw and/or imitated the behaviour of other monkeys. The excitement was enormous. The neuroscientist V. S. Ramachandran saw in them a basis for 'the extraordinary human capacity for empathy' even though they had been found in monkeys and not humans, and called them 'Gandhi neurons' because they broke down the barrier between human beings. The hype has largely dissipated since then, but we do know that, like many other mammals (notably elephants, whales and dolphins), humans have a strong tendency towards sympathy and fellow feeling. We also know that although this may be manifest unconsciously or consciously in the brain, it is also expressed in the heart. David Hume's musical metaphor – that humans resonate with each other like strings of the same length wound to the same tension – turns out to be almost literally true.

'As in strings equally wound up, the motion of one communicates itself to the rest; so all the affections readily pass from one person to another, and beget correspondent movements in every human creature.'
David Hume

Every year on the night of the summer solstice,

following an ancient tradition, villagers in San Pedro Manrique in Spain walk barefoot across seven metres of red-hot embers. In 2010 researchers monitored the heart rates of the fire-walkers, and found these could approach two hundred beats per minute towards the climax of the walk – enough, in theory, to cause a heart attack, although individuals remained outwardly calm and showed no ill effects afterwards. The researchers also monitored the heart rates of the fire-walkers' friends and relatives, as well as those of visiting strangers. They found that these followed a near identical pattern to those of the fire-walkers throughout the event, spiking and dropping almost in synchrony. But, the researchers found, the closer the social ties between the fire-walker and another person were, the more their heart rhythms were synchronized. Those of a walker's spouse or close partner were closer than those of a friend, and those of a friend were closer than those of spectators who were strangers. The correlation was so strong that they could accurately predict people's social intimacy simply by looking at the similarities between their heart-rate patterns.

At the other end of the spectrum, social isolation, anxiety and depression often correlate closely with heart disease and heart failure. The disorder and economic dislocation following the collapse of the Soviet Union was followed by a huge rise in both conditions. In Britain today, people who live alone have double the risk of serious heart disease: you really can die of a lonely heart.

It may be that one day humans or post-humans will live without a heartbeat. Artificial hearts that pump with a smooth, continuous flow have already proven to be viable, at least for short periods. Whether we could really live with them in the long term looks less likely: the circulatory system has evolved in tandem with the surge and cease of blood over hundreds of millions of years, and there may be all kinds of subtle but important effects that depend upon it. But if we could be smoothly engineered would we miss the heartbeat? Would the absence, even if covered up with a thousand other noises, give the feeling that something very deep within had died? Perhaps designers will create a fake beat and vibration, rather as they contrive the expensive-sounding clunk of a car door or the haptic feedback on a mobile phone.

Interventions that already exist can achieve astonishing things – but also, inevitably, have shortcomings. In his poem 'The Halving', Robin Robertson describes an operation to replace a naturally defective aortic valve in his heart. Robertson tells how, coming round afterwards, delirium and pain gave way to 'a blackness'; a disturbance of mood and cognition that some call 'pump-head'. In *Adventures in Human Being*, the doctor and writer Gavin Francis explains that pump-head is seen in up to a third of patients whose blood has been pumped through a bypass machine while their hearts were stopped for surgery. Some become violent, others quiet and yet others disinhibited and foul-mouthed. The cause is not known. It may be that tiny fatty particles loosed from the arteries by surgery or bubbles introduced into the blood disturb the cerebral blood flow. It may result from inflammation in the brain following the trauma of having your chest prised open and your ribs wedged apart. It may be a byproduct of cooling the blood in the bypass machine. But, says Francis, there may be another reason: even the most modern machines cannot mimic the natural pulse from the heart, and it may be that the heart's internal rhythm is essential to our wellbeing.

The rhythm of life

One characteristic of music – and perhaps ultimately its most important one – is that it connects people. It started, we may presume, with vocalization, singing and rhythmic movements of the body ranging from beating time to full-blown dancing. Pitch and rhythm vibrate throughout the body, but particularly in the chest cavity containing the heart and lungs, and especially when sung. Harmonic intervals also resonate, either as overtones in a single voice, or as a chord when sung with others in polyphony. In such moments it can feel as if we are getting to the core of

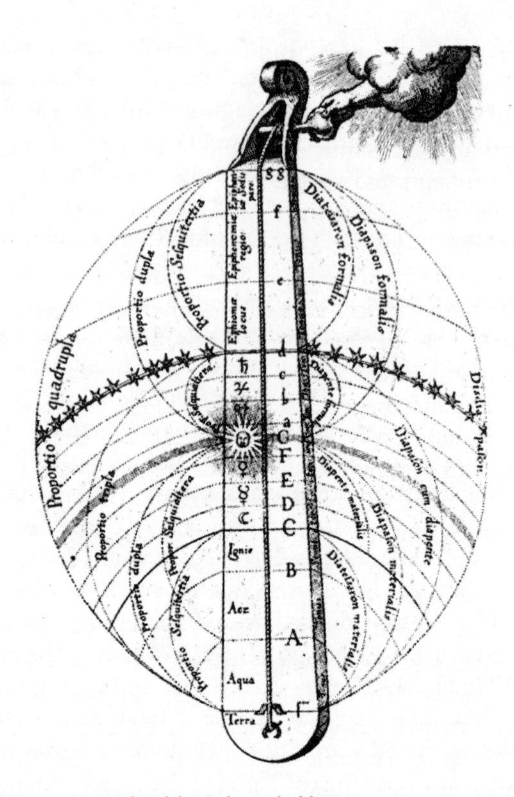

The Divine Monochord by Robert Fludd (1617).

'In every song there is distance . . . All songs are implicitly and often explicitly about journeys . . . In the sharing of the song, absence is shared and so becomes less acute, less solitary, less silent. And this reduction of the original absence . . . is collectively experienced as something triumphant.'
John Berger

existence. With music we can touch and be touched across distance.

In the early seventeenth century Robert Fludd used the image of a cosmic lyre to express the Pythagorean vision of a universe created according to mathematical ratios found in harmonic intervals. Today, psychologists, musicologists and others look, first, to findings in fields such as neuroscience or physiology. So, for example, if two tones form an octave, one set of auditory neurons may be firing twice as fast as another. Every firing of the slower one will have the same predictable relationship to the firing of the faster set, while neurons sensitive to both will get a repetitive pattern that is predictable and easy to interpret. The basis of harmony as we experience it, then, may be successful prediction in the early stages of perception – activating reward circuits in the brain. All this may be true, and yet more remains. Frank Wilczek, whose theory of harmony I have just described, also says that the equations for atoms and light are virtually the same equations that govern musical instruments and sound.

But whatever is the case with pitch, harmony and timbre, rhythm is essential to all or almost all the music that humans enjoy. It provides a frame against which the much higher-frequency vibrations of a single note or the multiple simultaneous pitches of harmony unfold within a musical narrative. Rhythm is vital and universal across traditions – whether in string quartets by Beethoven and Schubert, in dance anthems pumped out in nightclubs or in the songs of the pygmies of Central Africa. Music can be little more than rhythm, as in Steve Reich's *Clapping Music* or, with the addition of timbre, his piece *Drumming*. Even a performance of *4'33"* by John Cage includes rhythms such as the sound of breathing and, for the very attentive, the listener's own heartbeat.

Other rhythms of the body besides the heartbeat have an influence on music. Regular movements in dancing, running and walking can clearly play important roles. But music is never entirely divorced from heartbeat. Universally, musical tempo covers the same range as our heart rate, albeit sometimes pushing at the extremes of below thirty beats per minute (*larghissimo*) and above two hundred (*prestissimo*).

Repetition and, hence, predictability are important characteristics of almost all music, but we also enjoy variation and change that push at the boundaries of patterns, so long as they do not push too far. This is shown clearly in syncopation. A study in which participants heard funk drum-breaks with varying degrees of syncopation and rated how much they enjoyed them, and the extent to which the breaks made them want to move, found that syncopation is 'an important structural factor in embodied and affective responses to groove'. This sounds like a stilted way of saying the bleeding obvious – that groove makes people move and feel good. But the researchers had a serious point: medium degrees of syncopation elicit the most desire to move and the most pleasure. Syncopation excites us

A study of hundreds of musical recordings from around the world identified six rhythmic universals among eighteen features of music. These included a steady beat, two- or three-beat rhythms (like those in marches and waltzes), a preference for two-beat rhythms, regular weak and strong beats, a limited number of beat patterns per song, and the use of those patterns to create motifs, or riffs.

Timbre – the character or quality of a musical sound – has been well described by the writer and journalist Tim Falconer as its smell or *terroir*.

Syncopation: 'a disturbance or interruption of the regular flow of rhythm'; a 'placement of rhythmic stresses or accents where they wouldn't normally occur'. 'Syncope', the medical term for fainting, derives from a common root meaning to 'strike out' or 'cut off'.

by quickening us forward – in Earth, Wind & Fire's 'Boogie Wonderland', for instance, at an insistent 132 beats per minute – while maintaining a reference to the frame beat so we know where to anchor ourselves. But no less important, it seems, is the requirement that beats that are neither too simple nor too irregular and hard to predict. In Stevie Wonder's 'Superstition', where several harmonic lines are syncopated on top of one another, the beat is smooth and predictable overall, but convoluted and highly unpredictable in detail: there is 'controlled freedom', and the effect is tremendously energizing.

A heartbeat that is too simple and too rhythmic can be a warning sign of congestive heart failure.

Flow

Music can sometimes be a lifesaver, but there are times when it is not enough. We need to connect as directly as possible to rhythms that are larger than human. Among the things that have helped me to keep going has been walking and running in wildness and wet – places where few people go and where, as a result, the land is more alive to itself and what was peripheral can become apparent. Moving over uneven ground, heart and lungs working hard, every stride can bring the risk of tripping and falling, but there is also a sense of connection – a conversation between the body and the land without the intervention of the conscious mind. As each foot strikes, a vibration passes up through the legs: a beat shared with the land that is often regular but sometimes varies according to the length of stride, the qualities of the rock or soil and the angle of descent or ascent. In such circumstances, my own version of pumphead – a nauseous, disembodied condition arising from too much time in front of a screen manipulating symbols, or contemplating much of what goes on in what we call 'the world' – gradually begins to dissipate and, despite the exertion, I am at peace.

'Every landscape has its own particular soul.' Christian Morgenstern

When we breathe fresh air the lungs connect the world directly to the heart. Some winds – föhn, levante, Santa Ana, sharav and sirocco among them – bring unrest, either when they are imminent or upon

'Attention: mark this caressing zephyr, this mild breeze; it is the gaze of death constantly watching us, entranced from afar – a premonition of the strong gale that is to follow, and which will bear us away in its bosom, like motes in the wind.' Francisco González-Crussi

arrival. But I have encountered others that are benign. Nearly twenty years ago I helped sail a small boat across the North Atlantic in winter. From Bermuda eastwards towards the Azores we had a steady wind for days, and the boat slid on a straight course through the water like a wet bar of soap from between grasping hands. The wind continued at night and, surging forward under the stars, it felt like perfect freedom. For years afterwards the wind blew in my happiest dreams.

'. . . the wind,
filling entirely all
movements;
eternal colours and
eternal lights,
sea colours and sea
lights.'
Juan Ramón Jiménez

I did not have another such experience until a decade later, when I travelled to the Philippines to meet Antonio Oposa, a lawyer working to save the remnants of once glorious rainforest and coral reefs so that future generations might have something from which to rebuild. One night I stayed over in a simple building which had once been the holiday home of Oposa's parents. The house, which stood on the shore of a small island far from the city, was open at the eaves, and all night a steady sea breeze blew through, bringing relief from the heat and humidity of the day, and carrying dreams, as if from childhood, of a boat about to set out on a journey. Now, many years later, I still sometimes catch a hint of a dream-breeze like that even in the town where I live. From time to time, and especially on warmer nights in spring, the air here carries the smells of the soil and the woods, and even of the sea more than fifty miles away. All are breathing and, sensing this, I feel my heart beat just right and no longer feel as if I am suffocating.

In the 1990s Oposa won a legal battle in the Philippine Supreme Court for the right of children and future generations to inherit the rainforest intact. The case was celebrated internationally.

'Petrichor' (from the Greek *petra*: stone, and *ichor*: divine fluid) is the smell in the air before or as rain falls on hot, dry, stony ground. Following this, we might speak of 'chomichor' (*choma*: soil) and 'dasichor' (*dasos*: woodland).

Robert Fludd and other mystics claimed that the

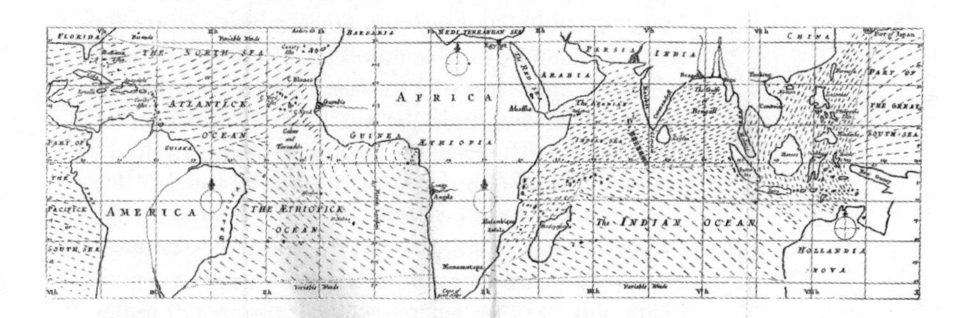

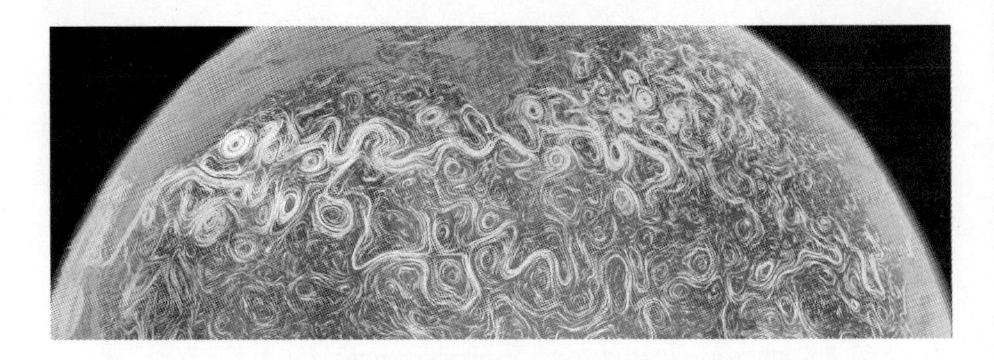

human body contains in microcosm all the features of the larger world, or macrocosm. You don't have to accept this as literally true to still see similarities and resonances across scales. During his study of the heart, Leonardo da Vinci discovered that blood swirls in vortices as it passes through the sinuses of Valsalva in the aorta, and, with the aid of the transparent glass models he built, saw that the turbulence helped in the smooth functioning of the aortic valve. In this he built on earlier observations of the effects of turbulent flow in rivers. Today we can also appreciate such flows on a vastly larger scale. Nature is, as Isaac Newton wrote, 'a perpetuall circulatory worker'. A map of the trade winds created by his contemporary Edmund Halley in 1686 shows one of the ways in which this is literally true. Today, such processes are understood in much greater detail. There are, for example, swirls and vortices in wind-blown currents on the ocean surface, made apparent to our eyes in beautiful dynamic maps such as NASA's Perpetual Ocean visualization.

Ocean currents differ in many ways from the circulation of blood. But there are similarities too. The forms of some surface currents resemble the flows that Leonardo saw, and the mixing and movement of seawater is vital to the continuance of abundant life. Such flows as we can see on the surface – from the smoke-ring shapes driven westward by the Agulhas Current off southern Africa to the billows of the Kuroshio Current extending off Japan into the enormity of the Pacific – are linked to others less readily apparent. In certain places on the planet, water rich in oxygen plunges to the bottom of the ocean and, over a cycle that lasts a thousand years or so, circulates around the planet before it surfaces again. At present, most of the additional heat of global warming is being soaked up in the oceans. The consequences (possibly foreshadowed by the rapid deoxygenation taking place at present) are hard to predict but are likely to be disruptive. For as long as a small, powerful beat continues inside us we can be witnesses and actors in the changes to come.

4

A HYPEROBJECT IN THE HEAD

Brain

The brain is a world consisting of a number of unexplored continents and great stretches of unknown territory.

Santiago Ramón y Cajal

The possibility suggests itself that no dreams, however absurd or senseless, are wasted in the universe.

Bruno Schultz

As if the sea should part –
And show a further sea –

Emily Dickinson

When I asked friends and acquaintances to recommend wonderful things that were relatively close by and cheap or free to access, several mentioned the great flocks of birds that feed on coastal mudflats and fly onshore in multitudes when the tide races in. The writer Robert Macfarlane talked in particular about knots: a species of stocky wader in the sandpiper family. In winter plumage they are silver-grey on top and white underneath, he explained, and when the light hits them in flight they seem to 'ping' brightly like flecks of snow or ice. Then they turn as a group and vanish – almost as if they've slipped out of our dimension into another. And then they turn again and they're in our world, visible once more. 'It's absolutely mesmerizing to watch.'

So just before dawn on the spring equinox I found myself at RSPB Snettisham, a nature reserve on the forehead of the skull-shaped curve of the Norfolk coastline which overlooks a large shallow bay on the east coast of England known as the Wash. Here, when the tide is high, knots and other birds come onshore in large numbers to roost beside a lagoon created by the extraction of gravel for runways in the Second World War. I had needed little persuading to come here: a murmuration, in which thousands of starlings wheel and morph like a single organism in the sky (and which may have evolved to deter and confuse predators), is one of the most beautiful things I know. The wonder is enhanced by learning that the same mathematics describe both this movement and the movement of superfluid helium, where the gas is cooled to near absolute zero and all the atoms are in the same quantum state. It may even be that dark matter, one of the great enigmas of cosmology, behaves in a similar way. Like music, a murmuration seems to be what Thomas

Browne called 'an Hieroglyphicall and shadowed lesson of the whole world'.

That the sea came in over the mudflats so fast was the first surprise. This was the highest tide of the cycle – Sun and Moon aligning to curve space-time just that little bit more – and the sheet of water tumbled in over itself like a river in spate. There was danger in this beast, heard before it was seen; but out on the mud the birds at its edge took to the air as blithe as the fingers of a harpist, only to loop down again before the oncoming water a little nearer to the shore. They repeated this movement several times until finally there was no mud left to land on, and only then did they start to fly in over the sandbar on which I stood towards the lagoon.

And here was the second surprise: the thrumming sound of hundreds and thousands of knots as they barrelled in, squad after squad, just a few feet above my head. I cannot find the words to say exactly what this sound was like. It was not unlike the roar of an aeroplane propeller absent the noise of the engine driving the propeller. And it was not unlike a bullroarer – one of those ancient musical instruments, sometimes known as aerophones, that have sacred associations in some traditions. But it was not very like them either. It was loud but it was discernibly composed of lots of smaller sounds – the fluttering of individual pairs of wings arriving at the ear fractions of a second apart with, perhaps, fractionally distinct timbres and microtones. Taken as a whole, the noise was massive and stirring, but there was gentleness in it. A little later I glimpsed the flash, disappearance and flash of knots turning in the Sun, and it was marvellous too, but the sound of their flight resonated within me longer and stronger than the vision of it.

The smallest time differences between sounds that we can detect are ten- to thirty-millionths of a second apart.

Sensing wonder

In *Wonder, the Rainbow, and the Aesthetics of Rare Experiences*, the scholar Philip Fisher writes that wonder is a relation to the visible world because 'only the visual is instantaneous, the entire object and all its details present at once'. Intuitively, this seems right. But the evidence, and a little reflection, show the truth to be larger and more complex. Certainly, vision is the richest human sense, capable of providing us with impressions of great detail and subtlety, but it is not single or instantaneous as it seems (and is it not even the fastest of the senses). Vision may seem effortless and direct but it requires as much as a third of the cortex to process. And what we see is also continuously shaped by what we hear, touch or sense in other ways, as well as by our mood, by what we remember and by what we already understand. In this sense we don't perceive objects as they are. We perceive them as we are.

The wonder for me on the shore at Snettisham that day was auraculous as much as it was miraculous. Light may travel almost a million times faster than sound, but because the pathways in the brain that process sound are shorter than those that process vision, we hear before we see. This immediacy (from the Latin for 'nothing intervening') can create a sense of immanence ('remaining within'), and may be part of the reason that unusual sounds and resonances are often a feature of sacred places. Ancient peoples recognized this well: prehistoric rock paintings are often located precisely where echoes are strongest. When flint of the kind that they used to make tools is knapped in a reverberative setting like a cave or near a rock face, it can create a high-pitched flutter very like a bird flying by. The musicologist Ian Cross and the anthropologist Ezra Zubrow, rationalists to the core, report a re-creation of the effect as 'quite unearthly . . . [as if it had] awoken some real yet invisible entity'.

When we hear, sound transmitted through the eardrum and three tiny bones reaches the cochlea, where different frequencies reverberate at different points along a membrane inside its snail-like curl. The vibration of the membrane brushes against tiny hairs linked to nerves that reach into the brain, where the signal is processed. The sensitivity of this system can be astonishing. In the right circumstances we can hear vibrations of less than the diameter of a hydrogen atom.

In everyday life, five senses are readily available to our conscious awareness: we *see* that we see, *hear* that we hear, *feel* that we touch, *taste* that we taste and *smell* that we smell. (In doing so we can, on occasion, feel vibrantly alive; sometimes, the mere act of noticing can bring about a state approaching the one that the Zen scholar D. T. Suzuki called enlightenment – a sensation 'like ordinary everyday experience, except about two inches off the ground'.) But these are not our only gateways to the world. By some counts we have as many as twenty, including the somatosensory system, which registers temperature and pain, as well as itch and tickle, and proprioception, which allows us to be aware of the position and movement of our head and limbs. (This is what allows you to, for example, touch the end of your nose with a finger when your eyes are closed.)

And fundamental among the senses enabling me to experience wonder in Snettisham that morning was one we seldom think about. As I looked up, tilting my head to listen and quickly rotating it to attend to the knots as they passed over, my vestibular system registered the movement and changing orientation of my head. The information it gathered and processed within a tenth of a second told my brain how my eyes needed to move to follow the birds' rapid motion and keep it in focus even as it continued to monitor the direction of gravity so that I could stay upright.

Two fingers' width in from the sides of the skull, next to the cochlea and set at right angles to each other like three adjacent sides of a cube, are three rings. Varying in size from a little smaller than a US one-cent coin to about the diameter of a British five-pence piece, these are the semicircular canals. Each is a hollow tube filled with fluid that swashes against tiny hairs on the inside surface of the tubes as the head moves. Nerves embedded in the sides of the tubes register the movement of these hairs within the fluid, enabling us to sense rotation and acceleration. And at the base of the tubes, where the canals meet, are two otoliths: tiny stones of calcium carbonate that rest on small sensing hairs within earlobe-shaped capsules. The pressure that the stones exert on the hairs under gravity tells us which way is down. Together, the canals and otoliths are our gyroscopes, accelerometers and sea anchors. They are always 'on', providing us with a continuous reference for the position of the head and its rate of change with respect to the Earth. They 'hear' where we are, and where we are going. Without them, I would not be able to attend fully to the flight of the knots or to much else, since if they malfunction, I will succumb to nausea, or fall over, or both.

Jellyfish, with whom we last shared an ancestor more than 540 million years ago, have statocysts, which are analogous to the otoliths of our inner ears.

The vestibular system is evolutionarily ancient. It developed in some of our earliest animal ancestors before the systems for auditory and visual processing, and it did so in tandem with the reticular core in the brainstem, another evolutionarily ancient system which today sends impulses throughout the entire brain to keep us awake and alert. Together with the reticular activating system, it has widespread connections across the rest of the brain, and as a consequence contributes to a wide range of brain functions from the most automatic reflexes to the highest levels of consciousness. It enables us to move through the world and attend to it as integrated, self-aware and balanced beings – helping me, for example, to get to where the knots are flying in the first place, follow their flight, and feel good about the experience. It is also sensitive to low frequency and infrasound below the threshold of about twenty hertz that can be registered by the cochlea, and for this reason the vestibular system is sometimes called the ear of the body. Conservation scientists working over many years with elephants report that they sometimes *just know* when animals who have been absent for hours, days or weeks are about to appear – predicting their subsequent return with near perfect accuracy. This is reckoned to be because the scientists 'hear' deep sounds from the approaching elephants through their vestibular apparatus without being consciously aware of it.

The mystery of the brain

The beautiful piece of engineering that is the vestibular system is entirely hidden from view, but it is not hard to get a sense of how it works, at least in general terms. The brain, by contrast, is dauntingly complex. Despite enormous progress over the hundred and twenty years or so since Santiago Ramón y Cajal and others showed it to be a network of neurons, much about the brain is still poorly understood and even utterly mysterious. Gary Marcus, a psychologist and neuroscientist, says that researchers still have yet to discover many of the organizing principles that govern its complexity, let alone develop a good theory as to how it actually works.

It is said often (probably too often) that the human brain is the most complex known object in the universe after the universe itself. The statement is so easy to make that the enormity of its claim can easily be overlooked. But every now and then a detail comes up that helps to drive it home. There are about a quadrillion – 1,000,000,000,000,000 – connections in a human brain. This number is staggering enough. Staggeringlier, at each synapse, where these connections take place, at least several hundred different

A quadrillion is, apparently, about the total number of ants alive on Earth at any one time, their total mass being approximately equal to the total mass of humans.

proteins self-organize to regulate the fusion, recycling and movements of the synaptic vesicles involved in neurotransmitter release between a reservoir pool, an active zone and a cell membrane terminal each time the synapse fires, which it may do hundreds of times per second. An image showing the large-scale organization of these proteins published in 2016 took researchers more than a year to create, and yet it represents a single instant in a process going on with almost inconceivable rapidity all the time. Reading a brief report about the image, I was reminded of the term 'hyperobject', which was coined by the literary theorist Timothy Morton to describe things that are so far outside normal categories of time and space that they defy perception and comprehension. Morton had in mind the likes of climate change, mass extinction and radioactive plutonium. But the brain also defies perception in its relation to time, space and dimensionality, and I think we can rightly call it a hyperobject in the head.

The human brain is implicated in more wonders, and horrors, than words can tell, but for all its complexities there are also things we can say about it that are both simple and true. It is, after all, just an organ of the body – a lump of tissue about the size and weight of a coconut with the texture of soft jelly or butter, and after death it rots just the same. In 'Going Home', Leonard Cohen sang of an alter ego (or a stage persona, or a true self) who knows he's really nothing but a brief elaboration of a tube. We don't have to think that this was *all* that Cohen (or any human being) was, but it is *one* of the things. And in that spirit, here are seven observations about the brain. Each is followed by an elaboration – some of them brief, some a little longer – of the substance behind the observation. Even the longest of them is sketchy (and may soon be dated) in comparison with what is known. I hope, however, that they are enough to act as starting points for curiosity and wonder.

1. The brain interacts, predicts and creates

'The cranium is a space-traveller's helmet,' wrote Vladimir Nabokov, but locked within that hard box of bone, and to all outward appearances an inert lump, the brain is in fact the most active organ in the body, requiring about twenty per cent of the energy we consume even though it accounts for less than two per cent of our mass. It is hungry because it is our most agile and outward-facing organ, responding within thousandths of a second to changes in its environment, as well as monitoring and controlling many inward bodily processes. 'The brain', writes the neuroscientist Antonio Damasio, 'is in the business of directly representing the organism and indirectly representing whatever the organism interacts with.' Sensations flow in and actions are initiated, and the brain distinguishes and manages interactions between the two. 'Somewhere in the middle is intelligence,' says Larry Abbott (yes, another neuroscientist); 'that's where the action is.'

'The human brain is the most public organ on the face of the earth, open to everything, sending out messages to everyone.'
Lewis Thomas

But rather than merely responding to outside events, networks of neurons are constantly stimulating each other. Whether you wake or sleep, the two hemispheres of the brain are continuously generating high-frequency electrical waves. Processing external stimuli – data from the world outside the head – accounts for a small proportion of the total energy consumed by the brain: perhaps less than five per cent. 'The nervous system primarily acts as an autonomous device that generates its own thought patterns,' says the cognitive psychologist Stanislas Dehaene.

The brain is vital but it is not everything. As the computer scientist Ben Medlock notes, each of the trillions of cells in the human body is a piece of networked machinery with as many components as a jumbo jet, and its ancestors evolved in encounter with the natural world over billions of years. Long before we were conscious, thinking beings, our cells were reading data from the environment and working together to mould us into robust, self-sustaining agents. We think with our whole body, not just with the brain.

Among the things that networks of neurons appear to be doing is creating predictions of what, based on your past experience, you will encounter next. They then update and correct these predictions in the light of new information. When, for example, we have the impression that we see something – whether it be a lemon or a lemming – the visual cortex is sending several times more information to the thalamus, an information hub at the centre of the brain, than the eyes are. Most of the time, our brain – a 'multi-layer probabilistic prediction machine' in the phrase used by the neuroscientist Karl Friston – takes information from the eyes to update a picture of what it already believes to be there. In this sense the brain makes a world first: it remembers, dreams and imagines before it sees.

2. The amazing architecture of the brain is not hard to understand

Santiago Ramón y Cajal used a similar image: 'We could say the [brain] is like a garden planted with innumerable trees . . . which, thanks to intelligent cultivation, can multiply their branches and sink their roots deeper, producing fruits and flowers of ever greater variety and quality.'

At least, it's not hard to understand in outline. One way to visualize the brain, the neuroscientist Christof Koch has suggested, is to compare it to a forest – though not just any forest. There are, says Koch, about as many neurons in your brain as there are trees in the Amazon basin, and their diversity of shapes – their distinct roots, branches and leaves covered with vines and creepers – is comparable. Imagine flying for hours over the Amazon in a small plane, he says, and you have some idea of the splendour and diversity hidden within the pinkish-grey lump of goo between your ears.

The comparison is beautiful, and apt so long as it is not taken as precise or literal. The numbers, for instance, are not quite right. The best estimate of the number of trees in the Amazon is 390 billion. That is more than four times the number of neurons in a human brain, currently reckoned to be about 86 billion. Still, even this is an enormous number, approaching the number of stars in the Milky Way. Translated into trees it would be enough to cover about 1.2 million square kilometres, which is almost twice the area of Texas or sixty times that of Wales. As for diversity, there are reckoned to be about sixteen thousand different species of trees in the Amazon, while the number of different kinds of neurons in the human brain is at least several hundred and perhaps more than a thousand. A typical neuron has about fifteen thousand dendritic spines, each a hive of activity, forming and re-forming – roughly comparable to the number of branches and twigs on a large tree.

The many kinds of neurons fall into three categories: sensory neurons carry information from sense organs to the brain; motor neurons send commands to muscles and organs; and interneurons relay information between neurons close together or over greater distances across different regions of the brain.

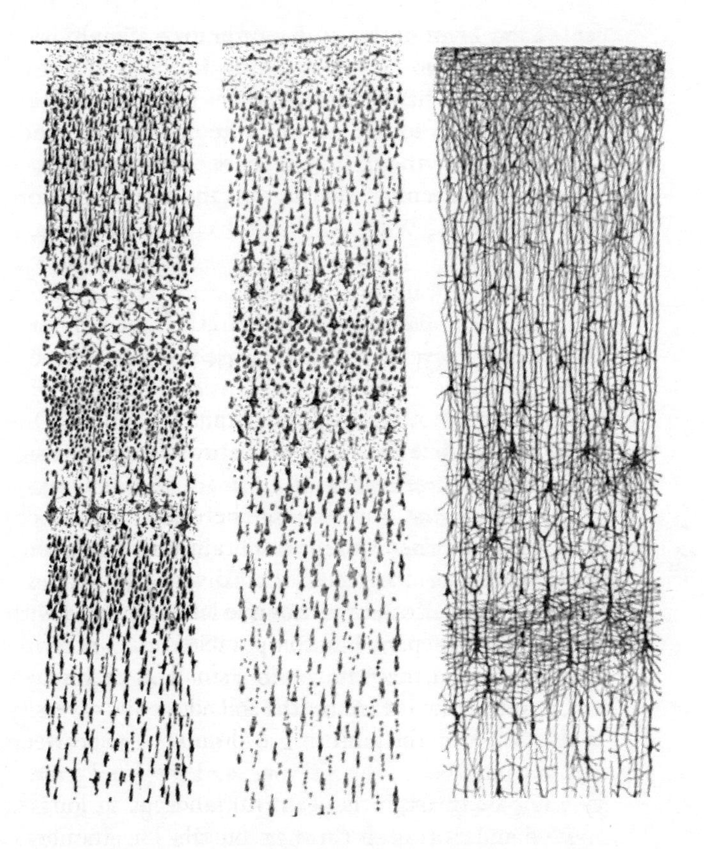

So although it is not quite on the scale of the Amazon, the brain does resemble a forest of tremendous dimensions, far bigger than even than Cosimo Piovasco, the nimble aristocrat in Italo Calvino's novel *The Baron in the Trees*, could explore in a lifetime. But we need to imagine it as a forest in which the trees and their roots are stacked (in the cerebrum at least six deep) into the sky as well as extending over the horizon: a Wistman's Wood or Black-a-Tor Copse reimagined by Piranesi or Escher. Within these thickets, the axons of some neurons carrying an electrical impulse away from the cell's central body extend like impossibly long roots or vines. If a pyramidal cell in the cerebral cortex, for example, were magnified so that its main cell body were the size of a human body, its axon would be a cable a few centimetres in diameter extending more than a kilometre.

Among the differences between the brain and a forest is, of course, in the way they communicate. Trees do 'talk' to each other, using chemical messages, vibrations and other signals in what has been called the Wood Wide Web. But, so far as we know, nothing within or between trees matches the rapidity of electrochemical transmission within and between neurons of the brain. It can take up to an hour for chemicals to travel to where they are needed in a tree to defend its leaves from attack. Inside a neuron, by contrast, signals are transmitted from the branch-like dendrites to the cell body, where those signals are processed, in thousandths of a second. The outcome of that processing is then sent with equal rapidity along the axons to synapses which communicate with the dendrites of other neurons. Because the whole sequence takes just a few thousandths of a second, neurons can fire many hundreds of such action potentials per second, encoding information in precisely timed sequences of impulses called spike trains. Typically, however, neurons fire as parts of networks that may number in the millions, and waves of activity will pass through these networks rather as gusts of wind and storms do through the trees – although these 'winds' are within the cells rather than above and around them, and some neurons are directly connected to others that are far away.

If you zoom out from a view of individual cells to the tissue as seen by a surgeon, the brain is breathtaking in a different way. Looking at its surface through the scopes of the neurosurgeon Henry Marsh, the author Karl Ove Knausgaard describes a landscape opening up in front of him:

> I felt as if I were standing on the top of a mountain, gazing out over a plain, covered by long, meandering rivers. On the horizon, more mountains rose up, between them there were valleys and one of the valleys was covered by an enormous white glacier. Everything gleamed and glittered. It was as if I had been transported to another world,

another part of the universe. One river was purple, the others were dark red, and the landscape they coursed through was full of unfamiliar colours. But it was the glacier that held my gaze the longest. It lay like a plateau above the valley, sharply white, like mountain snow on a sunny day. Suddenly a wave of red rose up and washed across the white surface. I had never seen anything quite as beautiful, and when I straightened up and moved aside to make room for the doctor, for a moment my eyes were glazed with tears.

Take away the surgeon's scopes and zoom out to the scale of everyday objects visible to the naked eye and none of this is apparent. What we can do, however, is appreciate something of the brain's overall architecture and organization.

First there's the cerebral cortex – the crinkly dome or fantastical bubble that, from Abby Normal in *Young Frankenstein* to the alien craniums in *Mars Attacks*, appears as a cartoon synecdoche for the whole brain. The cortex resembles a sheet, rather like a cloth that has been scrunched up into a ball with a deep fold (the medial longitudinal fissure) separating its two hemispheres, but joined beneath by a thick white band called the corpus callosum. Rotating it in the palm of one's hands much as one might turn a small globe, the maze of ridges (gyri) and furrows (sulci) on each hemisphere can be grouped into four regions, or lobes, each of which specializes in different functions. Broadly speaking, the frontal lobes (which sit behind the forehead) deal with reasoning and decision-making, plus planning and execution of voluntary movements. The parietal lobes, which sit under the crown of the head, mainly handle somatosensory and touch information from the body. The temporal lobes, which are on the sides, process hearing and speech, and play important roles in memory and spatial navigation. The occipital lobes, at the back, are responsible for vision.

Unfolded and smoothed out, the cortex would cover about 2,000 square centimetres, or about the same area as a car tyre. Its six layers are together 2–4 millimetres thick, or about half as thick as an iPhone.

But the brain is much more than the cortex. Four-fifths of its neurons are in other, no less essential, parts of it. Underneath the cortex – and partly separated from it by the ventricles (spaces filled with

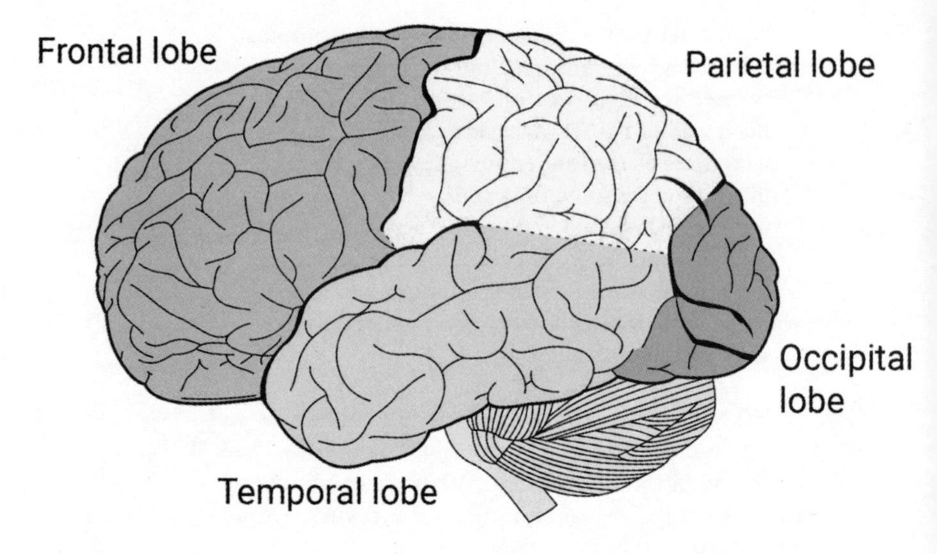

The cortex.

cerebrospinal fluid) – is an array of knobs, nodes and networks that make up the limbic system. Sometimes called the mammalian brain, this system plays key roles in the formation of memories and in shaping emotions. Seen in a graphic cross-section from the left or right side of the head, the limbic system looks a little like an index finger and thumb curled so that they are almost touching at the front, with the cingulate gyrus forming the index finger on top, and the hippocampus (where memories are formed) the thumb underneath. The almond-sized amygdala (which is involved with fear and aggression) is perched between where the two would meet.

Sitting in the core of this array is the thalamus (Latin for 'deep chamber'). This consists of two walnut-sized knobs that relay information from the sense organs to lobes in the cortex, and receive and integrate information sent back from those regions. Surrounding the thalamus are the basal ganglia, which are involved in the control of voluntary movement. Also adjacent is the midbrain – a relatively small region in humans but one with fundamental roles: it contains the reticular core, which (as noted in

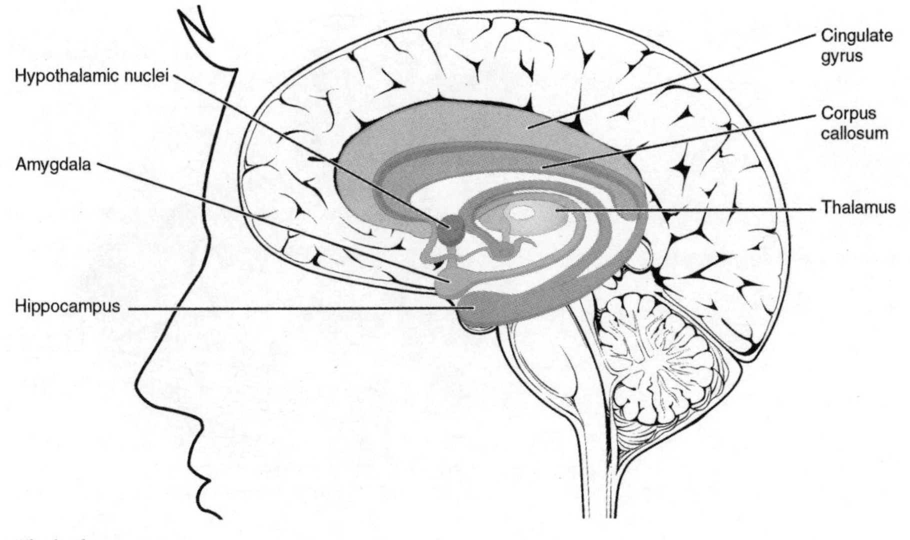

The limbic system.

connection with the vestibular system) relays information from the eyes and ears, and is also involved in basic behaviours such as sleeping, waking and walking. It also contains neuron clusters that produce dopamine, a neurotransmitter used in the brain's reward circuits.

Underneath the limbic system is the hindbrain, or brainstem, which, exposed to view, looks as strange as a fórcola, the non-intuitively shaped swerving rowlock of a Venetian gondola. It consists of three structures at the top of the spinal cord. The lower portion, called the medulla oblongata, controls involuntary functions such as breathing and heart rate, and is associated with our state of wakefulness. Above it is the pons, which relays signals between different parts of the brain. The third component is the cerebellum (Latin for 'little brain'), which sits at the back of the head just underneath the occipital lobe like a second brain that has sprouted somehow but not yet grown to full size. Its two fat, fleshy wings, a bit like those of a seriously overweight manta ray, span about ten centimetres. The pleats on its surface resemble the throat of a blue whale, but sliced in cross-section it looks more like a

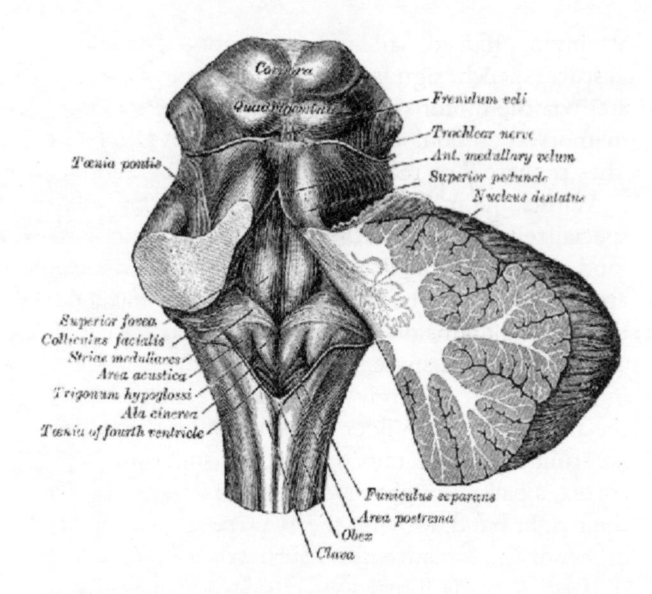

Hindbrain, with cerebellum in cross-section.

cauliflower. The cerebellum is involved in controlling balance, coordinating movement and learning new motor skills. Even though it only accounts for about ten per cent of the total volume of the brain, the cerebellum contains nearly four times as many neurons as the cortex (albeit more simply organized) and more neurons than the whole of the rest of the brain put together. This gives you some idea why computers can beat a human champion at chess or Go but cannot yet play with a ball as well as a five-year-old or a dog. All that movement takes huge amounts of brain power.

3. The brain is about movement and mapping

To say that the main purpose of the brain is to enable us to move around may sound like one oversimplification too many, but the neuroscientist Daniel Wolpert does not hesitate. 'To understand movement is to understand the whole brain,' he says. Movement is the only way we have of interacting with the world, whether foraging for food or attracting a waiter's attention. Indeed, all communication, including gesture, speech, sign language and writing, is mediated via the motor system. 'When you are studying memory, cognition, sensory processing,' says Wolpert, 'they're there for a reason, and that reason is action.'

With some major exceptions, such as those with specialized defences and those like corals to whom the food comes along by itself, animals that do not move are likely soon to be dead. To help them navigate, the brain contains or creates maps of both the body and the external space. The use of the word 'map' in this context is barely metaphorical. In the 1920s the neurosurgeon Wilder Penfield found that when he stimulated a certain part of the somatosensory cortex, a patient might report a tingling sensation in their right hand. Moving the electrode a centimetre higher on the somatosensory strip would evoke touch sensations in the forearm or elbow. A similar thing happened with the motor cortex. Stimulating one part of it would cause muscle twitches or small movements

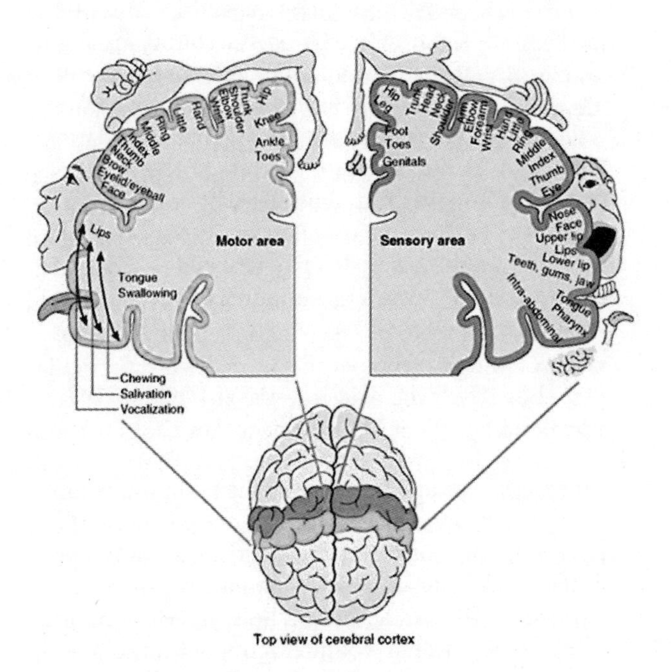

Body mapping in the somatosensory and motor cortex.

on the opposite side of the body, and moving the electrode to an adjacent area would cause the same movement in an adjacent body part. Penfield found that although there were small differences between individuals, the overall organization of these maps is essentially the same in everyone: the body is represented on the surface of the brain in a highly ordered fashion, with adjacent body parts mapping precisely onto adjacent areas of the brain. (These areas can remain active even when a body part is lost: a factor in phantom limb syndrome.)

Something very like a map is also created when the brain starts processing information it receives from the eyes. The pattern of light and dark that has struck cells in the retina is maintained in ganglion cells – individual strands in the thick cable that is the optic nerve – all the way to a region of the occipital lobe, and the exact layout of those cells is replicated so that this region has a fully laid-out image, in electrical activity, of the optical image that is present at the back of the eye.

But perhaps the most striking instance of mapping in the brain is found in what are known as place cells and grid cells. A combination of the former, first identified in the 1970s in the hippocampus of rats, fire when an animal is in a particular place and facing in a particular direction. The same combination of place cells fires when the animal visits the same location again, but a different combination will be active when it visits another site. Grid cells, which were first identified more recently in an area called the entorhinal cortex, differ from place cells in that they are active not according to whether the animal is in a particular place but according to where the animal moves on a grid that is, in effect, being projected by the brain onto external space. Imagine the room in front of you has a grid on the floor made up of tessellated triangles. One grid cell will fire whenever you reach the corner of any triangle in that grid. Shift the grid a little, and another will be responsible for firing every time you reach the corners of the new grid's triangles, and so on.

Place cells and grid cells suggest a deep evolutionary connection between mapping and memory, with

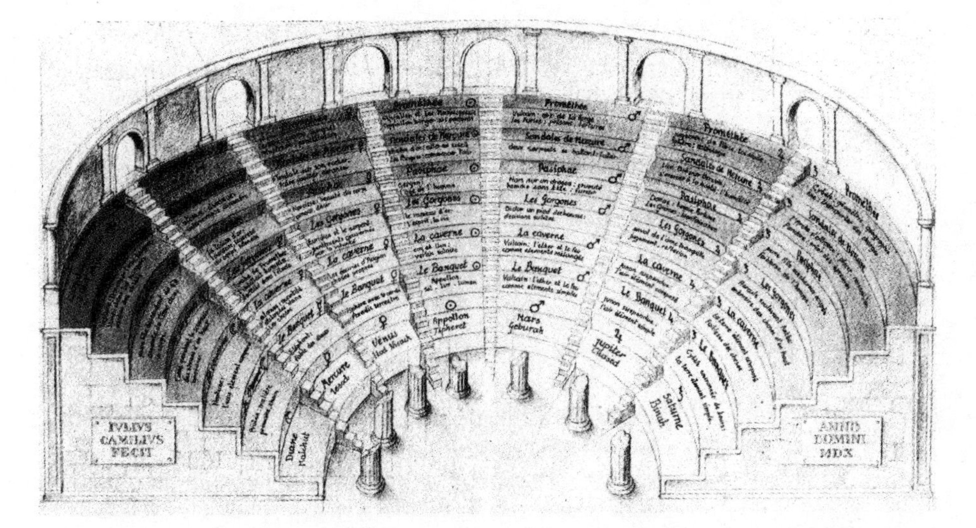

The memory theatre or palace – a technique developed to help the storage and ready recovery of memories by allocating each one a niche in an imaginary location – has deep roots in the way our brain evolved.

mapping having come first. 'Every animal that has to move to acquire food cares about the "where" above all things,' points out the neuroscientist Hugo Spiers. As our ancestors developed more intellectual lives, he suggests, the hippocampus storing those 'wheres' was co-opted into an all-purpose mechanism for anchoring important events into memories.

4. Brains are always changing

To the naked eye, a living brain looks inert, but within there is seething and almost constant transformation. Neurons, axons, dendrites and synapses are constantly adapting to their electrical and chemical 'experience', as part of learning, to maintain the ability to give different responses to different inputs, and to keep the brain stable. The states of all of these components are constantly affected by tides of chemicals sent out from the brainstem that determine such things as wakefulness, and by hormones from the body that help drive motivations. As we learn, the strength of synaptic connections between neurons changes and, with reinforcement over time, new pathways are created while old ones are lost. 'The brain is a vast dynamic network,' says the neuroscientist David Eagleman; 'instead of hardwired, it is live-wired.'

And although the brain is not physically moving, its architecture changes substantially over the arc of a lifetime. In early development there is massive growth in the number of synaptic connections followed by rapid refinement. During the first month of life after birth the number of connections increases twentyfold, from about fifty trillion to one quadrillion, and new connections continue to be made for at least the first three years. But as a child's brain develops, pathways that are little used are pruned, and during peak pruning periods as many as a hundred thousand synaptic connections may be eliminated every second. This is a crucial part of how the brain learns and becomes more able as irrelevant 'chatter' is edited out. Later, as a child's brain matures towards adulthood, a fatty white substance called myelin increasingly sheaths the axons of many neurons, insulating them and improving their function when they fire. With old age comes shrinkage and decline.

One of the ways in which the brain changes that has received most popular attention in recent years is brain plasticity, and in particular its ability to adapt, to recover from injury and to mitigate effects associated with ageing. An extraordinary example of adaptation is the instance of a young woman in China who was admitted to hospital complaining of dizziness and who was found to have no cerebellum – that is, the entire lower-back part of her brain was missing. Despite this, she had been living almost normally for twenty-four years, with only moderate motor deficiency and mild speech problems such as slightly slurred pronunciation. A dramatic case of coping with trauma occurred in the United States, where a young girl had an entire hemisphere in her brain removed in an operation to save her from cancer and has been living a normal life ever since. In both cases, other areas of the brain had stepped in to take over the essential functions that would otherwise have been lost.

But even when such extreme circumstances are

excluded, there is clearly a lot that can be done in the normal run of life by almost everybody to build new capabilities in the brain and to adapt to its eventual decline. Simple things such as a daily walk promote neurogenesis, or the generation of new neurons from stem cells in adults. And better brain function is also protected simply by remaining socially engaged. A study of elderly nuns in the United States found that while nearly a third of them had brains with physical signs of full-blown Alzheimer's, few of those who did showed symptoms. Having responsibilities and learning new skills appeared to be protecting them, and even as some parts of their brains were degenerating, other parts were finding ways to compensate and keep them functioning normally.

The plasticity of the brain is also evident in its relation to technology. With the help of tools, the mind becomes more than the brain on which it depends, and this in turn shapes the brain. 'We shape our tools,' observed the media theorist Marshall McLuhan, 'and afterwards they shape us.' Fire is one of the earliest and most important examples. Cooking allowed early humans to access more nutrients from food than their ancestors, enabling the evolution of much bigger brains. But tools come in countless forms. Consider the humble case of a stick for a blind man. Over time, his brain and mind reorganize, projecting tactile sensation onto the tip of the stick and treating it like part of the body. The man turns touch into sight. 'Mind-changing technology has a futuristic ring to it,' says the cognitive archaeologist Lambros Malouforis, 'but humans have used it since they first evolved.' Protean as it is, the human mind is always an unfinished project. Even if they live up to the hype, interfaces that transmit thoughts directly to and from brain to machine or another brain (which are said to be under development by Faceboook, Elon Musk and others) will be just one more extension of the mind–tool cerebellum.

'He'll . . . twist and turn / into every beast that moves across the earth / transforming himself into water, superhuman fire . . .' Description of Proteus in *The Odyssey*

5. Human brains are not the ones that are amazing

Evolution has created brains in wonderful variety. Some are quite differently organized from those of humans, yet able and intelligent. And some that are organized in similar ways to ours, such as those of elephants, dolphins and whales, are greatly underrated.

To begin, brains are not the only things that think; we just need to be a little less narrow-minded in what we mean by 'thinking'. Many organisms remember and process information about the world around them without a brain, and even without any neurons at all. Single-celled slime moulds cooperate to remember, make decisions and anticipate change. They can slow their movement in anticipation of an expected dry spell and avoid sticky areas where they have already travelled – this is a kind of externalized spatial memory that reminds the organism to explore somewhere new. Other creatures, such as sea slugs, embody, without any mental representation in their simple brains, complex forms of hyperbolic geometry that human mathematicians did not discover until the nineteenth century.

There is also a lot more going on in non-animal beings such as plants than meets the casual eye. Although they do not have anything we recognize as a nervous system, plants make exactly the same chemicals – such as serotonin, dopamine and glutamate – that serve as neurotransmitters and underlie mood and emotions in us. And plants have signalling systems that work in very much the same ways as those of humans and other animals. They feel, remember (possibly using prions, the proteins infamously linked to mad cow disease) and learn. We detect chemicals by smell and taste; plants sense and respond to chemicals in the air and soil and on themselves. Roots that are growing towards an obstacle or toxin sometimes alter course *before* contact. In 2016 it was discovered that plants may 'see' underground by channelling light from the surface all the way to the tips of their roots. Leaves turn to track the Sun.

'It is scarcely an exaggeration to say that the tip of the [root of a plant] . . . acts like the brain of one of the lower animals . . . receiving impression from the sense organs and directing the several movements.'
Charles Darwin

Plants attacked by insects and herbivores emit distress chemicals, causing adjacent leaves and neighbouring plants to mount chemical defences, and alerting insect-killing wasps to move in. At least one plant that eats insects, the Venus flytrap, can 'count' the number of times a struggling insect touches its trigger hairs, registering up to at least sixty touches and using the information to ramp up its digestion.

Plant cognition may be awesome, but animal brains can be even more so. A honeybee brain only has about a million neurons and the mass of a grain of sugar, but bees recognize patterns, scents and colours in flowers, remember their locations, and more. Famously, the waggle dance communicates the direction, distance and richness of nectar that a bee has found to members of its hive, but much less well known is that other bees will interrupt the dance if they have experienced trouble, such as a predator, at the same flower. And bee brains have more in common with human brains than you might expect. They contain, for instance, the same 'thrill-seeking' hormones that in human brains drive us to seek and derive pleasure from novelty. Nicotine-like pesticides may damage or kill bees but before doing so they tickle the reward centres in their brains in much the same way as nicotine acts on the brains of human addicts.

The brain of an octopus is even more astonishing. The last common ancestor of octopuses and humans – probably a worm-like creature with a few hundred or a few thousand neurons – lived over 550 million years ago, and since then we have evolved along entirely independent lines, with the consequence that our brains are organized in completely different ways. The brain of an octopus is not in its scrotum-like 'head' (which in fact houses its guts) but is partly (one third) wrapped around its oesophagus, or throat, while two thirds of it is distributed across its eight arms. The pianist Igor Levit playing a Bach partita, in which each hand dances with what seems like complete autonomy, is impressive, but an octopus can do up to eight different things at once, with each arm literally thinking for itself albeit coordinated by its central brain.

Octopus brains have only about five hundred million neurons – a twentieth of the number in a human brain – but they are quick and able learners. This is all the more amazing when you realize that they hatch out after their parents die and have to work everything out for themselves through observation, trial and error. They can solve novel problems that defeat dogs and small children – recognizing, for example, symbols placed on doors behind which food has been hidden even after the locations are switched. In the wild they have been known to board fishing boats and prise open holds to steal crabs. In captivity they are able to break out of aquariums that keepers had believed to be secure and shimmy across open spaces to neighbouring tanks in search of food. Some octopuses can recognize individual humans: experimenters have dressed people up in exactly the same clothes, but the octopuses could still distinguish between people they liked and those they didn't, and behaved differently towards them. Clearly, human camouflage doesn't fool a creature that can change the colour and texture of its own skin to match its surroundings and communicate with others of its kind.

Birds also show remarkable mental abilities with brains that are much smaller than and very different from ours. Some parrots can understand and correctly use a hundred words or more, but the cleverest birds of all are probably members of the crow family, or corvids. They are capable of multistep causal reasoning, flexibility, imagination and prospection, and they apply these abilities in tool manufacture, mental time travel and complex feats of social cognition. Crows in urban Japan, for example, drop hard-shelled nuts onto pedestrian crossings, having figured out that cars will run over and crack them but will also regularly stop so that people can cross, allowing the crows to retrieve the nuts safely. Ravens can imagine being spied on and refuse to cooperate with other birds that have cheated on them in the past. And yet the corvids, like other birds, have no neocortex – the part of the brain in humans and other mammals long thought to be essential for advanced mental function. Having diverged from mammals more than 300 million years ago, avian brains have instead evolved neurons densely packed into clusters called nuclei, which do the heavy lifting instead.

Few other bird families rival corvids for smarts, but the brains of many other species can do things that elude humans. The philosopher and author Charles Foster notes that when swifts hunt bees they select the stingless drones, identifying them by eye even though they are flying at fifty-five kilometres an hour. Contemplating this and other ways in which swifts live at high speed and over long distance, Foster concludes that 'wonder is a function of resolution' (that is, detail and focus). By comparison, 'we, at our very best, are snails'.

The brains of songbirds outmatch ours in a different way. Starlings and nightingales can sing or imitate some sixty different songs with far greater fidelity and rapidity than we can ever achieve or even hear. Wrens can sing over a hundred distinct notes in eight seconds. Mockingbirds can copy hundreds of birds' songs and other noises, and sing twenty of them in a minute. They do so with an elaborate set of seven pairs of muscles in the syrinx in which two membranes can vibrate independently to produce two different, harmonically unrelated notes at the same time, shifting in volume and frequency to create some of the most acoustically complex sounds in nature. All this is managed in dedicated areas of their tiny brains.

The syrinx is the lower larynx or voice organ in birds. It is named after the nymph transformed into a reed by her sisters in order to hide her from Pan, the god of fertility, and fashioned by him into a musical instrument.

Many of us are used to the idea that while human brains share the same basic architecture as that of other mammals, we have an exceptionally well-developed cortex, and that this supposedly supports a superior intelligence. So, for example, a recent study found that while the brains of African savanna elephants have around 257 billion neurons, or nearly three times as many as a human, nearly 98 per cent of them are in the cerebellum and only 5.6 billion are in the cortex. The human cortex, by comparison, contains about 16 billion neurons, and they are more densely packed than those in an elephant, which is likely to enable them to work together faster. On similar grounds, the larger brains of dolphins and whales (the brain of a sperm whale, for example, is six times the size of a human brain) are often viewed as being a function of the processing power necessary to control a much larger body rather than as being indicative of superior cognitive skills.

But the reality is richer than that. Elephants have a relatively larger and more elaborate hippocampus, a part of the brain associated with memory formation. And while whales and dolphins have only five layers in the cortex compared to our six, it is possible that their abundant glial cells play important roles in cognition. More to the point, however, the brains of other mammals mean they can do some things humans cannot do, and cannot do some things that humans can. An elephant's brain controls a trunk that can pick a peach without bruising it as well as knock down a wall. It can initiate and process signals that allow it to talk with its fellows across ten octaves (most humans can manage about three). It can manage and almost always avoid dangerous conflict with other elephants. A sperm whale's brain prepares it for dives deeper than a mile, managing the pumping and distribution of blood and oxygen while the whale stops breathing for up to two hours, and enabling it to hunt giant squid in total darkness. The whale's brain devotes huge resources to creating, detecting and analysing sound, both at very close distances and across hundreds of miles. It allows the animal to navigate over thousands of miles of open ocean and to keep track of family and friends over decades.

Orcas can, and do, kill a seal heavier than a grown man in an instant, but there is not a single recorded case of hostility by orcas towards humans in the wild. Dolphins have been observed stopping their exuberant behaviour and quietly accompanying a boat with a dead man aboard. Other dolphins have abandoned their food to surround and support a suicidal woman at sea.

In Orcas and other dolphin species, the limbic and paralimbic cortex, which process emotion, are much more developed than they are in humans, and this may account for their extremely strong family bonds and their almost uncanny ability to read human intentions. And the playfulness, compassion and forbearance these animals sometimes show towards humans is just one thing that their brains make possible. For decades humans have been puzzling over how to communicate with bottlenose dolphins, and during this time dolphins have not only learned how to follow our instructions through various symbolic means but also how to communicate some of their thoughts to us. We, however, are still largely blind to what they are saying to each other. It was only in 2016 that researchers published evidence that dolphins chatter with each other to solve difficult problems such as how to open a food canister. In the case of elephants, the many remarkable qualities of their intelligence include a marked tendency to gentleness (when they have not been traumatized by humans) – a wisdom that humans often lack.

So we should not underestimate the brain power of non-human animals. As the behavioural ecologist Peter Tyack says, it is seldom meaningful to attempt to fit different species along a linear scale designed to measure human intelligence. The radiologist and poet Amit Majmudar observes that the brains of other animals enable them to be 'routinely superhuman' in one way or another. 'They outstrip us in this or that perceptual or physical ability, and we think nothing of it. It is only our kind of superiority (in the use of tools, basically) that we foolishly select as the marker of "real" superiority.'

Tools, in turn, are facilitated by two capabilities: our dextrous hands, and language. In regard to the latter, human brains, uniquely among those of primates, have a direct connection between the part of the motor cortex involved in voluntary movements and the brain area that facilitates motor control of the larynx. This is enabled by a tweak of just two base pairs in *FOX2P*, a gene we share with apes and even with mice.

And for as long as we are still human, how we process and express feelings matters at least as much as how we manipulate objects in the world. In *Beyond Words: What Animals Think and Feel*, Carl Safina documents the lives of elephants, wolves and orcas (as well as the humans trying to save them from extinction), and marshals compelling evidence that these creatures largely experience the same range of emotions as humans. 'Fear, aggression, wellbeing, anxiety and pleasure are the emotions of shared brain structures and shared chemistries, originating in shared ancestry. They are the shared feelings of a shared world.'

6. Brain science has come a long way in a short time

To every schoolchild's delight, the ancient Egyptians considered the brains to be useless for the afterlife, and would pull those of the freshly dead out through the nose with a hook and throw them away before turning the remains into mummies. Given this unpromising start, brain science has made considerable strides.

The Greek physician Hippocrates, who was born in the fifth century BC, is the first on record, at least in the West, to argue that the brain is the seat of thought, emotion and sensation. Compared with the Egyptian belief this was a step in the right direction, but for a long time afterwards little that was true was added to it. Aristotle, a generation after Hippocrates, said the purpose of the brain was to cool the blood. And Galen, some six centuries later, ignored the solid tissue of the brain and said that three fluid-filled ventricles inside were responsible for the faculties of imagination, reason and memory. For well over fifteen hundred years Christian and Muslim physicians largely repeated Galen. In the early 1490s Leonardo da Vinci depicted these ventricles as three bubbles set in a row behind the eyes. His sketch in no way resembles the actual shape of ventricles of the brain, and although Leonardo had previously drawn the exterior and interior of the skull with such care that in the latter the impressions of the veins and arteries on the bone are visible in exquisite detail, there is no evidence that he studied the structure of the brain itself with any of the attention he later brought to the heart.

René Descartes was still following an essentially Galenic conception of the brain in the early seventeenth century and described the brain as controlling the body's movements by pumping fluid from the ventricles through the nerves to other organs and limbs. It is a hydraulic vision something like the steampunk mechanical elephant built by Les Machines de l'île in Nantes. Descartes recognized, however, that two major mental functions lay beyond the capacity of such a machine – namely, the generation of an infinite variety of thoughts, and the capacity to report them. His solution was the soul, a sort of secret sauce intervening from beyond the material realm by way of the pineal gland.

The most important breakthroughs since Hippocrates were made by Thomas Willis, a

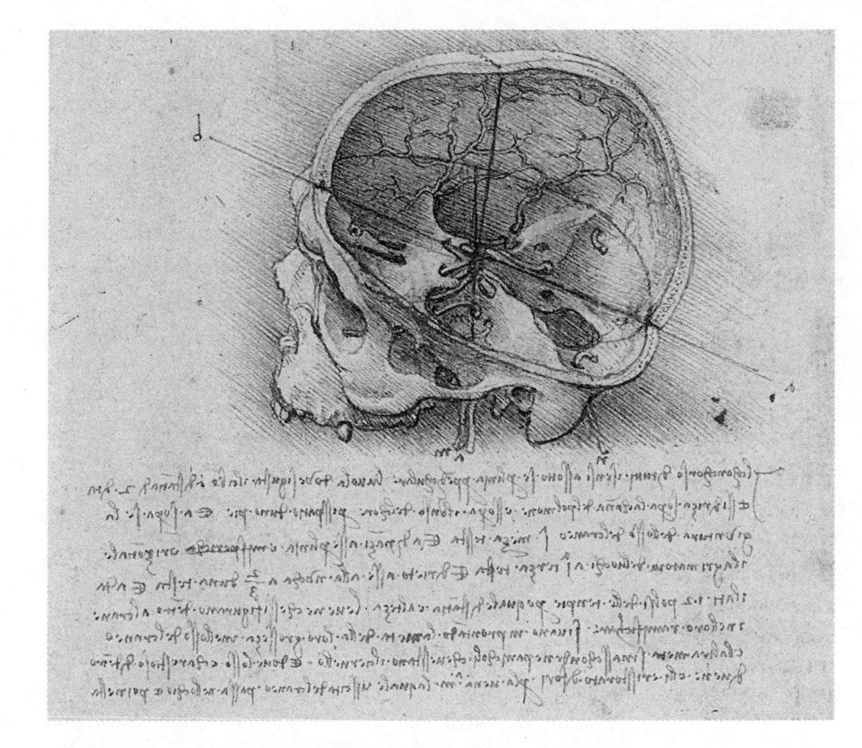

physician of the generation after Descartes who lived and worked in Oxford. In Willis's time autopsies on the brain tended to break through the top of the skull, but this often damaged and deformed the brain inside, rather as a heavy blow to a soft-boiled egg can smoosh its contents and squirt breakfast on your shirt. Willis and his collaborators took a different tack, coming from below and lifting the brain whole and entire from the skull. This allowed them to see it not as a gloopy mass sticking to the inside of the skull but as an independent organ divided into distinct parts.

Examining more closely, Willis found that the two carotid arteries that pump blood through the neck join in a circle at the base of the brain before entering the brain itself. With an experiment less gruesome but no less revelatory than those performed by William Harvey on the hearts of dogs forty years before, Willis cut open a spaniel's neck, and tied one of the arteries

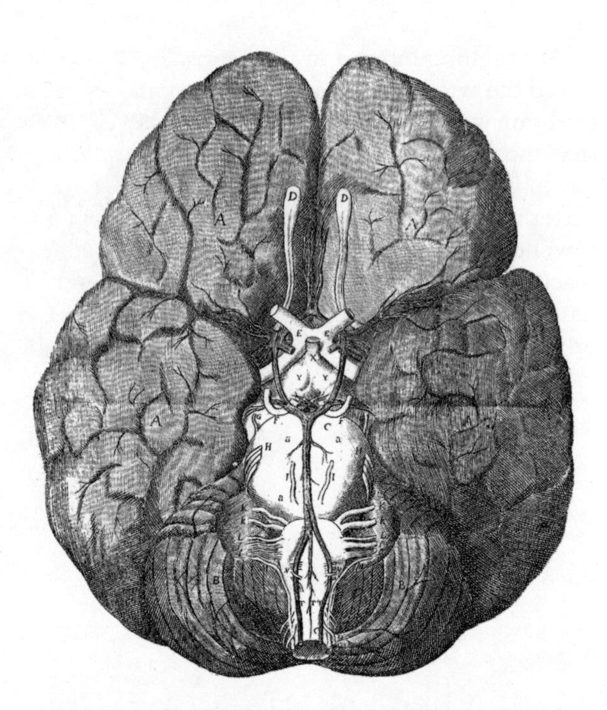

'I addicted myself to the opening of heads.' The brain as seen from below. Thomas Willis, 1664.

off. 'The dog was not altered by it,' he wrote, 'but continued very lively and brisk, and was so far from taking unkindly what was done to him that within a quarter of hour after, he got loose and followed the doctor [Willis] into town as he visited his patients.' He had discovered a structure, now known as the circle of Willis, that allows the entire brain to be supplied with blood even when part of the circle is damaged. Smaller arteries branching off the circle spread across the brain 'like the serpentine channels of an alembic', carrying what Willis supposed to be a 'chymical elixir' of animal spirits into the brain. But when he injected dye into these arteries Willis found that none of it ended up in the ventricles. The conclusion was clear: the ventricles had nothing to do with the mind but were just 'a complication of the brain infoldings'. Galen and the generations who followed him had mistaken background for foreground.

Abandoning interest in the ventricles, Willis followed the arteries and found that each major part of the brain was supplied by separate vessels – evidence that the different segments of the brain did different jobs. He also discovered grey structures with patterns of stripes buried deep in the cerebrum that travelled up from the brainstem and into the higher reaches of the brain. And he compared the brains of humans and animals and found that those of a fish or a cow had the same basic architecture – medulla

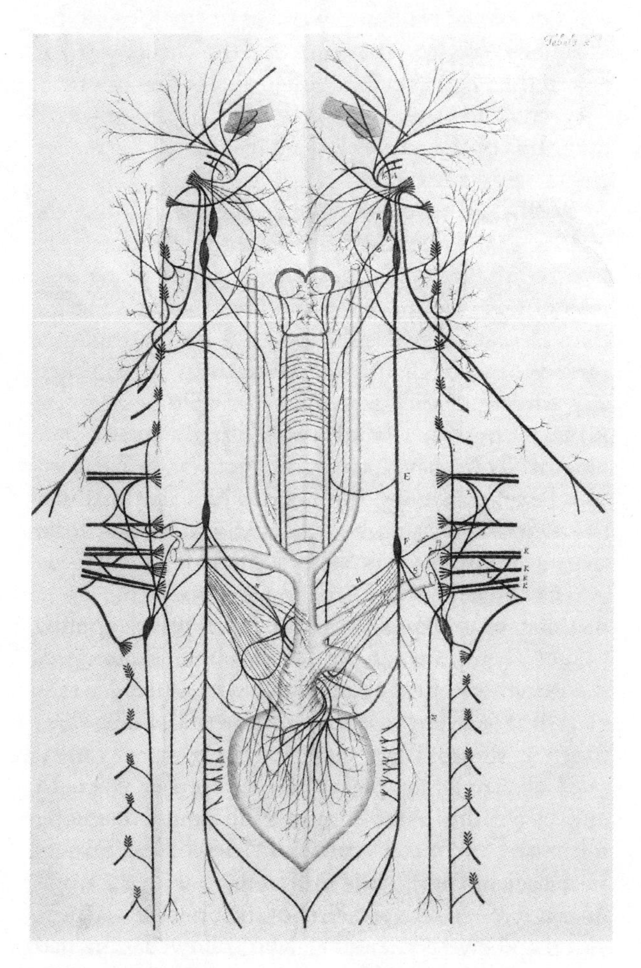

Willis's 'map' of nerves.

oblongata, cerebellum and cerebrum. Noting that humans had a gigantic cerebrum compared with animals such as dogs and cows, Wills conjectured that this played a role in our superior mental abilities. But he also saw that humans and animals had practically identical cerebellums, with a comparatively simple texture, which suggested to him that they worked like simple machines, creating 'spirits' that travelled to the heart and other organs without any supervision by the higher faculties. Willis also followed the nervous system from the brain into the body. He traced cranial nerves that slipped through holes in the side of the skull and wended to the tongue, the voice box, the lips, the teeth and the cheeks. He followed the vagus nerve along its fan of fine branches that entwine the heart and the diaphragm. His diagrams of the nerves looked like maps of streams, rivers and watersheds.

'The Spirits inhabiting the Cerebel perform unperceivedly and silently their works of Nature, without our knowledge or care.' Thomas Willis

Willis had revealed a whole new world, but the 'spirits' that somehow transmitted signals from the brain to the body and back again remained entirely mysterious – no more satisfactory as an explanation than Descartes' hydraulic forces. With hindsight we can see hints of their true nature in an ancient therapy. Already in the first century AD a physician to the Roman Emperor Claudius was using a torpedo – a kind of electric ray that can deliver a 200-volt shock – to relieve headaches. But nobody had any idea why this worked, and it was only in the late 1700s, more than a hundred years after Willis's death, that scientists began to formulate an idea of electricity as a distinct natural phenomenon. Working separately, Luigi Galvani and Alessandro Volta then showed that the nervous system operates under the influence of electrical activity, making the severed legs of a frog move by the application of external current. (Others used electricity to make human corpses convulse and rise from a table, inspiring shudders among witnesses and Mary Shelley to write *Frankenstein*.) In the second half of the nineteenth century Emil du Bois-Reymond was able to show that nerves and muscles themselves generate electrical impulses. Within decades researchers found they could measure tiny

In 1880 Du Bois-Reymond outlined seven 'world riddles' beyond current scientific understanding. Three of them, he said, would never be solved: the ultimate nature of matter and force; the origin of motion; and the origin of simple sensations.

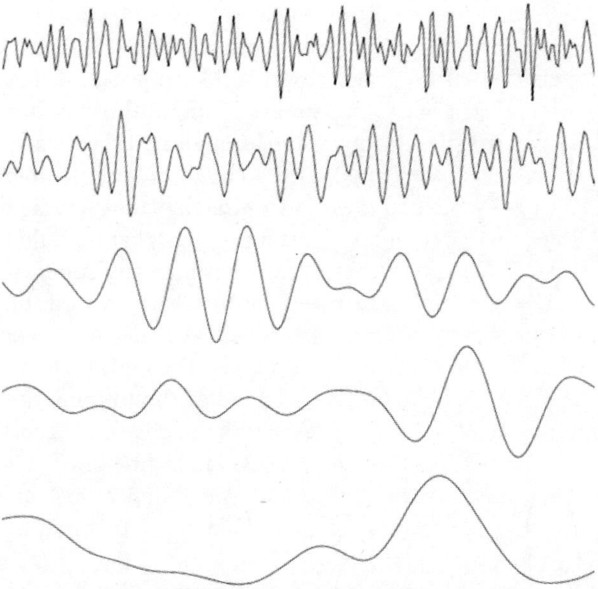

Gamma Waves

Beta Waves

Alpha Waves

Theta Waves

Delta Waves

The lowest frequency waves in the brain, delta waves at 1–4 hertz (Hz), are associated with deep, dreamless sleep. Theta waves at 4–8Hz are associated with light sleep or meditation. Alpha waves (the first to be discovered) at 8–13Hz are present during relaxed wakefulness. Beta, at 13–30Hz, are associated with normal waking consciousness. Gamma waves, at 30–70Hz (and approaching the lower end of the range of pitches played by instruments such as a cello), are associated with the formation of language, ideas and memory processing. 'As we make our way through the day,' writes Marcus du Sautoy, 'the brain behaves like an orchestra playing a symphony, moving between fast and slow movements with the occasional scherzo when we generate new ideas or encounter new situations.'

electrical currents arising in animal brains directly exposed to electrodes, and in 1924 Hans Berger recorded the first electroencephalogram, or EEG, of a human brain through sensors placed on the outside of the skull. The 'brain waves' that Berger recorded result from the synchronized activity of large numbers of neurons in various parts of the brain. It has since been found that they occur across a spectrum of frequencies.

But modern neuroscience had really begun a generation earlier with the work of Santiago Ramón y Cajal. He was not the first to suggest the importance of the recently discovered neurons in brain function but, using staining techniques developed by Camillo Golgi (who would share the 1906 Nobel Prize with

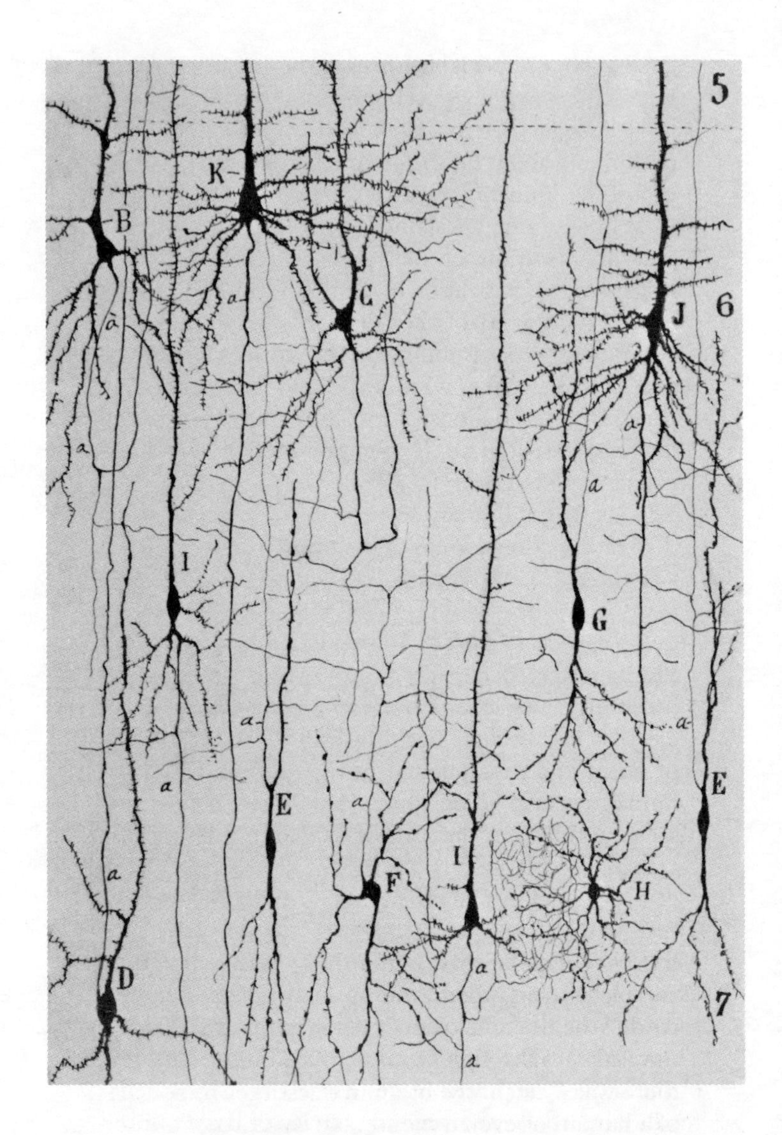

him), Cajal studied them in unprecedented detail and recorded what he saw in drawings as fine as Leonardo's best anatomical work. Ramón y Cajal showed neurons to be much more intricate and diverse in shape than any other cell types. He also found that individual neurons in insects matched and sometimes exceeded the complexity of those in humans, but that in humans some neurons connected over (comparatively) long distances. This suggested that mental capabilities depend not on special features of the cells themselves but on the way they are connected. This 'connectionist' view gave rise to the neuron doctrine and a way of thinking about information processing in the brain that still dominates research today.

The neuron doctrine:

(1) the neuron is the fundamental structural and functional unit of the nervous system;
(2) neurons are discrete cells which are not continuous with other cells;
(3) neurons are composed of three parts: dendrites, axon and cell body;
(4) information flows along neurons in one direction, from the dendrites to the axon via the cell body.

While Ramón y Cajal was exploring the structure and organization of neurons, others investigated larger-scale organization in the brain and in particular how certain areas of the cortex are associated with certain functions. Building on the work of Paul Broca and Carl Wernicke, who in the mid-nineteenth century identified regions of the cortex associated with language, Korbinian Brodmann used a technique called the Nissl stain to identify forty-three discrete regions that still today serve as a guide for studying distinct functional areas of the brain.

Among many remarkable achievements in the twentieth century was the first complete map of every neuron in a brain. In 1970 a team led by Sidney Brenner started cutting a brain into thousands of ultrathin slices and photographing them under an electron microscope. Each slice was floated on a bead of water and manipulated using a toothpick with a human eyelash glued to the end. The slices, which were so thin as to be almost invisible, were so delicate that accidentally touching them with the eyelash destroyed them. The researchers would gently stroke the surrounding water to nudge them into place. After fourteen years of this sort of thing they published a complete wiring diagram in a 446-page paper in the world's oldest and most revered scientific journal, the *Philosophical Transactions of the Royal Society*. But there was a catch: the brain in question was not that of a human or even a mouse, but instead came from a worm less than a millimetre long. *C. elegans*, which likes to live in the film of water between soil particles and rotting vegetation, has just 302 neurons (less than 0.0000004 per cent of the number in a human being) and nine thousand synapses (less than 0.00000001 per cent). Working at the same pace, it would take millions of years and billions of pages to create a wiring diagram of a human brain.

From the 1980s and 1990s onwards new non-invasive

Brain-visualization technologies include magnetic resonance imaging (MRI), functional MRI, diffusion MRI, magnetoencephalography (MEG), electroencephalography (EEG), and positron emission tomography (PET).

imaging technologies helped to revolutionize brain science. One of the most revealing is functional Magnetic Resonance Imaging, or fMRI, which uses powerful magnets to detect changes in blood flow in the brains of subjects who are exposed to various stimuli – images, sounds, thoughts. Activated regions can be presented on a screen as luminous blobs of colour. But fMRI has significant limitations. There is a time lag, and different neuronal events that happen a second or more apart can blur together when the excited area appears onscreen. This is a real liability when studying an organ that works at millisecond speed. Nor can fMRI reveal what brain cells are actually doing. The technique registers activity only at the scale of hundreds of thousands of neurons, and a lit-up area might represent a large number of neural processes. Given this lack of precision, it is remarkable what has been achieved.

One recent example is an atlas that shows how the meanings of words are arranged across different brain regions in a colourful 'quilt'. Investigators have found that no region holds one word or concept, that a given location can be associated with several related words, and that each single word lights up many different brain spots. Together they make up networks that represent the meanings of each word we use. Using the same data, researchers are working on atlases to show how the brain holds information such as narrative structure.

The proteins in our cells and tissues are responsible for everything from repair and maintenance of the body to the production of signalling chemicals. The brain has about three hundred unique proteins; the testicles have nearly a thousand.

The first decades of the twenty-first century have seen huge efforts to map the structure of the brain at every conceivable scale, with the hope that this will unlock ever more of its secrets. One source of inspiration for this approach is the perceived success of earlier giant mapping projects. In 2000, following a decade of international collaboration, the Human Genome Project published an initial rough draft of all the genes in the human body. Then US President Bill Clinton called it 'the most wondrous map ever produced by humankind', with enormous potential to increase human wellbeing. The achievement inspired other large-scale efforts such as one to map the human proteome (the entire set of proteins expressed by a genome), and researchers began to call for a similar project for the brain. After all, the 'mind' of the worm *C. elegans* had been charted twenty years earlier, and there had been many advances in understanding and imaging technology since them. Why not map all the neural connections – the 'connectome' – in, first, relatively simple animals such as the fruit fly, zebra fish and mouse, and, ultimately, the human brain itself? Following the neuron doctrine, an analogy was made to the wiring diagram of an electronic machine or computer, on the principle that function and behaviour are consequences of the way components are wired together. (Many researchers believe that faulty 'wiring' in the brain is the cause of disorders such as autism and schizophrenia. Better maps, they hope, will help researchers understand and treat these conditions as well as others such as Alzheimer's and Parkinson's.)

New technologies have produced striking results. Among them is Brainbow. Using four fluorescent proteins derived from a jellyfish, scientists created a palette of about a hundred different colours with which they could label individual neurons in a mouse brain in unprecedented detail. Images published in 2007 had a luminescent strangeness comparable to the paintings of Paul Klee, with dabs of multiple colour like leaves of a psychedelic tree falling across purple star space. And although the practical applications of Brainbow may prove to be limited, the beauty of these images may have helped stimulate increased awareness of and support for research.

At the time of writing there are over a dozen major public and private brain map, atlas and database projects, with a combined budget of billions of dollars. Notable among them are the Brain Activity Map Project of the US National Institutes of Health, which aims to map every neuron in the human brain over ten years from 2013, and the EU's Human Brain Project, a ten-year programme that was begun in the belief that it would be possible to simulate the human brain with supercomputers within a decade or two, but which has since run into trouble. Sebastian Seung, of the Princeton Neuroscience Institute, inspired tens of thousands of 'citizen scientists' to join EyeWire, an online game in which they help map neurons in the retina of a mouse. With BrainSpan, an atlas of the developing human brain published in 2013, researchers funded by the US National Institute of Mental Health found that 95 per cent of genes are active during foetal brain development – significantly more than in adult brains – and confirmed the significance of brain areas thought to be important in the development of autism. In 2014 researchers at the Allen Institute for Brain Science, where mapping projects had been under way for ten years, published a detailed atlas of the brain of a mouse, showing mid-level connections between 295 brain regions. It was an impressive piece of work but very far from the complete picture, and a mouse brain, with just seventy-one million neurons, is less than a thousandth of the size of a human brain. Nevertheless, as a remarkable discovery in 2017 described in the next section of this chapter suggests, such painstaking work is starting to yield remarkable results.

'A mouse is miracle enough to stagger sextillions of infidels.'
Walt Whitman

Some researchers hope that a complete map of the human brain will make it possible to record and store a person's very essence and 'replay' it, effectively conferring immortality. The idea seems far-fetched, but consider the case of Anna Bågenholm, who, in a skiing accident in 1999, fell into freezing water and became trapped under the ice. Her heart stopped for well over two hours, and her core temperature fell to around 13.7°C. By any usual definition she was dead. But because her brain had cooled down before cardiac arrest and therefore required much less oxygen than usual it was largely undamaged, and after doctors restarted her heart she gradually recovered full health. If the brain can survive virtual stasis like this, it is argued, then why should it not be possible one day to bring it back to life or even to copy its essential architecture in another medium?

Other researchers are sceptical about the power of brain mapping, at least in the near future, and point to the case of the Human Genome Project. Wondrous as it is, that map turns out to have been only a very small step towards an account of what makes us human.

The project identified about twenty thousand genes, but we now know that there are at least several hundred thousand switches in our DNA that interact in complex ways, turning genes on and off at different stages in ways that are fundamental to development and bodily function. It is likely that the organization of the brain is no less complex. Eve Marder, a prominent neuroscientist, cautions: 'If we want to understand the brain, the connectome is absolutely necessary and completely insufficient.'

This is epigenetics: an international consortium is now trying to map the epigenome, and no one can say when it will be finished.

Meanwhile, new techniques are casting light in places that were unthinkable just a few years ago. In the case of something called optogenetics, an approach developed by Gero Miesenböck, Karl Deisseroth and other neuroscientists in the early 2000s, this is literally true. Optogenetics makes it possible to engineer neurons inside the brain of a living animal that can be switched on and off with a laser and which light up when active.

The origins of optogenetics are as remarkable as its potential applications. Speculating in 1979 on the future of brain science, Francis Crick wondered if a tool could be developed that would switch some neurons on or off while leaving others untouched. 'It is conceivable', he wrote, 'that ... biologists could engineer a particular cell type to be sensitive to light.' And it turned out that a candidate had already been discovered. In the early 1970s the biochemist Dieter Oesterhelt had described a microbial opsin – a light-sensitive protein in a bacterium that thrives in salty lakes in Africa by converting light into energy. No one supposed at first that genes from the bacteria could be made to manufacture this protein in a mammalian brain. Amazingly, however, this is precisely what Miesenböck, Deisseroth and their collaborators eventually did.

in 1953 Francis Crick was one of the co-discoverers of DNA. From 1979 until his death in 2002 he sought to advance the understanding of the neural basis of consciousness.

Christof Koch has called optogenetics the fourth great breakthrough in neuroscience after the dye-staining of cell types by Golgi and Ramón y Cajal in the late nineteenth century, the application of microelectrodes from the 1950s onwards, and the advent of fMRI in the 1990s. Optogenetics makes it possible, Koch says, 'to intervene in the network of the brain in a very delicate, deliberate, and specific way'. It enables observation of the brain in action at the level of the elementary parts that are thought to underlie sensation, learning, memory, metabolism, hunger, sleep, reward and motivation. And there may prove to be therapeutic applications. Clinical trials under way at the time of writing are exploring a way to treat retinitis pigmentosa, an inherited degenerative eye disease that causes severe vision impairment.

Clear Lipid-exchanged Anatomically Rigid Imaging/immunostaining-compatible Tissue hYdrogel.

Another new technique, also developed by Deisseroth, is called CLARITY. This replaces the fat between brain cells, which is opaque, with a clear hydrogel without disturbing the arrangement of the neurons. The post-mortem brain remains intact but is now, amazingly, transparent. Videos of the result, such as one published by the journal *Nature* in 2013, show networks of incredibly fine cells and nerve fibres in a clarified mouse brain, visible in unprecedented detail and glowing green against a black background, or coloured differently according to type and function. The detail and beauty of the image is extraordinary.

There are many unknowns ahead. One concerns the role of signalling chemicals called neuromodulators. Even in the relatively simple systems of thirty or so neurons that control the stomachs of crabs and lobsters, neuromodulators can fundamentally change how a circuit functions. And what is true for the stomach of a crustacean is likely be true for the brain, where neuromodulators influence vastly larger systems. Another case in point are transposons – wandering snippets of DNA that hide in genomes, copying and pasting themselves at random. These have been found to be active in the development of adult brain cells from stem cells. It may be that they generate novelty and complexity in brain cells in a way that our genes alone cannot, even though fifty per cent of our genes are used to develop this two per cent of our selves. If this is true, then our brains are hosting a kind of evolution in miniature, unique in every single case.

The brain surely hides many other surprises. Some researchers argue that the currently popular model of the brain as a computer-like information processor is fundamentally wrong: a metaphor no less misguided than those of hydraulics or clockwork in earlier centuries. They stress that even though we have what is supposedly a comprehensive map of the simple worm brain of *C. elegans*, we still don't understand many basic things about how it works. Eric Kandel, who won a Nobel Prize for identifying chemical changes that take place in a marine snail after it learns something, has suggested that it may be decades before we understand how human memory works, while Kenneth Miller suggests it may take even longer just to figure out basic neuronal connectivity.

7. Consciousness is astonishing but not entirely mysterious

Consciousness has acquired a kind of aura. We are fascinated both by our experience of the world and the fact that we *can* experience the world. 'If [my] head were to try to encapsulate in a few words everything that is most amazing about itself,' says physician and philosopher Raymond Tallis, 'those few words would be: it has a world.' For in consciousness, an extraordinary creation (or re-creation) takes place. All sensory stimuli enter the brain in more or less undifferentiated form as streams of electrical pulses created by chains of neurons firing certain routes; and yet we feel that we experience actual sunlight reflected off water, the actual sound of distant bells, the actual touch of a breeze or the smell of wood smoke. 'Who could ever tire of this radiant transition,' writes Annie Dillard, 'this surfacing to awareness and this deliberate plunging to oblivion – the theatre curtain rising and falling? Who could tire of it when the sum of those moments at the edge – the conscious life we so dread losing – is all we have, the gift at the moment of opening it?'

The essayist Walter Benjamin described aura as 'a form of perception that invests or endows a phenomenon with the ability to look back at us'.

It would be no less easy and true to bring obnoxious memories to mind. But as Pierre in Tolstoy's *War and Peace* learns when taken prisoner and dragged through cold, hunger and filth, the simplest sensations – eating when hungry, drinking when thirsty, sleeping when sleepy – can bring unmitigated joy.

The riddle of how this happens is often referred to as 'the hard problem' – or even, by some particularly imaginative types, as 'the really hard problem' – of consciousness. The phrase was coined in 1995 by the philosopher David Chalmers, who asked why physical processing should give rise to what we sometimes experience as a rich inner life. But the

The journalist Ambrose Bierce (1842–c.1912) was dubious, and tweaked Descartes' formulation to 'I think that I think, therefore I think that I am.'

nub of the question dates back at least three hundred years. Descartes had concluded that his conscious experience was the one thing of which he could be absolutely sure: 'I think, therefore I am.' And yet he recognized that this experience, which seemed to him to exist beyond matter and time, must somehow be connected to a world that was entirely material. The mathematician and philosopher Gottfried Leibniz captured the essence of the puzzle in 1714 with an analogy to the high-tech machinery of his day, the mill. 'If we imagine a machine whose structure makes it think, sense and be conscious,' he wrote, 'we can conceive of it being enlarged in such a way that we can go inside it like a mill. Suppose we do: visiting its insides, we will never find anything but parts pushing each other – never anything that could explain a conscious state.'

Generations have wrestled with this riddle, with little result. In 1989, writing in the *International Dictionary of Psychology*, the psychologist Stuart Sutherland concluded that 'it is impossible to specify what [consciousness] is, what it does, or why it evolved. Nothing worth reading has been written on it.' The philosopher Evan Thompson is more patient. Consciousness, he wrote in 2014, is not subject to analysis in the way that many other things are because there is no way to step outside it and measure it against something else. It cannot be explained in non-experiential terms. 'Perhaps, therefore, the best we can do is fall back on analogies or metaphors. According to Indian yogic traditions, for example, consciousness is that which is *luminous*: it has the power to reveal, like a light.'

'The attempt at introspective analysis in this case is . . . like seizing a spinning top to catch its motion, or trying to turn up the [light] quickly enough to see how the darkness looks.' William James

But research over the last quarter century has led many who study the brain to believe that it is possible to do more than create analogies and metaphors. We can, if nothing else, get a better sense of what it is that we do and don't know. The neuroscientist Anil Seth has suggested that this be called 'the real problem' of consciousness.

With brain-imaging techniques it is possible to observe which neurons are most active when we are conscious – to see which parts of Leibniz's mill are whirring, and are therefore at least a part of what Francis Crick called the 'neural correlates of consciousness'. For the most part, the relevant neurons seem to be in certain regions of the cortex, where the number of local connections is very large, and greatly exceeds the number of entry and exit points to other brain regions. This suggests that those neurons communicate with each other more than with the sensory organs and motor apparatus (although they do have strong connections to centres of emotion and memory). And one model – drawing on ideas proposed by Francis Crick and Christof Koch, who pioneered research in this field in the 1990s – is that consciousness arises during the synchronization of many of these neurons, through trillions of synapses, oscillating together and all the while influenced by the reticular formation, thalamus, hippocampus and limbic systems. The results of new mapping by Christof Koch and others at the Allen Institute published early in 2017, however, hint that long-distance connections may also play an important role in making consciousness possible, perhaps by enabling the sharing of certain states across the whole brain. The researchers found three neurons in a mouse that branch extensively throughout the brain, including one that wraps around its entire outer layer of both hemispheres like a crown of thorns. The neurons originate from a small, thin sheet of cells called the claustrum – an area that Koch believes acts as the seat of consciousness in both mice and humans.

Research suggests that interoceptive awareness – a sense of the physiological condition of your body – depends heavily on an area called the anterior insula, while the cognitive self, which situates us in the world, is largely in the medial prefrontal cortex.

We need to consider the possibility that (almost) all brain functions (not just consciousness) are distributed across the brain and that all parts of the brain contribute to those functions, argues the neuroscientist Henrik Jörntell. 'Neuroscience needs to start investigating how network configurations arise from the brain's lifelong attempts to make sense of the world. We also need to get a clear picture of how the cortex, brainstem and cerebellum interact together with the muscles and the tens of thousands of optical and mechanical sensors of our bodies to create one, integrated picture.'

In 1989 the mathematician Roger Penrose proposed that classical physics could never fully explain how the brain produces thought and conscious experience, and that quantum processes must be involved somehow. Penrose's conjecture attracted a lot of attention, but his critics countered that the key quantum effects of superposition and entanglement are extremely fragile phenomena at anything except very low temperatures. The neurophilosopher Patricia Churchland summed up what came to be the mainstream view: 'Pixie dust in the synapses is about as explanatorily powerful as quantum coherence in the microtubules.'

Still, the idea that there might be quantum effects

in the brain has not been completely dismissed by serious scientists, and it has come to seem a little more plausible since the discovery in the 2000s that other living systems take advantage of them. Plants, for example, exploit a quantum property of photons (whereby they are probabilistically 'smeared' over space) to make photosynthesis more efficient. In a paper published in 2015, the physicist Matthew Fisher suggested that even if electrical impulses among neurons in the brain (which are well described by classical physics) are the immediate basis of thought and memory, a hidden quantum layer may still determine, in part, how those neurons correlate and fire.

Other proposals to better understand consciousness draw on information science. Integrated information theory, first outlined by Giulio Tononi in 2008, is one. Noting that whatever information you are conscious of is wholly and completely present to your mind, and that to be conscious you need to be a single, integrated entity with a large repertoire of highly differentiated states, Tononi proposes that consciousness can be expressed in mathematical terms. He claims that both the quantity and quality of consciousness in a particular brain can in principle be calculated, though the calculation may, for now, be intractable.

This can all begin quite quickly to become abstruse, if not incomprehensible to non-specialists. Approaches that don't try to solve the whole problem may be helpful. Michael Graziano, for example, asks us to consider consciousness as a representation of attention in the brain. Attention – in which some signals are enhanced at the expense of others – is, he says, a well-understood, mechanistic phenomenon that can be measured in animal brains and can even be programmed into a computer chip. What if awareness, which seems to be closely associated with consciousness, is just a distorted account of attention? Consciousness, Graziano suggests, may be a cartoonish reconstruction analogous to a schematic city map: physically inaccurate but nevertheless useful because it can be used to make predictions, try out simulations, and plan actions.

Following a similar line, the cognitive neuroscientist

Stanislas Dehaene seeks to characterize consciousness according to what it does. Consciousness, Dehaene argues, is inordinately open and yet very selective, and it chooses, amplifies and propagates thoughts that are likely to help an individual survive and thrive. Thanks to consciousness, he observes, we can keep important data in mind for a long time, and pass it on to any other mental process. This can give us enormous flexibility and all the advantages that can come with it.

'Though the non-logical, instinctive subconscious mind must play its part in [the artist's] work, he also has a conscious mind which . . . resolves conflicts, organizes memories, and prevents him from trying to walk in two directions at the same time.' Henry Moore

But, Dehaene adds, to understand consciousness you also have to look under the hood. Consciousness rides upon a much larger sea of preconscious and unconscious processes that perform vast numbers of integrative tasks essential to our continued existence, from monitoring and controlling basic body functions to processing complex external information. The information that reaches the conscious mind is the outcome of the very complex sieve that we call attention selecting from myriad unconscious and preconscious mental processes. The brain ruthlessly discards the irrelevant information and isolates a single conscious object or ensemble based on its salience and its relevance to our current goals. A classic example, known as the cocktail party effect, is when we suddenly become conscious of hearing our name spoken in a noisy room by someone to whom we were not previously attending. Multi-sensory information and meaning are bound together in the brain unconsciously (we have no introspection, for example, into how a sign evokes meaning), and we dramatically underestimate how much vision, language and attention can occur outside our awareness. 'We constantly overestimate our awareness,' says Dehaene, 'even when we are aware of the glaring gaps in our awareness.'

'Rational thinking is not necessarily our greatest property, and although we prize it, it [can] be a handicap. We have to recognize that in addition to conscious rational thinking our minds are capable of other, more powerful mental processes that lead us by intuition to grasp a tiny sparkling fragment of reality.' James Lovelock

A tiny sparkling fragment

Some doubt that consciousness can be 'just' an emergent property of the brain. Nature, they argue, doesn't make jumps, so there can be no point at which something that does not have consciousness suddenly acquires it. At least one prominent researcher – Christof Koch – has embraced a version of this idea, suggesting that consciousness may be a fundamental, elementary property of matter.

'Physics is mathematical, not because we know so much about the physical world, but because we know so little: it is only its mathematical properties that we can discover. For the rest, our knowledge is negative.' Bertrand Russell

This idea, which is sometimes labelled panpsychism, has some appeal. Like Baruch Spinoza's vision, where God and nature are not separate, it seems to unite everything. According to the philosopher Galen Strawson, an awareness of our profound ignorance of the nature of matter should warn us not to dismiss such ideas out of hand. Citing W. H. Auden, Strawson suggests that 'Matter', like love, is 'much / Odder than we thought'. As the philosopher Philip Goff puts it, 'the only thing we know about the intrinsic nature of matter is that some of it – the stuff in brains – involves experience'.

It is certainly the case that the ultimate nature of matter (and, indeed, the universe as a whole) is beyond our present understanding. Physicists tell us that at the deepest level of explanation they can offer so far, the world is weird: the wave function of quantum mechanics does not allow for a measure of both the location and the velocity of an elementary particle at the same time, only probabilities. The same goes for electric charge, energy and spin. Further, elementary particles flash in and out of existence and are entangled instantaneously across distance in defiance of the speed of light.

Towards the end of his life, Richard Feynman is said to have become fascinated by the question of whether particles such as electrons forming a cloud around atomic nuclei are in some sense 'thinking'. And perhaps the question goes back at least as far as the ancient Greek philosopher Parmenides, who said that to think and to be are one and the same. Perhaps, as Wisława Szymborska wrote in her poem 'Conversation with a Stone', even small pieces of rock are bursting with laughter.

'"I will write a book on leaves of flowers, . . . I'll sing to you on this soft lute, and show you all alive The World, where every particle of dust breathes forth its joy."' William Blake

But maybe panpsychists are conflating things that are not alike. The fact that matter is wonderful and strange may not necessarily mean it is conscious in the sense we apply the term to ourselves or to some other animals. Rather, something genuinely new may be emerging in certain arrangements of matter and energy, including (but not necessarily limited to) brains. Maybe consciousness arises at one of those tipping points in nature – analogous to when water boils and turns into a gas, or when electromagnetic waves arise from the movement of charged particles – when something different, or even fundamentally new, emerges. Could there be a field of consciousness that is activated by critical states of matter, in the way that the Higgs field gives mass to matter? 'It sounds like a crazy idea,' says the mathematician Marcus du Sautoy, 'but perhaps in order to get to grips with such a slippery concept [as consciousness] we need crazy ideas.'

The Higgs field is an energy field thought to exist everywhere in the universe. As particles pass through it they are 'given' mass and – like objects passing through treacle (or molasses) – become slower, and cannot travel at the speed of light because they have mass.

Whatever the larger truth, humans sometimes perceive the world *as if it were* conscious, or at least part of consciousness. In the Zen Buddhist tradition, the thirteenth-century sage Dōgen said that fences, walls, tiles and pebbles are also 'mind' (心, *shin*). And seven hundred years later the philosopher Maurice Merleau-Ponty recorded the painter André Marchand as saying, 'In a forest, I have felt many times over that it was not I who looked at the forest. Some days I felt that the trees were looking at me, were speaking to me . . . I was there listening.' Or, as the nature writer Barry Lopez puts it after observing and simply being present in the Arctic landscape over decades, 'One must wait for the moment when the thing – the hill, the tarn, the lunette . . . ceases to be a thing and becomes something that knows we are there.'

You don't have to go this far to accept that consciousness is not located solely in the brain. 'We have the impression . . . that consciousness is in our heads,' says the psychologist Riccardo Manzotti; 'yet nothing we have found in the brain warrants this.' Consider the case of an apple. 'What,' he asks, 'if the experience is *the apple itself*? After all, the apple is definitely the most applish thing around, and the only thing that has the properties of an experience of an apple. It's round, it's red, it's shiny. So why can't the subject, consciousness, be identical with the apple, out there where the apple is?' Consciousness is not *about* the

'The mind is like a rainbow. The rainbow is outside of our body, but it is a piece of our mind.' Riccardo Manzotti

apple, suggests Manzotti, it *is* the apple – or at least the relation between me and the apple.

However our views diverge, perhaps we can agree on the importance of paying attention with intensity and commitment. This can only be a good thing if, as Einstein said, our sense of separateness from what we call the universe is only a kind of optical delusion of consciousness. For among the most important challenges facing you and me, surely, is the need to develop awareness worthy of the complexity and beauty of creation.

'Wrens cut across, or rather scramble under, the imaginary boundaries which we are accustomed to draw between different types of country.' Max Nicholson

The door to the shed where I write is usually open in the warmer months. Outside – in my tiny garden, and from my neighbour's tree or beyond – birdsong trickles through and colours everything. Sometimes, when I am far away in my head, a wren zips right past the open door faster than I can consciously register it. Only afterwards, in the moment constructed in my brain, does the flight of the tiny bird reveal itself.

5

EDGE OF THE ORISON

Self

... like leaves are the generations of men –
old leaves, cast on the ground by the wind,
young leaves
the greening forest bears when spring comes in.

Homer

Life gives everything to everyone, but
most men are unaware of it.

Jorge Luis Borges

How beautiful can a being be!

Caetano Veloso

Titian's *The Three Ages of Man*, which was painted in 1514, the same year that Dürer engraved *Melencolia*, is a sensuous and colourful work. The almost naked figures glow with vitality in an Arcadian landscape of deep greens and blues. But the message of the painting is simple, almost stark: humans are born, grow and flourish, and then age and die. The details of any given life are, of course, a lot more complicated. From the cellular level to our most complex thoughts and emotions, different parts us are born, flourish or die at any given moment. But the basic idea of a life in three parts – which is an old one, and appears in various forms, including the riddle of the Sphinx – contains an essential truth. The arc of a life resembles the trajectory of a day, and each of us has a morning, an afternoon and an evening. This chapter follows that arc in an exploration of the wonder that it is, simply, to live, and how that can change over a lifetime.

The Sphinx devours all travellers who cannot not answer this riddle: 'What is the creature that walks on four legs in the morning, two legs at noon and three in the evening?' Oedipus answers 'Man', and the Sphinx dies. But for Oedipus, the story is just beginning.

1. Morning

It is good to know that a man can make his living by studying the laughter of babies, and when I met Caspar Addyman, a psychologist and self-described infantologist at the University of London, he was charm itself. Addyman's hair is turquoise and he likes to quote the musician Victor Borge: 'Laughter is the shortest distance between two people.' He readily concedes that some of the findings of his research may seem like common sense. Babies, for instance, laugh before they can talk. Less obviously, he suggests, baby laughter is, in a sense, the opposite of crying: it says something like 'let's do more of this' and is important, even vital, to the wellbeing of both a baby and its carers. And, he claims, the research he and other psychologists are doing is leading to some genuine surprises as well as a deeper understanding of matters that were previously only the stuff of hunch or intuition.

Only recently, for example, have researchers begun to understand the extent to which, from infancy, humans are programmed for fun. Ultrasound scans show that foetuses 'practise' smiling and laughing (as well as crying) in the womb. A global study by Addyman and his colleagues has shown that, the world over, almost all babies smile in the first three months of life. By the time they are two to four months old most healthy babies have mastered the trick of social smiling, which actively engages the parent and is used to elicit a response. Babies generally start laughing at around three months. Addyman's research shows that peekaboo is the most popular way to make babies laugh everywhere in the world. The game works throughout the first years of life, he says, although it gets more sophisticated with age. In the first six months, babies are genuinely surprised by your return. Then they learn to anticipate it and are pleased that their prediction comes true. By the time they're toddlers, they realize they can make parents and other carers laugh, a development that coincides with the start of a true sense of self. They are often playing to humour you, though this is also the age when tantrums start. Despite these changes, however, the basic nature of the game remains the same: it is all about eye contact, says Addyman – pure social interaction, stripped down to its barest elements.

Infants as young as eight months can use a specific type of humour: teasing. A baby may, for example, willingly hand over the car keys she's been allowed to play with, but whip her hand back quickly, just before allowing her father to take them, all the while looking at him with a cheeky grin.

Sigmund Freud asserted that most infant laughter is a form of *Schadenfreude*, or joy in the misfortune of others. But all the research, Addyman insists, shows that Freud was wrong: young children and even babies are empathic and moral. For instance, babies empathize with the distress of others – crying in response to the cries of other babies and stroking or offering toys to those who seem upset.

Psychologists have shown that babies have a clear

preference for kindly figures, and learn better with parents and carers who laugh often. Addyman says the converse is also true: mirthful babies encourage adults to teach them things they are interested in. Either way, however, the really important point seems to be that laughter encourages interaction and cooperation. Jokes help – and both infants and adults laugh more at them in company than alone – but they are secondary to the serious purpose of fun, which is to build closer bonds and help us learn. And, for all its potential scientific and social value, there is a large element of fun in Addyman's research. 'We are connecting babies to heart monitors and seeing what happens when we tickle them,' he told me. Tickling, it seems, may be one of several forms of touch that, like hugs, are vital to healthy development.

Since we met, Addyman and other psychologists have worked with the composer Imogen Heap to create music intended to please babies as much as possible. They tested diverse samples of tunes and sounds on a cohort of babies, and asked thousands of parents to vote on the silly sounds that made their babies happy – the top ten included 'Boo!' (66 per cent), raspberries (57 per cent), sneezing (51 per cent), animal sounds (23 per cent) and baby laughter (28 per cent). Plosive 'pop' sounds are also especially pleasing. The resulting 'Happy Song' is in a major key, and has 163 beats per minute, a simple and repetitive main melody with drum rolls, key changes and rising pitch glides to provide opportunities for anticipation and surprise, and a familiar repeating chorus.

Anyone seeking hope for the human race may take some comfort from babies. At the same time, anyone who has spent time with babies knows it's not always an easy ride being one, let alone looking after one. To truly understand how they experience the world, says the psychologist Alison Gopnik, smoke four packs of cigarettes and drink four double espressos, or go to Paris and fall in love. This is (apparently) a fantastic state to be in, but it does mean you wake up at three o'clock in the morning crying. The point is that babies are paying lots of attention, but in the wide-eyed, indiscriminate sort of way that adults do when they are in a hyper state brought about by certain drugs, exotic travel or romance. Gopnik makes another comparison: being a baby is like being immersed in an engrossing movie; you are not in control, your consciousness is not planning and your self seems to disappear, and yet the events in the movie are very vivid in your awareness.

The mental world of babies may also, on occasion, have a lot in common with being on a psychedelic trip. A team led by Robin Carhart-Harris at Imperial College London has studied the effects of psilocybin – the active ingredient in psychedelic mushrooms – on states of consciousness in adults, and has shown that it deactivates 'hubs' in the brain (notably the posterior cingulate cortex and medial prefrontal cortex), thereby reducing long-range connectivity between brain regions. These hubs are like conductors of an orchestra, says Carhart-Harris; introduce psilocybin and the conductors leave the room. Participants in studies often report that they start to feel as if they are melting into everything around them. It's as if the brain and mind move back to an earlier stage of development. Cognition becomes less constrained and less analytical, more influenced by imagination and wishes, but also fears. Psilocybin makes patients more childlike. They become emotionally volatile, and they tend to giggle a lot.

Parents and others may have observed that after feeding on breast milk, a baby's facial expressions can look very like those of an adult who is way high.

Curiouser and curiouser

The very young may experience a sense of confusion, revelation and delight, but do they have a sense of wonder? Annie Dillard suggests not: 'They bewilder well, but few things surprise them. [Everything] is new to [them], after all, and equally gratuitous.' Over time, however, infants do become increasingly aware of regularities in experience and begin to develop a sense that there is a world and there are people distinct from them. With loving and supportive carers and fellows, they certainly become increasingly curious about the outside world as they grow.

Curiosity is not, of course, peculiar to human beings. Many animals are curious, especially when circumstances are benign enough for them not to be preoccupied by the immediate challenge of survival. Even *Caenorhabditis elegans*, the humble worm with 302 neurons mentioned in Chapter 4, will, when placed in a big patch of its favourite food and plenty of potential mates, often go looking elsewhere. Some worms can't get no satisfaction. But highly intelligent animals such as New Caledonian crows are especially curious. Through trial, error and observation as they grow, baby crows learn how to fashion sticks into barbed hooks with which to extract grubs from holes. Their curiosity enables them to develop a deep understanding of how physical objects work and interact, and to learn to use that knowledge to accomplish their goals.

But humans are more curious than almost any other animals, and even very young humans are better than New Caledonian crows at noticing the novel, the accidental and the serendipitous, and at using that experience to imagine new opportunities. And once sparked, curiosity in humans seems to be hard to stop. When, through curious behaviour, we light upon a new discovery, a dopamine 'hit' is delivered to reward centres in the brain, increasing the likelihood that we will remember the discovery. Way beyond babyhood and throughout our lives, humans are often insatiably inquisitive, and our curiosity peaks in circumstances where we think we have a good idea of an answer or outcome but don't know for sure. Hence, the more we know, the more scope there is for us to be curious.

By the age of three or four children start to make mental 'maps' of causal relations between things and people. They may decide, for example, that because eating makes you grow, therefore eating more will make you grow indefinitely. (This is what happens to Alice in Wonderland when she eats the cake marked 'Eat me'.) And as children mature, these causal maps become increasingly complex and accurate, extending to desires and beliefs, emotions and actions – their own and other people's. They also use these maps

When love is withheld or absent very early in life – as was so dramatically the case for many Romanian orphans until the 1990s – brain development can be seriously impaired, and individuals may never become healthy adults. On other hand, children who have had the opportunity to flourish before they are subject to trauma can be astonishingly resilient. In a memoir of internment at Auschwitz, the historian Otto Dov Kulka describes a moment when he got snared on the electrified wire. Recalling his childish mind, in which he believed himself dead, Kulka says he felt 'the boundless curiosity a human being possesses from the moment he first becomes aware of his mortality; curiosity that transcends death'.

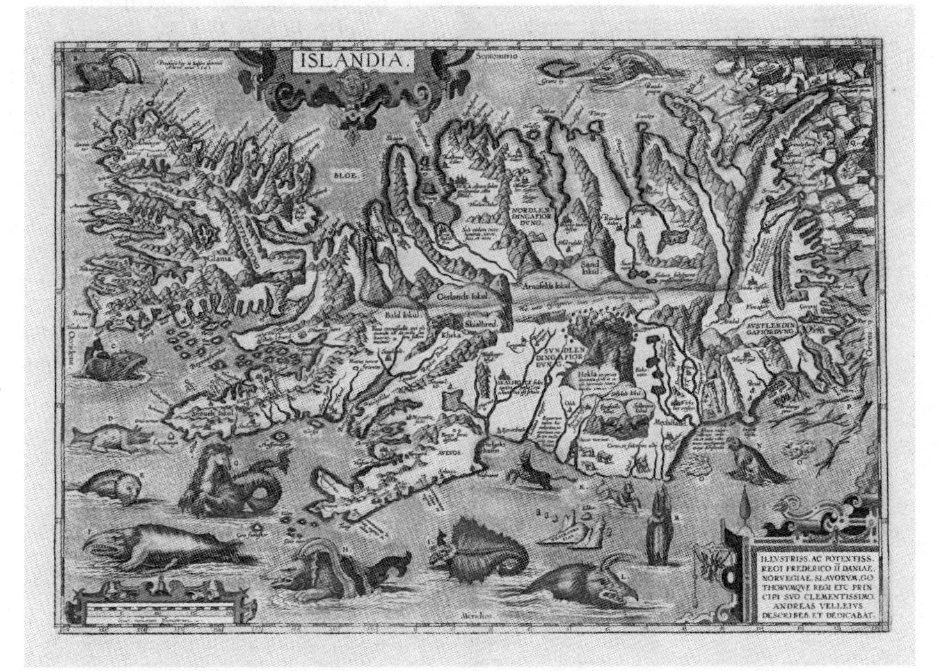

With its sea monsters and picture-book volcano, this map of Iceland by Abraham Ortelius (1590) speaks the language of an atlas from childhood.

to imagine different ways that the world could be, with an ever-increasing number of possibilities and counterfactuals. This, says Alison Gopnik, is the foundation of the great flowering of pretend play in early childhood in which fantasy and imagination transform the world around young children. Underpinning it all, she stresses, is love: just as children can learn freely because they are protected by adults, they can imagine freely because they are loved.

'To young children,' writes Robert Macfarlane, 'nature is full of doors – is nothing but doors, really – and they swing open at every step. A hollow in a tree is the gateway to a castle. An ant hole in dry soil leads to the other side of the world. A stick-den is a palace. A puddle is a portal to an undersea realm.' Macfarlane is writing about a project that encourages children to explore the semi-wild space of Hinchingbrooke

Adventure for children doesn't have to be in the countryside. For Walter Benjamin, growing up in turn-of-the-century Berlin, everything in the courtyard of his family's building was 'a sign or hint'. The anarchist and writer Colin Ward notes the miraculous capacity of children for wringing happiness out of the most disheartening circumstances, in cooperation with 'the equally miraculous transforming power of time'.

Country Park in Cambridgeshire and then make maps of their discoveries. The project leaders, Deb Wilenski and Caroline Wending, are themselves wonderstruck at how, in the children's imaginations, those maps seem to emerge from the ground and the water themselves, rather than being constructed, drawn or laid over the land by a human hand.

A scratchier and no less true account than Dillard's is given by Abel Magwitch, the convict in Dickens's *Great Expectations*: 'I first become aware of myself, down in Essex, a thieving turnips for my living. Summun had run away from me – a man – a tinker – and he'd took the fire with him, and left me wery cold.'

As the wider world comes into focus, children also acquire a more defined sense of self. Annie Dillard has a graceful image for her own memory of this: 'Like any child, I slid into myself perfectly fitted, as a diver meets her reflection in a pool. Her fingertips enter the fingertips on the water, her wrists slide up her arms. The diver wraps herself in her reflection wholly, sealing it at the toes, and wears it as she climbs rising from the pool, and ever after.'

Secure and healthy children are emboldened to explore and, in doing so, engage in what Maurice Merleau-Ponty called the first task of true philosophy: to see the world anew. In his *Autobiography*, the poet John Clare, who was born in 1793, describes how, as a young boy, one summer morning he wandered alone far from his home village:

> I had imagind that the worlds end was at the edge of the orison & that a days journey was able to find it So I went on with my heart full of hopes pleasures & discoverys expecting when I got to the brink of the world that I could look down like looking into a large pit & see into its secrets the same as I believd I could see heaven by looking into the water So I eagerly wanderd on & rambled among the [trees] the whole day till I got out of my knowledge when the very wild flowers & birds seemd to forget me & I imagind they were the inhabitants of new countrys the very sun seemd to be a new one & shining in a different quarter of the sky still I felt no fear my wonder seeking happiness had no room for it I was finding new wonders every minute & was walking in a new world often wondering to my self that I had not found the end of the old one

The point of the child, wrote the socialist thinker Alexander Herzen, is not simply that he or she will grow into an adult. 'The purpose of the child is to play, to enjoy himself, to be a child, because if we follow any other line of reasoning, then the purpose of all life is death.' And few have captured the wonder of their untrammelled play better than the writer and artist Bruno Schulz looking back on his childhood in Poland at the beginning of the twentieth century:

> In those far-off days our gang of boys first hit on the outlandish and impossible notion of straying even farther . . . [to] where boundary lines petered out and the compass rose of winds skittered erratically under a high arching sky. There we meant to dig in, raise ramparts around us, make ourselves independent of the grown-ups, pass completely out of the realm of their authority, proclaim the Republic of the Young . . . It was to be a life under the aegis of poetry and adventure, never-ending signs and portents. All we needed to do, or so it seemed to us, was push apart the . . . old markers imprisoning the course of human affairs, for our lives to be invaded by an elemental power, a great inundation of the unforeseen . . .

The idea that our sense of wonder fades as we put away childish things and grow towards adulthood is widespread. 'It is not now as it hath been of yore,' wrote Wordsworth in his 1804 poem 'Intimations of Immortality from Recollections of Early Childhood'. And this is not only the case for the heirs of the Romantic tradition in the West. In the 2016 documentary film *Beyond Clouds and Dreams*, Phil Agland vividly captures the revelatory wonder experienced by nine- to eleven-year-olds across China at natural beauty in an increasingly poisoned and endangered world. And yet, Agland told me, the common assumption in China is that as these children move on into their teenage years they will quickly forget.

'There was a time when meadow, grove and stream, / The earth and every common sight, / To me did seem / Apparelled in celestial light, / The glory and the freshness of a dream.'

But this is not the whole story. Some cultures and ways of approaching the world seem to hold on to a sense of deep wonder and connectedness to

For the Potawatomi, all beings are persons, and elders of the tribe advise Kimmerer to go and spend time with the Standing people (the trees), the Beaver people and even the Rock people. Their different perspectives will remind her that humans don't have to figure out everything by themselves – that there are teachers all around them.

non-human as well as human beings well beyond childhood. Robin Wall Kimmerer, a biologist and a member of the Potawatomi Nation of North America, suggests this is evident in the language of her people. English is a noun-based language in which about thirty per cent of words are verbs; but – like some real-world cousin of the *Ursprache* of Tlön in Jorge Luis Borges's fiction – Potawatomi is about seventy per cent verbs: things happen rather than are. Kimmerer takes the example of a body of water that in English we call a bay.

> A bay is a noun only if water is dead. When bay is a noun, it is defined by humans, trapped between its shores and contained by the word. But the verb *wiikwegamaa* – to be a bay – releases the water from bondage and lets it live. 'To be a bay' holds the wonder that, for this moment, the living water has decided to shelter itself between these shores, conversing with cedar roots and a flock of baby mergansers. Because it could do otherwise – become a stream or an ocean or a waterfall, and there are verbs for that, too.

This approach is not mystical but is rather a more generous and insightful kind of realism that respects and pays attention to process, cycle and change. Aligning with Kimmerer, Robert Macfarlane imagines 'landscape' as a noun containing a hidden verb:

> landscape *scapes*, it is dynamic and commotion-causing, it sculpts and shapes us not only over the courses of our lives but also instant by instant, incident by incident. It can deplete as well as supplement; leaving our bodies feeling hard and consolidated as crystal – or dispersed and mobile as a flock of birds.

2. Afternoon

As we approach adulthood, new varieties of intense experience become available to us. These can cover a huge range, which, if we're fortunate, can include anything from sheer relief to ecstasy. They do not all have the combination of emotional and intellectual engagement that I suggested in the introduction make for full-on wonder, but many of the people who have these experiences describe them as wonderful.

Take privation followed by relief. After trekking out and back across Antarctica by himself for eighty-six days, the all-but-starving young explorer Aleksander Gamme came upon a food cache he had set aside for his return journey. It included, gloriously, a bag of Cheez Doodles. No prose description could do justice to Gamme's joy on finding this snack, and his reaction is worth watching online. 'It's hard to describe how hungry [I was] at that moment,' he helpfully explains.

Sometimes the cause of intense experience is elusive even though it is grounded in the fabric and functioning of body and brain. 'For several instants I experience a happiness that is impossible in an ordinary state, and of which other people have no conception,' wrote the novelist Fyodor Dostoyevsky. 'I feel full harmony in myself and in the whole world, and the feeling is so strong and sweet that for a few seconds of such bliss one could give up ten years of life, perhaps all of life ... I [feel] that heaven [descends] to Earth ... I really [attain] God and [am] imbued with him.' The language is mystical but Dostoyevsky is describing his state of mind at the onset of a phenomenon with an entirely material cause: he was subject to epileptic seizures.

The most common form of epilepsy causes generalized seizures in which synchronized electrical discharges overwhelm the whole cortex. This often leads to loss of consciousness and after-effects that are extremely unpleasant. But there is another type in which the electrical 'storm' is focused in a small region of the brain, and the person usually remains conscious. Dostoyevsky was subject to this second

Such seizures may be focused in the insula cortex, the main function of which seems to be to integrate 'interoceptive' signals from inside the body such as heartbeat with 'exteroceptive' signals such as touch. (The front of the insula is also thought to be responsible for how we feel about our body and ourselves, helping to create a conscious feeling of 'being'.)

type, which is known as focal epilepsy, and recorded more than a hundred episodes in his diaries and letters over two decades.

Fabienne Picard, a neurologist, has recently found that focal epilepsy is not as rare as was once thought. She has also found that the feelings of those who experience it fall into three broad categories: heightened self-awareness; a strong a sense of physical wellbeing; and intense positive emotions. 'The immense joy that fills me is above physical sensations,' one patient told Picard. 'It is a feeling of total presence ... of unbelievable harmony of my whole body and myself with life, with the world, with the "All".'

The transfiguring joy felt by Dostoyevsky and others with focal epilepsy results from a brain disorder, and is an involuntary and solitary experience. But young adults in perfect health also have ecstatic experiences that they have actively sought out, and very frequently such experiences are shared. Often, one or more of the trinity of sex, drugs and rock and roll are involved – and more about them in a minute – but a vital arena for many people is religious faith, in which being together with others is central.

Evangelical Christianity is a good example. 'There's a depth of community in the church that comes from the fact that you're ... brothers and sisters in Christ, not just a random bunch of people trying to have a community,' Nicky Gumbel, a charismatic in the Anglican communion, tells the philosopher and author Jules Evans. Gumbel manages the Alpha Course, an introduction to Christianity that has engaged more than 27 million people. 'There's a different level of trust and intimacy,' he says, adding, 'Of course that can be abused, but rightly used it can be an amazing thing.' Believers report often feeling completely loved and at peace.

Some volunteers for Islamic State and other militant groups are reported to experience an ecstatic sense of shared purpose built on strong bonds of trust. The anthropologist Scott Atran, who has traced the social networks and life trajectories of militants, especially those from the margins of European societies, suggests that they are 'surfing for the sublime'. Though they may support or commit atrocities, their emotional focus is the poetry of brotherhood and sisterhood in a cause that they believe to be transformative. They find a sense of belonging and redemption in what they see as a righteous struggle against the injustices and deficiencies of consumer societies.

Of course, one doesn't have to subscribe to a religious doctrine to understand the appeal of feeling part of something far larger than oneself, and being forcefully reminded of our relationship to the bigger world tends to have a benign effect. Some psychologists claim to have found that experiences of awe – 'in the upper reaches of pleasure and on the boundary of fear' – can lead to significant positive changes of behaviour. The psychologists Paul Piff and Dacher Keltner monitored people on whitewater rafting trips and visits to groves of giant trees, as well as in more familiar environments. They found that, compared with a control group, these people afterwards made more ethical decisions and showed greater generosity and compassion. 'Even brief experiences of awe ... lead people to feel less narcissistic and entitled, and more attuned to the common humanity [we] share.' Piff and Keltner are firm advocates of what they call 'everyday awe', and encourage people to actively seek out whatever gives them goosebumps, be it looking at trees, night skies, patterns of wind on water or the quotidian nobility of others.

'Awe is the perception of something so physically or conceptually vast that it transcends your view of the world and you need to find ways to accommodate it. It's a basic sense that what you have experienced doesn't fit in with your expectations of the world, so you have to recalibrate.'
Paul Piff

These moments of 'everyday awe' and the sense of awakening that typically comes with them appear to be widespread across time and culture. Spiritual and religious traditions in many parts of the world recognize their importance and incorporate them into systems of belief. In Zen Buddhism there is *kenshō*: an insight or awakening. In Hinduism there is *darśana* or *darshan*: a vision of the divine. Christians speak of grace, and for Sufis there is a vision of divine light. But according to the psychologist Guy Claxton, such experiences need not necessarily be framed within a religious context, and also occur for those who do not profess a faith. In Britain, where the majority of the population do not take part in organized acts of worship, almost two-thirds of both teenagers and adults report having had at least one experience of this kind, and it may well be that many more do but keep quiet about it.

According to research by Claxton and others, these moments, or 'glimpses', are usually fleeting and elusive, seldom lasting more than a few minutes and sometimes only a few seconds. And they vary in their intensity, from merely pleasant to seeming to strike at the very foundations of normal ways of perceiving and feeling. They tend appear out of the blue and disappear again of their own accord, and cannot be held on to by an act of will. Despite their evanescence, however, they are regularly felt not to be illusions but revelations of a deeper reality, accurately and intensely perceived.

Two other things characterize these states. First, they are typically very physical experiences: there is a burst of vitality and aliveness – often described as brightness, energy and warmth – and an intensification of perception. Second, there is a shift from feeling separate to feeling connected – 'from longing to belonging'. One of their most attractive characteristics, says Claxton, is a liberation of affection: 'Instead of being too busy to care, you notice what needs doing to look after the people and the environment around us, and naturally do it . . . There is a feeling of ease, as a complex weight of considerations and concerns seems to drop away, and life appears radically simplified.' You experience the world as if everything matters.

Glimpses of this kind tend to be felt in the body as much as the mind. But they do not, for the most part, have a sexual dimension: they tend more to *agape* than *eros*. (The novelist Henry James recognized this in the paintings of Fra Angelico: 'a passionate pious tenderness . . . a perpetual sense of sacredly loving and being loved'.) This is not to say that erotic experiences are necessarily without a spiritual dimension. Sex where love is present can be holy, and what is holy can also be expressed in erotic ways as it is in the Song of Songs:

'As they lay together . . . it was as if they were at the very centre of all the slow wheeling of space and the rapid agitation of life deep, deep inside them all, at the centre where there is utter radiance.'
D. H. Lawrence

> How much better is thy love than wine!
> and the smell of thine ointments
> than all spices!

Thy lips, O *my* spouse, drop *as* the honeycomb: honey and milk *are* under thy tongue; and the smell of thy garments *is* like the smell of Lebanon.

Sex

Copulation between non-human animals can be startling, and startlingly enthusiastic, in many ways. Duelling hermaphrodite flatworms seek to inseminate each other with a stab from one of the two penis-like organs they both possess. Some male bush crickets have a pair of curved rods shaped like coat hooks extending from their genital openings that they use to stimulate a sensitive area of the female's genitals. You can find a lot more of this kind of stuff on the Internet, including on pages maintained by one of the world's most respected public service broadcasters. BBC Earth explores 'The Twisted World of Animal Sex Organs' and explains 'Why Animals Are Kinkier Than You' – with 'Why Seals Have Sex with Penguins' being just one instance. But in general it is not a rabbit hole that I would recommend. In any case, in contemplating the sexual behaviours of other members of the animal kingdom we should sometimes put the snickering aside and simply marvel at the beauty. The courtship dance of Japanese cranes is one of the great glories of life on Earth. The mating of bonobo chimps may not be so physically prepossessing, but their culture of love not war has a lot to recommend it.

The drive to copulate is hardwired across the animal kingdom – including, obviously, humans. Sometimes the wiring can get confused, with rather wonderful results. In a manifestation of synaesthesia, some people (and they seem to be mostly women) see intense colours during orgasm. For some individuals, only one colour is present at a time, but the colour changes from one orgasm to the next. What prompts a particular colour to manifest is a good question, and there is surely an Ig Nobel Prize in there somewhere.

But not all mis-wiring, or 'cross-talk', is so pleasant.

Parodies of the Nobel Prizes, the Ig Nobels are awarded for research that 'first make people laugh, and then make them think'. In 2016 the Ig Nobel Prize for chemistry went to Volkswagen for solving the problem of excessive automobile pollution emissions by automatically producing fewer emissions whenever their cars were being tested. The prize for medicine was given for discovering that if you have an itch on the left side of your body you can relieve it by looking into a mirror and scratching the right side of your body (and vice versa).

In *Bonk*, Mary Roach describes the case of a woman who had an orgasm whenever she brushed her teeth. It sounds hilarious, but she had a horrible time of it. The joke wore out long before the toothbrush did (though the inventor of the first electric toothbrush, a Dr Woog, did develop a champion vibrator). Another puzzling case concerned a man with severe and intractable hiccups. He tried every conceivable way to stop them until – in what was, presumably, a desperate last resort – he had sex with his wife. At the moment of ejaculation the hiccups stopped.

Orgasm has been defined as a reflex of the autonomic nervous system that can be either facilitated or inhibited by cerebral input – that is, by what you are thinking and feeling. This is a rather depressing definition, but it does say something important about the nature of the beast. We are talking about a behaviour that, whatever else we attach to it, is in large part automatic and primitive. The autonomic nervous system – which also regulates heart rate, digestion and respiratory rate, and reflexes such as coughing, sneezing and vomiting – is largely outside conscious control. It is different and distinct from the somatic nervous system, which transmits skin sensations and enables voluntary movements. (Unlike the pathways of the somatic system, those of the autonomic system do not run exclusively through the spinal column. This means that some people who are paralysed from the neck down as a result of spinal injury are still capable of experiencing orgasm.) Brain scans show that many areas in women's brains become quiescent during the actual moments of orgasm – including, rather surprisingly, those involved in emotion. The effect is reported to be less marked in men but this may be because male orgasms are shorter, and harder to detect (in the brain) in the first place.

This is not to say that what we think and feel during intercourse is unimportant. All the evidence suggests that the most exquisite and intense sexual pleasure in humans depends on prior psychogenic arousal by a partner who is imaginatively attuned to the loved one's sensations and desires. A study by William Masters and Virginia Johnson published in the 1970s found that homosexual couples were often more adept at bringing their partners to the heights of ecstasy than heterosexuals were. This was because they (and women in particular) 'tended to move slowly . . . and to linger at [each] stage of stimulative response, making each step in tension increment something to be appreciated'. Often demonstrating more empathy for their partners than many heterosexuals did, and prolonging each tease, lesbian and gay couples tended to be less fixated on getting the job over and done with. Clearly, there is a lot here from which many of us could learn if we want to make the sexual act a more wondrous experience. The philosopher Luce Irigaray argues that men and women can find great wonder when they open themselves fully to respect the irreducible difference of the other and respond with generosity. Wonder, she writes, is 'indispensable not only to life but also . . . to the creation of an ethics, notably of and through sexual difference. This other, male or female, should surprise us again and again, appear to us as new, very different from what we knew or what we thought he or she should be.'

Drugs

Love is the drug, sang Roxy Music. But it's not the only one associated with the search for intense, life-enhancing experience. The most widely consumed, at least in the West, is alcohol. Unlike orgasm, alcohol operates almost entirely in the brain alone, where it works by binding to at least two types of chemicals. With the first, called GABA, alcohol acts like a sedative, calming the brain and making its owner feel more relaxed, while with the other, called glutamate, alcohol excites the brain and makes it more active. The two effects would seem to be at odds – thus Lady Macbeth: 'That which hath made them drunk hath made me bold; / What hath quench'd them hath given me fire.' At the same time, alcohol triggers the release of dopamine in brain regions involved in pleasure and reward. Not bad for a chemical that is basically yeast wee.

In many countries alcohol abuse damages and destroys huge numbers of lives. The abuse of other mind-altering chemicals can inflict grave harm too, but there is evidence that moderate and careful use of certain of them can, in some instances, be beneficial. A few researchers make special claims for psychedelics such as LSD, psilocybin and ayahuasca. When administered responsibly, they say, these drugs can contribute significantly to human wellbeing and open the doors to extraordinary wonders.

In the years up to the mid-1960s the US federal government funded hundreds of tests with psychedelics on alcoholics, people struggling with obsessive-compulsive disorder, depressives, autistic children, schizophrenics, terminal cancer patients and convicts, as well as on healthy artists and scientists to study creativity, and on divinity students to study spirituality. Although many of the studies were poorly designed by modern standards, positive results were reported frequently. Then LSD escaped from the lab and swept through the counterculture. Increasingly, the recreational use of psychedelics became associated with instances of psychosis, flashback and suicide. In 1970 the US government prohibited their use for any purpose, and research came to a halt.

In the last decade there has been a revival of interest in the possible benefits of psychedelic drugs, with a few clinical trials run by universities in the United States and Europe. In 2006, the psychopharmacologist Roland Griffiths published a paper on the administration of psilocybin to healthy volunteers in carefully controlled conditions and reported that it could give rise to mystical experiences with 'substantial and [enduring] personal meaning and spiritual significance', including feelings of unity, sacredness, ineffability, peace and joy, the impression of having transcended space and time, and the sense that the experience has disclosed some objective truth about reality. A third of those who ingested psilocybin in the study ranked the experience as the most spiritually significant of their lives, while two-thirds of participants rated it among their top five, comparable to the birth of a child or the death of a parent. Not all experiences were positive, however. About one in five reported negative feelings such as anxiety.

In a 2015 article on the revival of research into psychedelics, the journalist Michael Pollan reported that no adverse effects had been reported in the new wave of trials, many of which were being conducted on patients with terminal cancer. This is probably because these patients were carefully screened, prepared for the experience in advance, and guided through it by therapists trained to manage the episodes of fear and anxiety that sometimes occur. Every trip was different, Pollan noted, but there were recurring themes. Several of the patients he interviewed described an experience of either giving birth or being born. Great secrets of the universe – such as 'We are all one' or 'Love is all that matters' – become clear to them. 'The usual ratio of wonder to banality in the adult mind is overturned,' Pollan writes, 'and such ideas acquire the force of revealed truth. The overall result is a kind of conversion experience, and the researchers believe that this is what is responsible for the therapeutic effect.' Patients also revelled in a

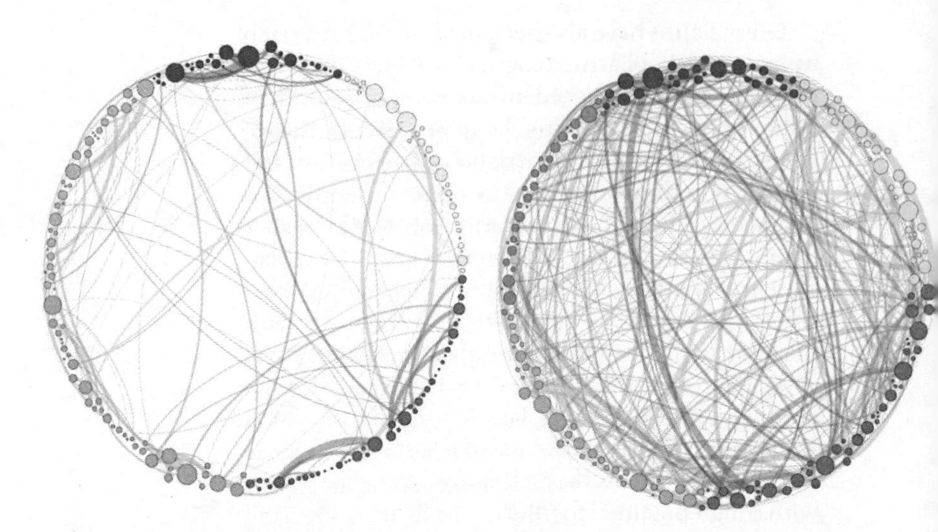

A visualization of increased connectivity in the brain under psilocybin (right).

'I am nothing; I see all; the currents of the Universal Being circulate through me; I am part or particle of God . . . From the earth, as a shore, I look out into that silent sea. I seem to partake its rapid transformations; the active enchantment reaches my dust, and I dilate and conspire with the morning wind.' Ralph Waldo Emerson

sense of being able to travel at will through space and time. It sounds a bit like Emerson's (non drug-induced) transcendent experience of becoming a transparent eyeball, with the body dissolving even as a perceiving and recording 'I' continues to exist.

Roland Griffiths has described the therapeutic experience of psilocybin as a kind of 'inverse PTSD – a discrete event that produces persisting positive changes [lasting years or decades] in attitudes, moods, and behaviour, and presumably the brain'. He stresses, however, that the drug is no replacement for the mental health benefits of continuous personal reflection and discipline: 'There's all the difference in the world between a spiritual experience and a spiritual life.' Griffiths also stresses that substantial risks could easily follow from the use of psilocybin without the appropriate psychiatric screening, preparation and monitoring. Still, results in carefully controlled settings continue to appear positive. In one trial, eighty per cent of smokers gave up smoking after three doses of psilocybin, and had still given up six months later. The most successful anti-smoking programmes to date have a success rate of well under half this.

Great claims have also been made for LSD. A study by neuropsychopharmacologist David Nutt and Robin Carhart-Harris published in 2016 showed that the brains of those on the drug 'lit up' spectacularly, indicating that they were experiencing images through information drawn from many parts of their brains, and not just the visual cortex. Another study claimed to find that LSD produced feelings of happiness, trust, closeness to others, and enhanced explicit and implicit emotional empathy. For now, however, the great majority in the medical community remain cautious about and/or sceptical of the alleged benefits, and psilocybin and LSD remain Class A drugs in the United Kingdom and Schedule 1 substances in the United States, which means they are judged to be dangerous with a high potential for abuse.

Some of those championing a revival in the use of psychedelics are more interested in what they believe to be their spiritual potential than they are in their therapeutic potential. A favoured term is entheogen, or 'God-facilitating', rather than psychedelic, which means simply 'mind-manifesting'. The Santo Daime Church, which operates in the Netherlands and Brazil, uses ayahuasca, an Amazonian vine employed by indigenous shamans to induce extraordinary states of consciousness, in its ceremonies. A spokesman for the Santo Daime Church suggests that ayahuasca can help adults become better children, and it sounds like a verse from the Gospel of St Matthew. Taking ayahuasca has now become the subject of mockery, with satirists reporting in 2016 that a South American shaman was dreading yet another session of guiding tech CEOs to spiritual oneness. But jokes aside, consider this from an account by a healthy young adult male working in a prestigious profession who attended a Santo Daime ceremony:

'Except ye . . . become as little children, ye shall not enter into the kingdom of heaven.' Matthew 18:3

> The music and the space of the room took on a dark resonance as the [ayahuasca] took hold, and I started to feel a deep and grinding fear that took form in visions. As I shut my eyes, I could see demonic beings moving towards me out of a dark realm . . . The strong and centred presence

[of a ceremony assistant] next to me was a big help through a horrible twenty minutes or so, during which I felt like I was going to be dragged down in the very pits of darkness . . .

The demons had taken on a horrifying clarity and objectivity, but at some point I realized . . . that my only escape from this was through love . . . I focused on the love I have for my wife, imagined holding her close, and this appeared in the vision as a light in my chest. The light grew and grew, and as it did so it transmuted the hell-like realms around me, and the colours of the visions became lighter and lighter . . . The love I was feeling in my chest expanded into an even more intense form, as I was shown my unborn child (my wife was 14 weeks pregnant at the time) as a baby girl in my arms. I sobbed out loud with love for this baby for what seemed like hours, and with my tears I was finally released from darkness into light . . .

Now that love and peace were present as the core of the experience, far more order and symmetry was manifest in the visions. The higher worlds that I had emerged into were spectacular [with] constantly shifting spaces of symmetry-laced geometrical surfaces and structures, like part-mechanical, part-organic crystalline cities, interlaced with animals, birds, semi-human forms and deity-type figures. This was no hallucination . . . I was *in* these perfectly formed hyper-complex worlds for hours, repeating to myself over and over 'never forget this, never forget this'.

Rock and roll

Taking an ayahuasca trip in the Amazon, New York or the Netherlands may be a long way to go for a mood-altering or life-enhancing experience. A more common and accessible method of altering mood is through music, and it has probably been enjoyed in this way for as long as we have been human. The uses of drugs and music are not, of course, exclusive of each other, but music can be remarkably powerful on its own.

'Whites Invent "Rock and Roll",' declares the headline of the satirical online 'newspaper' *The Onion* for 21 November 1955 – 'Authorities Assure Public that Negroes Had Nothing to Do with Popular Music Form'. The joke is, of course, mainly about racism, but it also speaks to the difficulty that some white Americans had with the rebirth of music at its most Dionysiac and filled with *Rausch* – giving us a rush.

Rausch means 'intoxication' in modern German, but the word has a long history. For Goethe it was an acceleration of movement leading to flowing joy. For Nietzsche it was excitement, rapture, primal desire, a feeling of power and of unity with the cosmos.

An unease with the ecstatic side of human nature evoked by music is not new in European culture. In October 1557, a small group of French Calvinists who had been promised religious freedom in France Antarctique, a colony on a small island in the bay of what is now Rio de Janeiro, were expelled by their Catholic overlord. Desperate, they took refuge with the Tupinambá Indians on the mainland. In his chronicle of their misadventures, Jean de Léry records his fear and wonder at a ceremony performed by their new hosts:

> ... we began to hear in the men's house ... a very low murmur, like the muttering of someone reciting his hours. Upon hearing this, the women (about two hundred of them) all stood up and clustered together, listening intently. The men [slowly] raised their voices and were distinctly heard singing all together and repeating this syllable of exhortation, *Heu, heu, heu, heu*; the women, to our amazement, answered them from their side, and with a trembling voice ... let out such cries for more than a quarter of an hour, that as we watched them we were utterly disconcerted. Not only did they howl, but also, leaping violently into the air, they made their breasts shake and they foamed at the mouth ... I can only believe the devil entered their bodies and they fell into a fit of madness.

But then the mood changes:

> After these chaotic noises and howls had ended ... we heard them once again singing and making

> their voices resound in a harmony [of] marvellous, ... sweet and gracious sounds. At the beginning ... I had been somewhat afraid; now I received in recompense such joy, hearing the measured harmonies of such a multitude, and especially in the cadence and refrain of the song [that] I stood there transported with delight. Whenever I remember it, my heart trembles, and it seems their voices are still in my ears.

In an utterly alien culture and in dire straits, de Léry found brief solace in music. But the sounds that brought him such sweet joy were, for their Tupinambá creators, part of a larger story – a dramatic arc in music and dance that also contained more turbulent emotions, trance-like states and, presumably, a kind of catharsis. Maybe the Tupinambá, unlike these early modern Europeans, understood the value of entering fully into both the dark and the ecstatic aspects in their religious practice and art.

For a recent study of 'strong experiences with music', the music psychologist Alf Gabrielsson asked Swedes to describe the nature of their experiences. The variety is remarkable:

> Feelings of weightlessness may be experienced when listening to Lumbye's 'Champagne Galop', Bob Marley's 'Exodus', or Bach's *St Matthew Passion* ... Metaphors employed by the informants include snowflakes (Bach), stone pillars (Mozart), an incense holder (Beethoven), and a red-hot toaster (Wagner). Listening experiences are compared to having lemonade in one's legs, being cut by a piercing laser beam, going on a roller-coaster, or being nailed to the bench. Other kinds of feelings include satisfaction, gratitude, perfection, love, solemnity, humility, admiration, patriotism, and sexual arousal. Music is experienced as particularly powerful in connection with illness and death (for example at funerals), as well as with love. It can prompt epiphanies that suddenly change a situation completely. Those performing often experience the strongest feelings, including

total involvement, no fear, no time, no self, trance, collective touch, universal humanity.

John Sloboda, another music psychologist, suggests a way to categorize these diverse experiences of music. When I met him at the Guildhall School of Music in London he described how (at least in industrial societies) they tend to be of three kinds. At the highest level, he told me, are peak experiences. These are very intense, memorable and often life-changing moments that people talk about for years afterwards. They tend to be rare and cannot be engineered: they just happen. The next level are thrills: visceral responses to music that are often associated with a physical manifestation of emotion such as a lump in the throat, tears in the eyes or a shiver down the spine. Thrills are much more frequent than peak experiences, and tend to be predictable. Most people can describe, and tend to agree on, the places in a piece of music that cause them – typically something such as a change in volume, rhythm or key or a voice singing particular, emotionally charged words or phrases. They tend not to last very long. The third kind of experience in Sloboda's category is associated with music in everyday experience. People tend to play or listen to music quite deliberately to change their mood – typically to energize themselves for minutes or even hours. Some people use music this way many times a day.

One of things I take from this is that peak moments and thrills in music are well and good, but we should not underestimate the benefits of everyday music and dance – not least because they may sometimes trip us almost by accident into a moment of transcendent significance and celebration. In *Dancing in the Streets*, her history of collective joy, Barbara Ehrenreich ends with a description of a street celebration in Rio de Janeiro. There was, she writes, no 'point' to it – 'no religious overtones, ideological message or money to be made – just the chance, much needed on a crowded planet, to acknowledge the miracle of our simultaneous existence with some sort of celebration'. Perhaps the first part of her observation is not entirely right: the point to Rio's celebratory culture may in fact be that it

'We need to worship less, consume less and play more.' Jules Evans

'The pleasure I get from playing is not the pleasure of expressing myself, but of subjugating myself to the music and experiencing the ecstasy that comes from being a part of it.' Steve Reich

flourishes not just despite the injustice and violence of life in the city, but precisely as a way of defying it. But Ehrenreich's account of this moment on the streets of Rio illustrates the key to the joy of music and dance: it is about taking part.

I try, in a small way, to do something like this when I sing in a community choir. We are not exactly brilliant, but we're not absolutely terrible either, and the sense of achievement and joy we get from the shared effort is disproportionate to the quality of our singing. What we do is a small but significant wonder in our own lives. We are an odd-looking, unlikely bunch – any citizen of Rio would laugh at the way we move – but we are mostly harmless, and I like to think that we have something in common with the Mbenzélé pygmies of the Congo rainforest, who don't just sit around consuming music but are always creating it, in order (as they say) to 'make the forest happy'.

3. Evening

The psychologist Daniel Kahneman identifies System 1 and System 2 thinking, with System 1 as fast, instinctive and emotional, and System 2 as slow, deliberative and logical. 'True' wonder, I would argue, allows for the full play of both.

'Behind the most efficient-seeming adult exterior,' wrote Ted Hughes, 'the whole world of the person's childhood is being carefully held like a glass of water bulging above the brim.'

Wondrous experiences in our early years and young adulthood like the ones explored in the first two parts of this chapter are often very immediate, and associated with strong physical sensations. But there is another dimension to what I would like to see as 'true' wonder, and that is strong cognitive engagement – an intensity of thought as well as feeling, or at least the beginnings of it. This deeper and larger wonder, as I would characterize it, takes account of what is not immediately present or apparent as well as what is. It integrates the experience of the moment (or a strong memory of it) with a larger perspective to create a sense of meaning. This kind of wonder is not restricted to older adults (and there are older adults who may seldom if ever experience it). It can be seen on the faces of the nine- to eleven-year-olds in Phil Agland's documentary of China's environmental crisis. But it can be particularly salient for those who have been alive longer, and for that reason have more days to look back and reflect upon.

The long view does not always offer reasons to be cheerful. There has never been a shortage of evidence that, as Edward Gibbon wrote, 'history is ... little more than the register of the crimes, follies, and misfortunes of mankind'. Indeed, there is a lot to suggest that all three extend far back and beyond the written record, and have been rife across societies of which Gibbon knew little or nothing.

There is also plenty to suggest that humans often tend to be 'wired for negativity'. In 2001 Roy Baumeister and other psychologists looked at the evidence from a large number of studies and found that bad emotions, bad parents and bad feedback have more impact than good ones, and that bad information is processed more thoroughly than good. 'The self is more motivated to avoid bad self-definitions than to pursue good ones,' they wrote. 'Bad impressions and bad stereotypes are quicker to form and more resistant to disconfirmation than good ones.' We want it darker. We kill the flame.

And we are, it would seem, above all creatures of fear. The novelist and essayist Marilynne Robinson suggests that fear may be a default posture of human beings. Her observation is informed by experience in her own country, the United States, but research suggests that frequent experiences of fear may be not just, or not only, a consequence of particular cultural and political conditions. The amygdala – a region that flags important, emotionally laden situations and that is critical for coding fear – acts extraordinarily quickly and unconsciously in the brain, and some researchers conjecture that our evolutionary past has given fear a dominant role in human psychology. Maybe our distant ancestors, vulnerable to terrifying animal predators and the predation of other humans, have bequeathed us a frequently hysterical being within.

'Most top predators are majestic creatures,' asserts the historian Yuval Noah Harari. 'Millions of years of dominion have filled them with self-confidence. Homo sapiens, by contrast, is more like a banana republic dictator. Having been so recently the underdogs of the savannah, we are full of fears and anxieties over our position, which makes us doubly cruel and dangerous.'

In addition to fear, it is supposed, our lives are dominated by appetite. Aristotle asserted that 'it is of the nature of desire not to be satisfied, and most men live only for the gratification of it'. Some scientists think they have found a neuroanatomical basis for this claim. Our brains seem to be wired in such a way that our wants always exceed our likes. Desire and pleasure are separate chemical systems in the brain. The desire system is vast and powerful; the pleasure system is much smaller and harder to trigger. We are therefore liable to becoming insatiable wanting machines – hungry ghosts. This condition can be unleashed altogether when the brain's normal control systems are impaired, as they are in Klüver-Bucy syndrome, which manifests in insatiable eating and a heightened sex drive, sometimes combined with irritability and distractibility. But even without physical damage to the brain, we are vulnerable, in a hyper-connected, hyper-capitalist world, to distraction and an endless chain of wants leading to what the philosopher Matthew Crawford calls the obesity of the mind. 'Human desires are indeterminate,' asserts another philosopher, Roberto Unger. 'They fail to exhibit the targeted and scripted quality of desire among other animals. Even when, as in addiction and obsession, they fix on particular objects, we make those particular objects serve as proxies for longings to which they have loose or arbitrary relation.'

'The United States suffers from pandemic attention deficit disorder, fake news, hackable everything, cyber war and social media bullies.' Rebecca Solnit

So too with habit. 'Remaining stuck in a rut is something to which we devote great effort,' observes the psychologist Vincent Deary. We do so because ruts are familiar and therefore comfortable. (This, incidentally, may be a major cause of procrastination, which the psychologist Seth Roberts describes as the tendency to do the same thing over and over in order to avoid change.) And recognizing this, entrepreneurs have worked out ways to reinforce and exploit our habit-forming natures. A really successful app, for example, creates a 'persistent routine' or 'ludic loop', triggering both a need and providing the momentary solution to it. 'Feelings of boredom, loneliness, frustration, confusion, and indecisiveness often instigate a slight pain or irritation and prompt an almost instantaneous and often mindless action to quell the negative sensation,' writes the former games designer Nir Eyal. 'Gradually, these bonds cement into a habit as users turn to your product when experiencing certain internal triggers.' The economists George Akerlof and Robert Shiller point to a particularly successful instance in slot machines in Las Vegas casinos. Ten years ago, deaths

In a ludic loop, we do something over and over again because every once in a while we get a reward. Like pigeons, we are more driven to seek feedback, when it isn't guaranteed.

due to cardiac arrest were an especially serious problem, but because of crowding the emergency crews could not get through. 'One surveillance video shows why . . . As a squad . . . defibrillates the heart arrest of a player, the surrounding players play on, their trance unperturbed, even though the victim is literally at their feet.' And more broadly in society, as we walk around staring at our screens, checking our newsfeeds rather than chatting with friends, pressing refresh to retrieve a new tweet, we are less attentive to the world around us, more likely to miss moments of wonder. Human desires may not necessarily be indeterminate as Unger asserts, but cocooned in the filter bubble of the Internet and, as the writer and journalist George Monbiot puts it, 'without a visceral knowledge of what it is to be hurt and healed, exhausted and resolute, freezing and ecstatic', we lose our essential reference points and become vulnerable to neuro-marketing and even worse forms of manipulation.

'To live in freedom, one must get used to a life full of agitation, change and danger.'
Alexis de Tocqueville

'In theory, freedom may be held in high regard,' says the writer on Buddhism Stephen Batchelor; 'in practice it is experienced as a dizzying loss of meaning and direction.' Some even dismiss its pursuit as contemptible. The philosopher and essayist Emil Cioran, author of books with titles like *The Trouble with Being Born*, believed he was fated to experience the world as weariness, boredom, meaninglessness and rebellious anger towards everything. 'It is not worth the bother of killing yourself,' he once wrote, 'since you always kill yourself too late.' The writer Michel Houellebecq – whose facial musculature, according to one report, 'seems evolved to express infinitely fine gradations of disenchantment' – thinks that Europeans made a wager that the more they extended human freedom the happier they would be, and have lost.

'For Cioran . . . thinking does not mean thanking . . . it means taking revenge.'
Peter Sloterdijk

But no generalities about cruelty, fear, addiction and the like tell the whole story. Time and again, many people find the resources and courage to overcome them. As the poet Rainer Maria Rilke said, 'If [the world] has terrors, they are our terrors; if it has abysses, these abysses belong to us; if there are dangers, we must try to love them.'

'The soul is a miraculous abyss of infinite abysses, an undrainable ocean, an inexhaustible fountain of endless oceans.'
Thomas Traherne

A fragment attributed to the Greek philosopher Heraclitus, who was born in the first half of the sixth century BC, says that all things come into being out of a conflict of opposites. The whole that arises depends upon opposed tensions held in balance, like the tips of a bow or a lyre held in equilibrium between the tension of the string pulling them together and the spring action of the limbs pushing them apart. Consider, for example, the trajectory of a human body over time. Unimpeded, many of its cells will tend to reproduce and grow endlessly, and may metastasize into cancer. It is only when the drive of individual cells to reproduce is restrained or held in balance through mechanisms such as programmed cell death that the whole will continue to flourish. We exist in a constant tension between living and dying.

Consider, too, the integrative nature of conscious experience, which holds different things together. In a short poem titled 'Encounter', Czesław Miłosz recalls a ride through frozen fields in a wagon at dawn. A hare suddenly runs across the road and one of the people in the wagon points to it. But then the poem turns: the sights it has brought to mind happened long ago and neither the hare nor the man who made the gesture is alive. The poem, whole and entire on the page in front of us, exists in the tension between what it recalls, which is irrevocably lost, and the fact that it is somehow remembered there.

'. . . where are they, where are they going / The flash of a hand, streak of movement, rustle of pebbles. / I ask not out of sorrow, but in wonder.'

The poet Les Murray notes that Carl Jung and many before him would call integrative experience of consciousness 'soul'. But not wanting to depend on a word he regards as almost worn out by overuse, Murray prefers to call it his 'poem-self', in which his three ordinary states of being – the waking conscious mind, the occult mind of dreams, and the body – are united:

> The fusion . . . produces an excitement frequently so intense that I can't bear it for too long at a stretch, but must get up and run outside for rests from it, then come back for some more. The poem I write during this experience will contain the experience, the more strongly the better the poem is, and will continue to contain it after the trance has left me.

In a good poem, the integrity experienced by its author is made available to the listener or reader. And, being fixed, it is something to which we can return, in wonder, for solace. The same holds true for the other arts, but also in the most compelling stories we tell ourselves about real life and the choices we have made, where incidents spread across time are brought together by a line of narrative – a story – that is charged with energy and meaning.

Take the case of Bryan Stevenson, a lawyer who campaigns in the United States against the lifelong imprisonment of children, for the reversal of convictions to death due to a miscarriage of justice, and for the abolition of the death penalty in all cases. Among his sources of motivation, he relates in his book *Just Mercy*, was a lesson taught him by his mother more than forty years before, when he was ten. On a Sunday morning after church, he and his friends met a small boy who was visiting from out of town. The boy had a severe speech impediment and stuttered so much that he couldn't even pronounce the name of the town where he lived. The young Stevenson, who had never before met someone who stuttered, thought the boy was joking, and laughed. Then he noticed his mother looking at him with horror, anger and shame. 'Don't you *ever* laugh at someone because they can't get their words out right,' she said. 'Don't you *ever* do that!' Stevenson was devastated by his mother's disapproval and then, to his mortification, she told him in front of his friends that he must hug the stranger and tell him that he loved him. The young Stevenson did so, hugging the other boy awkwardly at first, and finally, on his mother's insistence, mumbled to the boy that he loved him. The boy hugged him back and said 'I love you too' without a single stutter. 'There was such tenderness and earnestness in his voice,' writes Stevenson, 'and just like that, I thought I would start crying.' The compassion that Stevenson discovered in himself that day has informed the work of his life, which is to help give voice to those who are unheard.

Or consider a story told by the theologian Thomas Berry of his childhood in North Carolina in the 1920s. At the age of seventy-nine, Berry recalled moving with his family when he was about twelve to a new house being built by a creek. Across the creek was a meadow. 'It was an early afternoon in late May when I first wandered down the incline, crossed the creek, and looked down over the scene,' Berry wrote. 'The field was covered with white lilies rising above the thick grass. A magic moment, this experience gave to my life something that seems to explain my thinking at a more profound level than almost any other experience I can remember.' And this early experience, he wrote, became a standard for him throughout his subsequent life and thinking: 'Whatever preserves and enhances this meadow in the natural cycles of its transformation is good; whatever opposes this meadow or negates it is not good. My life orientation is that simple. It is also that pervasive. It applies in economics and political orientation as well as in education and religion.'

'The long-lost glittering hour that means more than age, more than logic, more than lore.' Amy Leach

In these autobiographical fragments, both Stevenson and Berry have reached an age when they have been able to step back, to glimpse their existence as a whole and to trace the thread of meaning that runs through it. This kind of reckoning may happen at almost any time but it may be particularly common when we are strongly aware of the proximity of death, and the nature of life can appear in stark and clarifying focus. Raymond Tallis writes, 'The sense of finitude animates a desperate desire to make a deeper, more coherent, sense of things, to seize hold of it in its greatness, to be equal in consciousness to the great world on which we find ourselves, of which we are conscious in a piecemeal, sequential, fragmented, small-world way. The idea of death is a threat, a goad, and an inspiration. And its power is available to all of us who aim to live abundantly.'

When, at the age of thirty-six, the neurosurgeon Paul Kalanithi was diagnosed with terminal lung cancer, he and his wife Lucy discussed whether or not to have a baby. 'Don't you think saying goodbye to your child would make your death more painful?' she asked him. 'Wouldn't it be great if it did?' he replied. Kalanithi chose to walk towards the pain, wagering that with it would also come something wonderful, and in his case it did. Addressing his healthy infant

'Sweet is the bitter sea.' Richard Jefferies

daughter Cady shortly before his death less than two years later, Kalanithi wrote, 'When you come to one of the many moments in life when you must give an account of yourself . . . do not discount that you filled a dying man's days with . . . a joy unknown to me in all my prior years, a joy that does not hunger for more and more, but rests, satisfied.'

Wisdom is difficult to define. One cross-cultural study identified six components to it: pragmatic knowledge of life; emotion regulation; pro-social behaviour; knowing one's strengths and limitations; decisiveness; and acceptance of uncertainty. There is no guarantee that ageing necessarily favours any of these, and a life does not have to be long to be wise and joyous. 'It is an inestimable joy that I was raised out of nothing to see and enjoy this glorious world,' wrote Thomas Traherne, who died of smallpox at the age of thirty-seven. But wisdom can and does come to some people in later years. Perhaps, with death a little closer but not so close that it is all-engulfing, it can be a little easier to detach from one's immediate self and become more of an observer of all that is. 'The most beautiful thing about our minds,' writes Carl Safina, 'might be the occasional triumphant moment when we see ourselves *not* in a mirror but from a distance.'

'The gift is to go on failing without being destroyed by it.'
Richard Holloway.

'Men that love wisdom must be acquainted with very many things indeed.'
Heraclitus

'With what strife and pains we came into the world we know not,' wrote Thomas Browne; 'but 'tis commonly no easy matter to get out of it.' Not many are as fortunate as the philosopher David Hume, who wrote shortly before his death in 1776 that notwithstanding the great decline of his person, he had never suffered a moment's abatement of his spirits, and possessed the same ardour as ever in study, and the same gaiety in company. The physician Oliver Sacks seems to have come close, writing shortly before his death in 2015 that he was able to see his life as if from a great altitude and with a deepening sense of the connection of all its parts. For those who have a harder time of it, the writer Jenny Diski, herself approaching death in 2016, suggested a modest aid. Just as we write books for the very young to help them learn names, actions and descriptions for the world they have suddenly found themselves in, so, she suggested, we should

Aún aprendo (I Am Still Learning) by Francisco Goya (*c.*1826). Ineluctably, our brains slow down as we get older. Our ability to carry out mental operations declines, as does our capacity for episodic memory and executive function. But two capabilities tend to deteriorate more slowly than others: reasoning and cognition based on past experience. These typically peak between the ages of about forty and fifty, and sometimes endure for as long as we stay mentally active and engaged.

make a book for those who are gradually losing their command of language and memory. 'The child's book is the book of becoming. The book for the elderly is the book of going. It had better be the most beautiful ever made.'

'I don't care how old you are. Do something beautiful.'
George Saunders

A study conducted at a hospice in Buffalo, New York, found that nearly all terminally ill patients reported having dreams or visions shortly before they died. Most of these were described as comforting, although about one in five was distressing. Other research also suggests that dreams express emotions that have been building within, and both positive dreams and troubled ones, which are often linked to unresolved issues in the individual's past, can erupt with excessive energy. Often, it appears, in the hours before a final delirium or unconsciousness takes hold, people try to find meaning. And coursing through these dreams and visions is the need for resolution and forgiveness and – given or withheld – love. Even for Oedipus, the most unfortunate of men, the end contains something wonderful (at least in Sophocles' imagination), as he is filled with serenity and love for his daughters.

The most wonderful thing in William Shakespeare's *The Tempest* is not Prospero's rough magic, which is associated with midnight mushrooms and creaky stage machinery, but the love between his daughter, Miranda, and Ferdinand, the son of his enemy. They are able to leave the ills of the past behind and begin afresh. The magician's achievement is that, recognizing his own mortality and the finitude of all things, he is able to forgive and to give, setting himself aside and nurturing the next generation.

'All of us are creatures of a day; the rememberer and the remembered alike. All is ephemeral – both memory and the object of memory'
Marcus Aurelius

Two of the everyday experiences that, as a parent, fill me most with wonder are running alongside my daughter in the woods and watching her sleep. The first is a moment of shared existence and exhilaration in play. And in the second, watching the gentle rise and fall of the breath in a new being, I am in awe that both this life, so completely beautiful, and the love I feel for it will pass away for ever.

6

OF MAPS AND DREAMS

World

You never enjoy the world aright, till the Sea itself floweth in your veins, till you are clothed with the heavens, and crowned with the stars.

Thomas Traherne

Our way back to the world involves us in an endless proliferation of detours.

Tim Robinson

There's no such place as home. And we live there, you and me.

Philip Hoare

When I was eight I loved to pore over an old atlas in my grandparents' house, imagining lands and stories far beyond my own little world. I also liked to go into the back garden, turn on an outside tap at the top of a gently sloping path and watch trickles of water snake down between the stones. I would imagine that I was looking down from a great height on a map of great rivers in wild country, and that one day I would travel to such places and encounter amazing things.

These childish daydreams share something with an urge that has also given rise to ancient and enduring visions of flight: the desire to know the world. Among the Eveny or 'reindeer people' of northeast Siberia, for instance, apprentice shamans develop the skills and intuitions that hunters acquire on the ground through dreams in which they soar effortlessly over the landscape, monitoring the movements of migratory animals. And among the Yanomami of the Amazon, initiates fly up into and beyond the tree canopy, where they learn about the forest spirits and their own place in the whole. '[My] thought spreads everywhere under the ground and under the water, beyond the sky and in the most distant regions,' says the elder Davi Kopenawa.

Shamans in pre-agricultural societies see the lands they need to know through methods akin to daydreaming or lucid dreaming. We in our highly technological culture often rely on maps – graphic representations that facilitate a spatial understanding of things, concepts, conditions, processes or events in the world. But there is no hard and fast line between maps and dreams because maps, however objective they may seem, are only interpretations. A map is always a way of groping through the unknown.

Maps can be ways of wondering at the world and ways of representing its wonders. They can also be wonders in themselves. For wonder, observes the scholar Philip Fisher, is often sparked by linking a local detail to a larger picture, setting the intimate and the celestial side by side. And this is exactly what maps do. Even the mundane act of searching for a route online allows us to fly in our minds wherever we wish, looking down as if from a god-like perspective that we can manipulate at will. 'The map's dissimulating brilliance,' writes the historian Jerry Brotton, 'is to make the viewer believe, just for a moment, that they are not still tethered to the Earth, looking at a map.'

Similarly, in a waking dream, the greater world is somehow represented in the mind. Part of the wonder here is the wonder of consciousness itself, which William James expressed so clearly when he asked, 'How can the [world] I am in be simultaneously out there and, as it were, inside my head, my experience?' Many people think there is still no good answer to this question that I know of – although I recently heard the mathematical physicist Roger Penrose identify a striking corollary of it. It is as if, he said, there are three distinct worlds, equally real, and yet each somehow encompassing the others. There is, first, the world of mathematics – unbounded and infinite, and something that Penrose, following Plato, believes really exists. Then, within the world of mathematics, there is the relatively small set of equations that, Penrose says, can explain all of physical reality. And finally, within and made possible by that physical reality there is the world of conscious beings and what they can experience. And yet somehow these conscious beings (or at least the ones who are good enough mathematicians) are capable of comprehending the mathematical world. Each world is therefore somehow nested in turn within another in an eternal loop, like the triangle devised by Penrose that has been called 'impossibility in its purest form'.

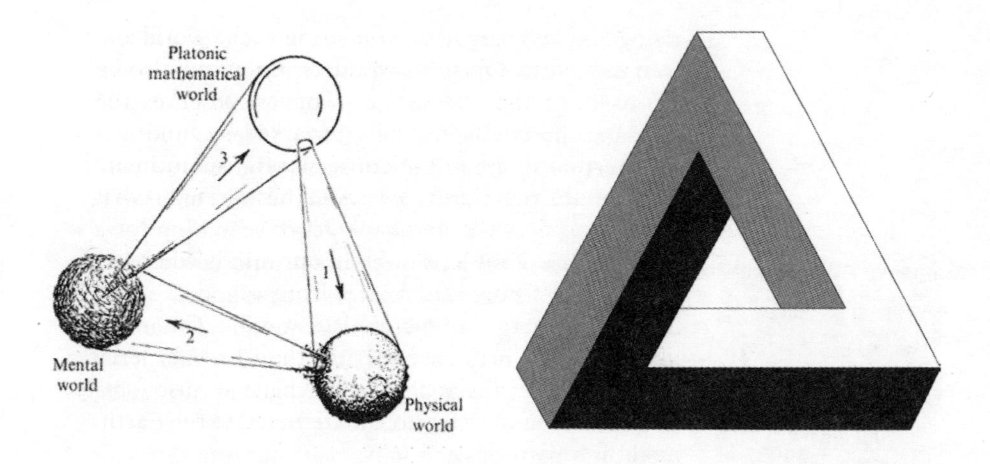

A distant mirror

Maps, like dreams, can be wildly wrong but still lead to remarkable discoveries. Setting sail in 1492, Christopher Columbus used a map based on one from antiquity that greatly underestimated the Earth's circumference and concluded he would need to sail just 3,700 kilometres west from Spain to reach China. The actual distance is 19,600 kilometres, and when Columbus thought he was in India he was in fact in Cuba.

Medieval European maps were often even more inaccurate than the one Columbus used, and yet they can reveal more than we may at first suppose. One of my favourites is the great mappa mundi or world map at Hereford Cathedral, which dates from about 1300 – just a few years before Dante wrote the *Divine Comedy*. A bright modern reproduction of it is stuck on the wall next to my desk, and I have made a pilgrimage to view the original, which is now browned with age and hangs in a dimly lit alcove. It is not, at first sight, especially beautiful, but it is a thicket of riddles and surprises.

On the Hereford map, an ocean encircles a world that looks very different from ours. In keeping with the custom of the time, east – the location of the

The Hereford map is flat, but medieval Europeans did not necessarily believe that the world was. In the third part of the *Divine Comedy*, Dante sees the heavens and the Earth as circles that surround and are surrounded by each other. The physicist Carlo Rovelli points out that this is a good description a 3-sphere, a higher-dimensional analogue of a sphere in which space is unbounded.

rising Sun and paradise – is at the top, and Jerusalem is in the centre. Outside and above this world, Christ sits in judgement. Within it, in the lower-left quadrant and directly below the centre, Europe and the Mediterranean are just about recognizable once you adjust to the orientation (or turn your head on its left side).

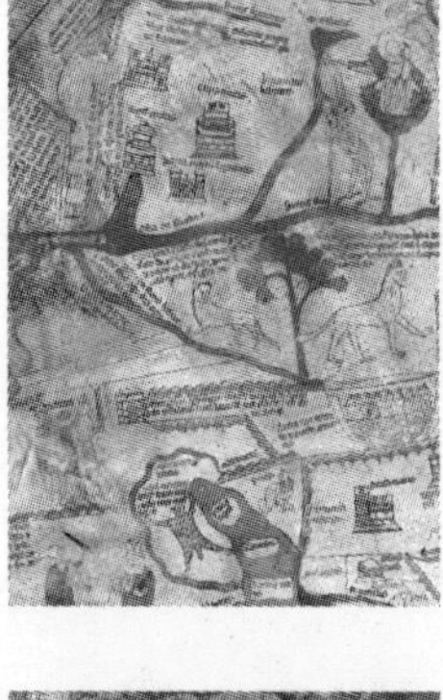

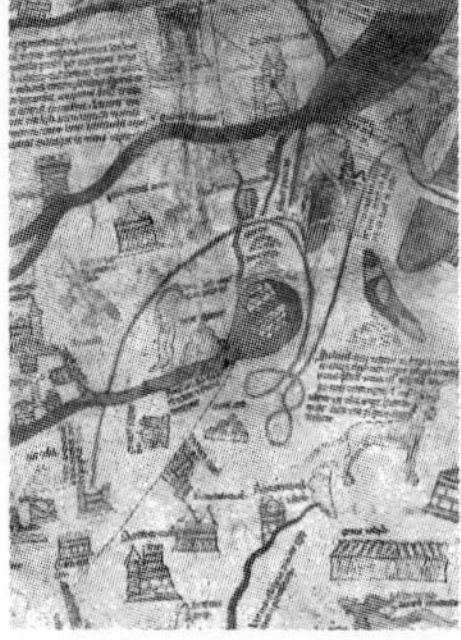

But much of the land on the map, especially Africa and Asia, is distorted beyond recognition, and populated with images of beasts, monsters and men. A legend on the map explains their stories. The lynx sees through walls and grows a valuable carbuncle in its secret parts. The manticore has a triple row of teeth in a man's face, a lion's body, a scorpion's tail and the voice of a siren. Semi-humans such as the Phanesii, a bat-like people with enormous drooping ears, live in Asia, as do the Spopodes, who have horses' feet. The Agriophani Ethiopes eat only the flesh of panthers and lions; their king has one eye in his forehead. The Gangines of India live on the scent of apples of the forest and die instantly if they perceive any other smell. The Arimaspians fight with griffins for diamonds. Fully human but utterly foreign, and terrifying, are the Scythians: they love war, drink the blood of their enemies from gushing wounds, and make cups from their skulls. The Hyperboreans, by contrast, are the happiest race of men: they live without quarrelling and without sickness for as long as they like, and only when they are tired of living do they throw themselves from a promontory into the sea.

The images and vignettes – wonders in their time – are bizarre to modern eyes. Taken as a whole, the map is fascinating and charming, but it seems little more use as a guide to reality than, say, the Map of the Square and Stationary Earth depicted by Orlando Ferguson, a flat-earther who flourished in South Dakota in the 1890s. The Hereford map can seem medieval in the pejorative sense: its creators were ignorant of, or had turned their backs on, the hard-won knowledge collated in the classical world, which had still been very much in evidence when the geographer Muhammad al-Idrisi made the *Entertainment*

The Hereford Mappa Mundi (*c.*1300).

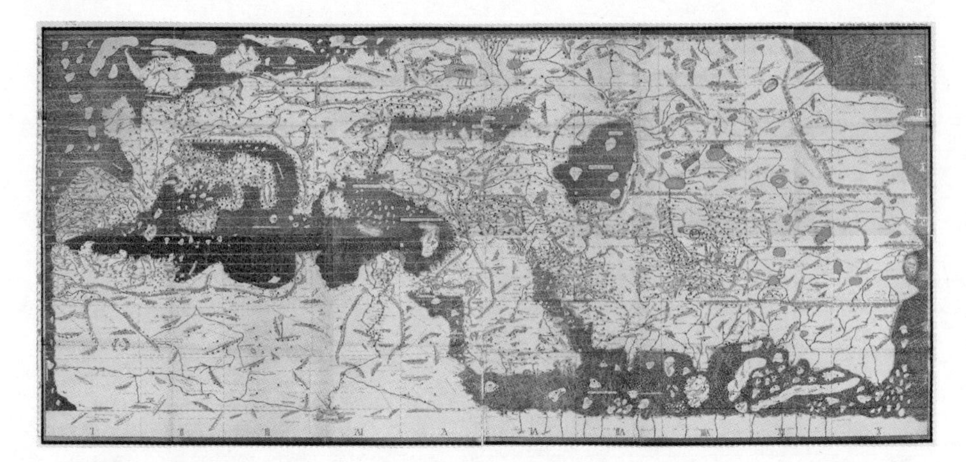

al-Idrisi's Tabula Rogeriana (1154).

for He Who Longs to Travel the World for the enlightened Norman king Roger II of Sicily in 1154. In the world map for that great work, which is based on an ancient Greek original, the shapes and dimensions of the Mediterranean lands, the Arabian Peninsula and much of western Asia are remarkably similar to those we recognize today.

But there are ways in which the Hereford map is more sophisticated than it appears. For, as Jerry Brotton notes, its flat unmoving surface charts the fourth dimension of time. Moving from east to west (top to bottom), the map lays out a history of the world, starting with creation and paradise, and followed by the succession of kingdoms from ancient Babylon through to contemporary western Europe in the time it was made. In some respects, then, it prefigures certain kinds of maps we find especially valuable today. For the most powerful and enduring maps are not only the ones that are proportioned and shaded like the ground they represent. They are often those that show what is not readily apparent, or highlight certain features in ways that have little to do with the visible appearance or actual proportions of land formations. William Smith's geological map of 1815, for instance, depicts rock strata in England and Wales and part of Scotland with waves of bright colour

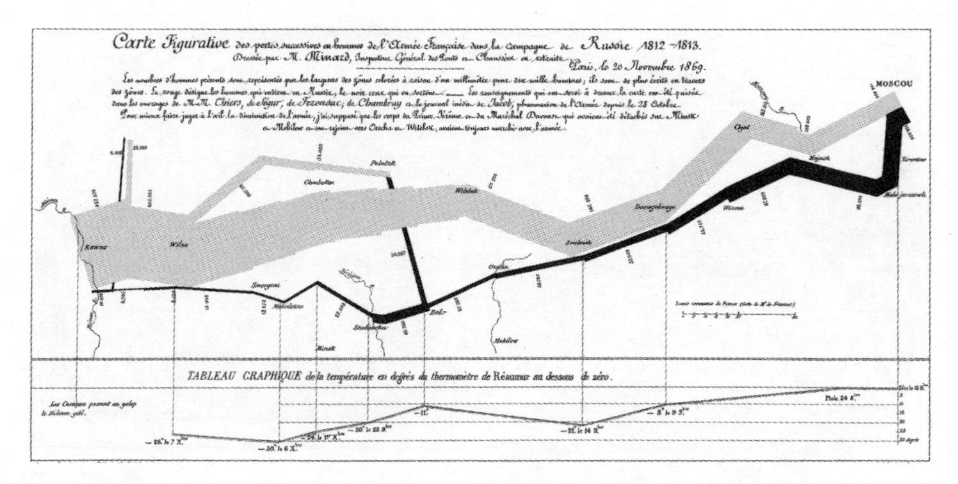

Minard's map (1869).

that resemble the marbling found in the endpapers of old books. Smith's colours bear no relation to what can be seen on the ground, but his map – signalling the vast hinterland of deep time within the land – helped kick-start a revolution in geology and hence our understanding of the nature and origin of life. Other innovations have extended and deepened the capability of maps to represent information. A map of Napoleon's 1812 Russian campaign made by the civil engineer Charles Joseph Minard in 1869, for example, displays six types of data (number of troops; distance travelled; temperature; latitude and longitude; direction of travel; and location relative to specific dates) in two dimensions. In its stark representation of (among other things) a criminal destruction of human life, it shows the potential of what are known as thematic and data maps to make visible what is not otherwise readily apparent. And maps that morph the shape or size of a region or country according to a variable such as population or economic activity also help us map a reality about the world that we cannot otherwise see. A 2015 map by Ben Hennig, which allocates land area according to population, bears a striking resemblance to a medieval mappa mundi.

Another way in which the Hereford world map speaks to our time is that our world is also full of

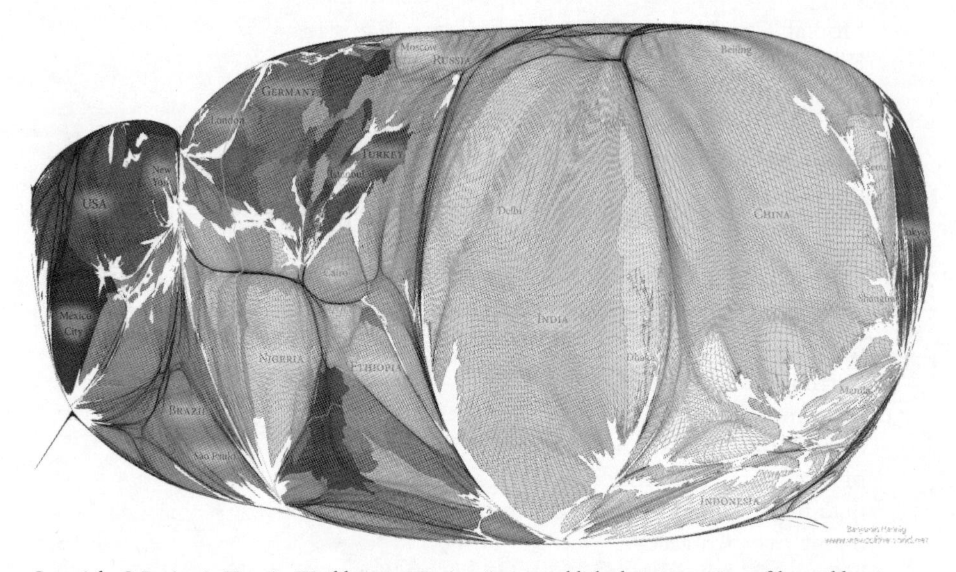

strange creatures, albeit ones that are real. For the living world overflows with astonishing things we have hardly begun to apprehend. Consider a few of the smaller examples from the animal kingdom. An insect called the branch-backed treehopper has grown a crescent moon on its back, the horns of which almost touch. Malaysian exploding ants contract their bodies when they see a threat, causing a poison inside them to boil and explode, drenching their antagonist. At least seven species of land snail are so small that they can literally pass through the eye of a needle.

In addition, misrepresenting and demonizing humans who are foreign to us, as the Hereford map does, isn't confined to the medieval past. In December 2001, as the US and its allies entered their third month of fighting in Afghanistan and began to prepare for the invasion of Iraq, the cover of the *New Yorker* magazine was a map showing New York City as New Yorkistan, divided into tribal territories such as you might find on a colonial map of central or southwest Asia. Manhattan teemed with Artsifarsis, Khouks, Bulimikhs, Psychobabylon,

Moolahs and Nudniks. Brooklyn had Fattushis, Khandibar and Fuggetabouditistan. Queens was home to Veryverybad, Kvetchnya, Irate and Irant. (In another century, in another context, Ambrose Bierce is said to have observed, 'War is God's way of teaching Americans geography.' As of 2017, there are more than eight hundred large US military bases in several dozen other countries, suggesting that the lesson is not yet done.)

And if the Hereford world map looks crude and misleading to us, then we should consider that many current attempts to map reality will probably seem little more impressive to future generations. We've already had at least one planetary lesson in the risks of not paying attention, when scientists ignored data-maps showing the depletion of stratospheric ozone over Antarctica. Maps of the brain often look impressive, but a vast amount remains unexplained and unknown. The universe, too, has been mapped at large scale, but this too may be hardly more than a beginning. We have little idea of much of the detail or of what may lie beyond its visible edge. Like those who created medieval maps, we are still surrounded by an ocean of ignorance.

The origins of maps, and dreams

Where do maps begin? Many animals have intuitive navigational abilities – and therefore access to a world – that leave humans looking very limited. Salmon can smell or taste a few molecules from their native river hundreds of miles out to sea, and follow the gradient back home. A range of creatures from insects to beluga whales are able to read the Earth's weak magnetic fields. Monarch butterflies make a 1500-kilometre, multigenerational flight from Mexico to the Rocky Mountains, calculating north using the position of the Sun by accounting for the time of day, the day of the year and the latitude. (This pattern of life, which has emerged through the subtle interplay of forces over thousands of years, is now threatened with extinction.) Loggerhead turtles navigate circular

Exactly how various animals are sensitive to Earth's magnetic field is poorly understood, but in some cases it may be thanks to a protein 'biocompass' with a nanoscale needle at its core.

ocean routes of more than ten thousand kilometres by reading local magnetic fields and return to exactly the same spot because they recognize the unique magnetic signature of that particular part of the coastline. (Loggerhead turtles are also threatened with extinction.) Robins maintain quantum entanglement in their eyes longer than the best laboratory system can do at temperatures near absolute zero and use their entanglement to 'see' the Earth's magnetic field. Clark's nutcracker, a bird that weighs 140 grams, keeps around ten thousand accurate maps in its head. Bar-tailed godwits hatch from their eggs in Alaska and fly to New Zealand through gales and hurricanes without ever stopping.

Our earliest human ancestors were most likely foragers, hunters and fishers. The more they knew about where and when plants would flourish, and where and when animal migrations would pass, the more likely they were to survive. Genetic mutations and learned behaviours that increased their ability to get to the right place at the right time in the search for roots and berries or on the track of prey were likely to ultimately increase the number of their offspring to reach reproductive age. They had to cover ground, remember and be ready for surprises. But their continued existence was also dependent upon passing on what they knew, enabling their fellows and offspring to create their own cognitive maps.

The ability to communicate to others the location of things that are not immediately present is not unique to humans. Honey bees describe the direction and distance to a valuable food source with their waggle dance (see page 165). And some of the wayfinding abilities of our ancestors were not so different from those of, say, elephants, who follow extended migrations between areas of forest, savannah and swamp. Elephant matriarchs must remember every year which route they have taken safely in the past and teach the young ones as they go. But at some point our ancestors learned to do something that, as far as we know, no other animal can, and that is to create material symbols that transmit complex information about place across distance and time. Enhancing our

Knowledge of ancient pathways has been little use to savanna elephants in many parts of Africa in the face of poaching. About sixty per cent of those in Tanzania were slaughtered in five years to 2015, and across the continent about a hundred individuals are killed every day. There is, however, evidence that some groups some groups have learned that they are relatively safe when they cross into countries such as Botswana, Namibia and Uganda. But Gabon, previously known to forest elephants as a safe haven, has been ravaged, with 25,000 – 80 per cent of the population – killed between 2004 and 2014.

capability with gesture and language, material symbols such as scratches and pigments on the ground or on rock gave us our first maps as well as new kinds of stories and dreams.

It is unlikely we will ever know when humans first created symbols out of material objects. The attribution of symbolic status to naturally occurring objects may predate *Homo sapiens*. It seems that three million years ago an archaic hominin in South Africa found a piece of red jasperite weathered in such a way as to resemble a face, and that he or she carried this object, known today as the Makapansgat pebble, or 'pebble of many faces', back to a home base several kilometres away. An extraordinary find in 2015, also in South Africa, suggests that *Homo naledi*, a small and distant cousin of anatomically modern humans, went to considerable trouble to place their dead in deep caves where access was difficult. This suggests a culture with the capacity for symbolic thought.

By about three hundred thousand years ago, almost perfectly symmetrical axeheads were being made with what archaeologists judge to be far more care than would have been necessary for strictly utilitarian ends. No less astonishing than the *naledi* find was a discovery in 2016 deep inside a cave in France: about 176,000 years ago Neanderthals had broken off and grouped stalactites into semi-circular walls and burned fires within them. Nobody knows why they did this, but symbolic or ritual behaviour must have been part of it. Finds in the Blombos cave on the southern coastline of South Africa dating to between 100,000 and 70,000 years ago show that anatomically modern humans were routinely ornamenting themselves. There are animal-bone paintbrushes and palettes: abalone shells in which they mixed pulverized red ochre, bone marrow and charcoal with water to form colourful pastes. The cave also contains an ochre slab with 75,000-year-old geometric engravings, and snail shells drilled through with holes so they can be strung as beads. While there are many uncertainties associated with evidence of early hominin activity (and the date of the naledi burials is not clear at the time of writing), it looks to be beyond

Michelangelo once said that he saw the angel in the marble and carved until he set him free. An early flint-knapper once saw a tool in flint and chipped until he or she set it free. Researchers recently found an Acheulean axehead more than 100,000 years old that was carefully shaped so as to display a delicate fossil shell at its centre.

Homo naledi resembled australopithecines – ancient upright apes that predate the homo genus – but also had characteristics like modern humans. Its skull was half the size of ours but housed a brain with frontal lobes like the ones in modern humans linked to speech and higher emotions such as empathy. *Naledi*, whose name means 'star' in the Sotho language and whose remains were first discovered in 2013 in *Dinaledi*, 'the chamber of stars', lived as recently as 300,000 to 200,000 years ago.

reasonable doubt that proto-modern humans had an eye for ceremony and beauty.

Star paths

The earliest-known map-like representation – depicting what may be a mountain, a river, valleys and routes around Pavlov in what is now the Czech Republic – is found in a cave painting dating from about 25,000 BC. A chunk of polished sandstone from a cave in Spanish Navarre dating from about 14,000 BC may represent similar features, as well as with animal etchings. But some of the oldest surviving maps may be of stars. Dots painted on the walls of caves at Lascaux in France some time around 16,500 BC, for example, appear to show Vega, Deneb, Altair and the Pleiades, while at the Cuevas de El Castillo in Spain a dot map made around 12,000 BC is thought to depict the constellation Corona Borealis.

There are good reasons for using stars to make maps. Though they rotate through the night and swing back and forth with the seasons, the stars remain in fixed positions in relation to each other as far as humans viewing with the naked eye are concerned, and pass through the same cycles of rise and set every year. A Navajo creation story may be a relatively recent expression of an ancient perception. 'When all the stars were ready to be placed in the sky, First Woman said: I will write [with them] the laws that are to govern mankind for all time. These laws cannot be written on the water as that is always changing its form, nor can they be written in the sand as the wind would soon erase them, but if they are written in the stars they can be read and remembered for ever.'

Some maps try to represent places found in dreams rather than those readily observable on Earth or in the sky. Take, for example, the San, also known as Bushmen, whose way of life endured in southern Africa for at least tens of thousands of years before they were all but exterminated in the nineteenth and twentieth centuries. According to the anthropologist

David Lewis-Williams, the San rock art known as the Linton panel depicts shamans in a state of trance. Their experience can be likened to being underwater, he says, so we see one of the human figures surrounded by fish and an eel. The panel also contains paths through this water dreamscape. San shamans alive today report seeing bright lines or 'threads of light' along which they can walk or simply glide, says Lewis-Williams. They also treat these lines as ropes that they can grasp and climb as they ascend to the spirit realm in the sky. Thus the panel may show a kind of spiritual sky map.

In traditional Aboriginal Australian belief, individual people are expressions of a sentient landscape. As the anthropologist Veronica Strang puts it, your spirit 'jumps up' from a given place such as a waterhole to enliven the foetus in your mother's womb (the ancestors having informed your parents of this quickening by some sign such as a strange birdcall). You grow up as part of a kin network with rights to your clan's estate, and you cannot be separated from your home because you are spiritually and materially composed of it, and your job is to relive the lives of the ancestors within certain largely fixed patterns of kinship. When you die, your spirit will be sung back into its home in the Earth.

The relationship may not always have been harmonious. Humans arrived in Australia around 50,000 years ago. In the following millennia more than 90 per cent of large mammals became extinct. It may be that both natural climate change and human action, such as the setting of large-scale fires, contributed to the extinctions.

The Aborigines do not use maps as Europeans understand them, but use dances and songs – in a phrase of the poet Les Murray, 'a vast map of song-poetry'. These teach how to move about the country,

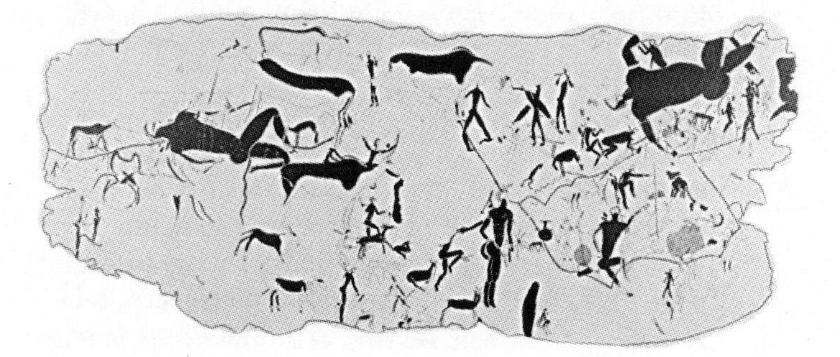

The Dreamtime, or Dreaming, is a generic term in English for a variety of Aboriginal Australian myths. These diverse stories tell of how the Ancestor Spirits travelled across the land creating the rivers, streams, water holes, rocks, plants, animals, people and sacred places.

The Greek myth of the Seven Sisters may be an independent creation with only coincidental similarity, but it is not impossible that both the Australian and the European versions derive from the same story, carried for tens of thousands of years as modern humans spread east and, later, west after leaving Africa 70,000 years ago.

The stars of the Pleiades, which are about 430 light years away, were born from the same cloud of gas and dust about 100 million years ago, and are still bound closely together by gravity. Only six or seven are visible to the naked eye, but the cluster contains several hundred stars, many of them hundreds of times brighter than the Sun.

learning its every feature and becoming a successful hunter-gatherer. They also teach about the law, and about obligations to the land itself and its non-human inhabitants. Stories and pictures of places are occasionally painted on rocks, bark and other materials, but they are not maps in the topographical sense. Rather, they are expressions of relationships and life lessons from the Dreamtime. In the Western Desert, for example, during a particular Dreaming called the *Tjukurrpa*, the ancestors left behind their songs, ritual designs and moral lessons for humans to follow. Often, the actions of these ancestors – full of jealousy, selfishness, violence and environmental degradation – show how humans should not act, with reminders embedded in rock forms and other features of the land.

Aboriginal Dreaming extends to the sky. In the *Tjukurrpa*, seven sisters flee unwelcome attention and take refuge in the heavens, becoming the constellation we know as the Pleiades. For the Yolngu people of Arnhem Land in the Northern Territory, a canoe takes the dead to Baralku, a spirit-island in the sky where their campfires can be seen burning along the edge of the great river of the Milky Way. The canoe is sent back to Earth as a shooting star to let survivors on Earth know that their loved one has arrived safely in the spirit-land.

But like those of other indigenous peoples, Aboriginal Australian traditions also recognize that stars can be useful simply as a means of orientation in space and time. Peoples of the Western Desert use the Pleiades to navigate at night, and date the start of winter from when, in May, the constellation rises at dawn. They are not alone in using the Pleiades in this way. The Pleiades are a particularly distinctive reference point – much closer together than the stars that make up other constellations – and because they always rise over the horizon in the northern and southern hemispheres at the same time each year, they have been seen as harbingers and markers of time. In the Andes, they were associated with abundance because they return each year at harvest time. The Berbers of North Africa noted the coming of the hot season or the cold and rainy season according to whether the Pleiades were in the west or the east. For the Blackfoot of North America, the setting of the Pleiades was a signal to prepare to travel to their buffalo hunting grounds. The Zuni of New Mexico call them the Seed Stars, because their disappearance in the evening sky every spring signals the seed-planting season. In prehistoric and ancient Europe they held special significance too. In the Lascaux cave, a line of dots is thought to represent quarter-moons, and counting thirteen dots along this line from the first winter rising of the Pleiades brings you to the time of year when horses are pregnant and easy to hunt. The Nebra sky disk, a fantastically rich object of hammered copper and gold that dates from around 1600 BC and was found in the Saxony-Anhalt region of Germany, depicts the Sun, the crescent Moon and a cluster of stars thought to be the Pleiades. In ancient Greece, the day that the Pleiades first appeared in the morning sky before sunrise announced the arrival of the safe season for sailing the Mediterranean. The name Pleiades itself is thought to derive from an ancient Greek term for sailing.

Navigators from many cultures have taken their cue from the stars. The open Pacific presented greater challenges than the Mediterranean, and early Polynesian voyagers developed extraordinary techniques to help find their way. During the daytime, they would observe distant clouds, which were likely to indicate the presence of islands over the horizon, the direction in which birds flew, and the direction and patterns of waves, which might be refracted by the shadow of a distant island. 'Stick charts', with shells or stones attached to a network of lines, would record some of this in a form that to Western eyes resembles a map. But for night-time voyaging they also learned to read the stars. Navigators would memorize a sequence of stars for a course sailed between certain islands. As each star rose above or set below the horizon, so the next one would be sought out. These 'star paths' would take the navigator right through the night. Since each star rose and set in a different place, the horizon was divided up into a sort of 'star compass'.

Back on land, in conditions where no fixed marks or features are apparent, such as during a whiteout in a snowstorm, people almost always end up walking in circles. No one knows quite why this is so, recognizing the problem, trekkers in open expanses have long made signs to guide others who follow. Among the Inuit, the solution is an inukshuk – a kind of cairn or stone man that means 'someone was here' or 'you

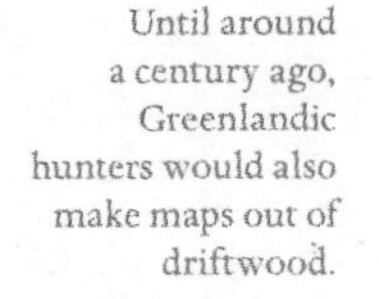

Until around a century ago, Greenlandic hunters would also make maps out of driftwood.

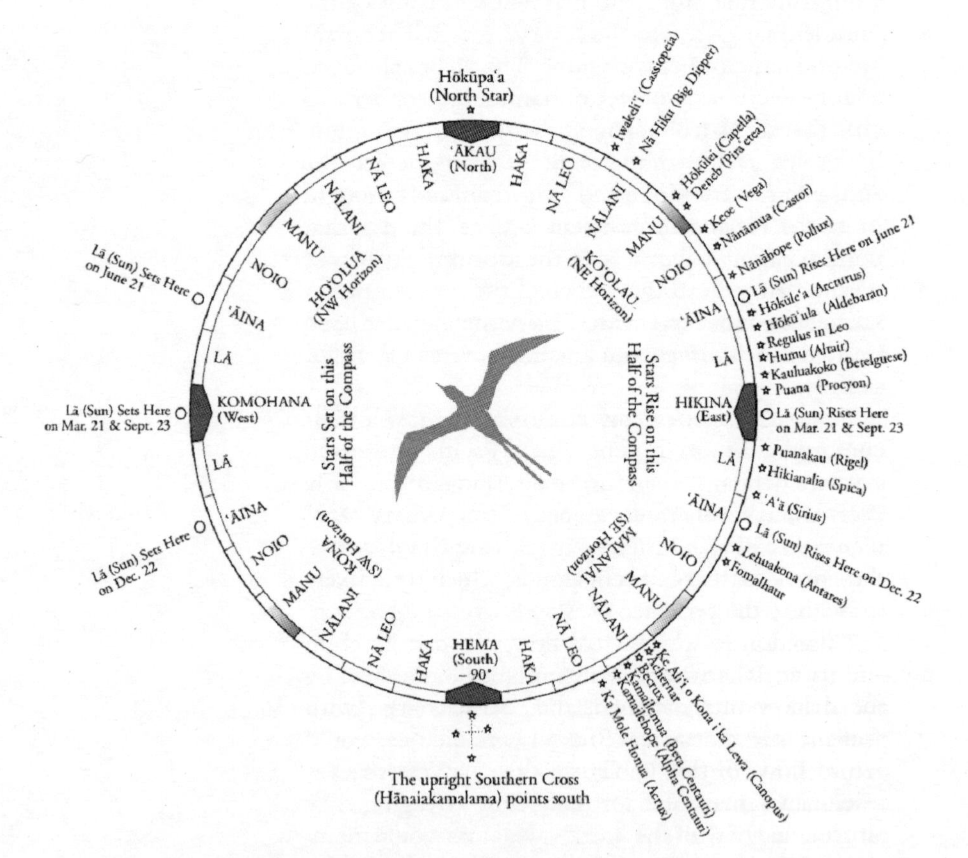

Nainoa Thompson, a Hawaiian navigator who is reviving traditional open-ocean canoe voyaging, describes the star compass as a mental construct for wayfinding framed around the places from which about 220 stars rise from and set back into the ocean. 'If you can identify the stars as they rise and set, and if you have memorized where they rise and set, you can find your direction,' he says. 'You have to constantly remember your speed, your direction and time. You don't have a speedometer. You don't have a compass. You don't have a watch. It all has to be done in your head. It is easy in principle, but it's hard to do.'

are on the right path'. At sea, this is not an option, and where currents and winds are unpredictable or hard to read and the sky is obscured, other options are needed. One such is fire. Hilltop beacons, we may presume, are ancient. But the first recorded lighthouse, the Pharos of Alexandria, was built between 280 and 247 BC. At nearly 140 metres high, it was half as tall again as the Flatiron Building in New York City and visible up to 50 kilometres from shore. As well as being a kind of inukshuk for seafarers, it was also an artificial star. In daytime a mirror reflected sunlight from its apex; at night a great fire was placed there to guide ships.

Charting the cosmos

Measuring and mapping the world was as important for economic and military purposes in the ancient world as it is in modern times. The Greek Alexandrine culture that built the Pharos was pre-eminent thanks to violent conquest. It wanted to project power and prestige and dominate trade routes in order to ensure access to resources, including slaves. In an account of wonder and wonders as starry-eyed as this one, it is important not to forget this. But improvements to measurement and mapping have seldom if ever been solely about control and exploitation. They have also been driven by sheer curiosity, and the ancient Greeks, perhaps more than any people before them, wanted to know the world.

For an enlightening account, see *Owning the Earth* by Andro Linklater.

This culture of curiosity dates from at least the sixth century BC, when the philosopher Anaximander introduced the gnomon – the pointer that enables accurate time measurement on a sundial – and the world map to his fellow Greeks. Though the map probably relied on a Babylonian model, it was distinctively Greek, centring its view of the world on the Greek holy site of Delphi or on Anaximander's home city of Miletus. With the continents of Europe, Asia and Africa gathered around the eastern Mediterranean, it shows essentially the same configuration as the Hereford map nearly two thousand years later.

'And so they tell us that Anaxagoras answered a man who was . . . asking why one should choose rather to be born than not – "for the sake of viewing the heavens and the whole order of the universe".' Aristotle

But no less important to Anaximander than measuring and representing time and space was locating the Earth within its larger context. Ultimately, he said, everything arose from and returned to *apeiron* – an indefinite or boundless whole. 'All things originate from one another,' he wrote, 'and vanish into one another, according to necessity . . . in conformity with the order of time.' The Earth, he believed, was a free-floating sphere at the centre of everything we could see, with all the other celestial bodies orbiting around it at different distances. This model was, as far as we know, unprecedented, and has earned its inventor the label the Father of Cosmology. But it had its flaws. Although Anaximander surmised correctly that the Sun was very large and farther from the Earth than planets such as Mars and Venus, he also supposed it was farther away than the stars.

Other cultures had conceived of the Earth as a sphere floating in space, but few if any took the further step of suggesting that the other heavenly bodies are spheres.

Two sets of calculations more than three hundred years after Anaximander's death brought precision (if not, at first, accuracy) to estimates of the sizes and distances of heavenly bodies. The first concerned the size

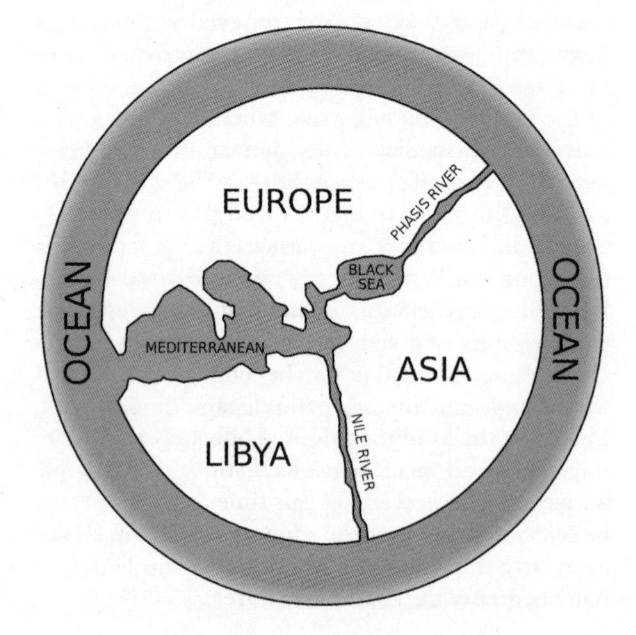

of the Earth itself. As soon as people had started to sail high-masted ships it would have become apparent to sharp-eyed observers that the hulls of these ships sank over the horizon before their sails did. This would have suggested that the surface of the ocean is curved. If you know the height of the mast and (more difficult) how far the ship has travelled, you can estimate the degree of this curve and hence the size of the Earth. But such a calculation is hard to do accurately, and in 240 BC the geographer Eratosthenes of Cyrene, who was head of the library of Alexandria, came up with a better method. Measuring the difference between the length of a shadow at Alexandria and one at the city of Syene five thousand stadia to the south on the same day at noon, he calculated the planet's circumference to within as little as two per cent of the true value. (Columbus ignored this estimate in favour of another, erroneous measure.)

A stadium is an ancient Greek unit of measurement equal to a little less than 200 metres or about a furlong. There is some doubt as to the exact length used by Eratosthenes, but his method was sound.

The second set of calculations regarded the sizes of the Moon and the Sun and their distances from the Earth. Aristarchus of Samos, an older contemporary of Eratosthenes, attempted both. Starting with the Moon, he observed that the curved shadow that the Earth casts across it during an eclipse of the Moon shows the Earth to be a sphere several times larger. From this he estimated the Moon's size. He also observed that a full moon covers about half a degree of the sky and, knowing that when an object covers half a degree its distance is about 115 times its size, determined how far away it is.

The Moon, which is 3,474 kilometres in diameter, is on average 382,500 kilometres away, or about thirty times the Earth's diameter (which is 12,742 kilometres). In a scale model where the Earth is the size of a basketball, the Moon is a tennis ball 7.2 metres away.

From this, Aristarchus realized that working out the size and distance of the Sun was a matter of simple trigonometry. When we can see exactly half of the lunar surface, the Sun, the Moon and the Earth form three corners of a right-angled triangle. Having the right angle and a value for the length of one side of the triangle (in this case, the distance between the Earth and the Moon), he needed only one more angle to calculate all its dimensions. Estimating the angle between Moon and Sun at this time to be 87 degrees, he calculated the Sun to be about twenty times farther away than the Moon. And because the Sun also covers half a degree of the sky it must, he reasoned, be twenty

The diameter of the Sun is over a hundred times that of the Earth. Its volume is more than a million times greater. If the Sun is represented by a basketball, the Earth is the size of a pea.

times bigger than the Moon, or five times bigger than the Earth. Eureka! Only not quite. We now know that Aristarchus misjudged the angle between Moon and Sun and as a result underestimated the distance to the Sun and its size by about twenty times. His logic was correct, however, and it led him to an extraordinary leap: it made no sense, he said, that a larger body should orbit around a smaller one, so the Earth must be revolving around the Sun.

The cosmologist Max Tegmark describes this work by Aristarchus as 'the moment when human imagination finally got off the ground and started conquering space'. Whatever you think of that assertion, it was certainly characteristic of a culture that was comfortable with the idea of measuring the heavens and the Earth with precision. By around 200 BC craftsmen were making analog 'computers' of the cosmos. In the case of the Antikythera mechanism, for example, a hand crank rotated some thirty interlocking gears to calculate simultaneously the positions of the Sun and Moon, the phase of the Moon, eclipses, calendar cycles and the locations of planets. Though knowledge of these computers was lost until modern times, the thinking that made them possible ultimately gave rise to our present mathematization of reality – the computed world in which we are ever more immersed.

Aristarchus' belief that the Earth went around the Sun was, however, not widely accepted, and not only because it was counterintuitive to anyone looking up in the sky. In the second century AD the geographer and astronomer Claudius Ptolemy rejected it on the grounds that the stars appear to be stationary. If the Earth really was moving then surely, he reasoned, the relative position of the stars would shift, rather as trees in a wood appear to move apart as you approach or slide behind one another as you walk by. This apparent flaw in Aristarchus' model was a major reason it fell out of favour and only began to be taken seriously again by natural philosophers after Copernicus made the case for a heliocentric universe early in the sixteenth century.

In 1687 Isaac Newton suggested a solution to the riddle of why, if the Earth was moving, the stars

The angel Ruh holding the eight celestial spheres of the Ptolemaic system, centred on Earth, from an edition of *The Wonders of Creation and the Oddities of Existence* by the thirteenth-century polymath Zakariya al-Qazwini dating to around AD 1600.

seemed to be standing still. He assumed, first, that (as the philosopher Giordano Bruno had intuited a hundred years before) other stars were bodies like the Sun, and appeared to be much dimmer only because they were much farther away. And he supposed, second, that (following the recently formulated inverse square law) the brightness of an object decreased by the square of its distance. Then, judging the star Sirius – the brightest star in the night sky – to be about as bright as the planet Saturn (whose distance from the

Earth was known and whose brightness in reflected sunlight Newton believed he could determine), he worked out how far away Sirius would have to be to appear only as bright as Saturn, even though it was actually as bright as the Sun. And from this he calculated that Sirius is about 800,000 times as far away from the Earth as the Sun is, or about 12.6 light years. This estimate was off by almost half: the actual distance to Sirius is about 8.6 light years (or 81.5 trillion kilometres). But it was in the right order of magnitude, and the idea that the stars were so far away that no apparent movement would be easily detected began to seem more credible.

The astronomer Friedrich Bessel settled the matter in 1838. Telescopes had improved a lot in the hundred and more years since Newton's time, and Bessel was able to measure the tiny apparent shift in the position of a star called 61 Cygni with respect to other stars when he made two observations of it six months apart (and thus when the Earth was on the opposite side of the Sun). Bessel calculated, just about correctly, that 61 Cygni was eleven light years away. The same principle is at work here as in the walk in the woods. It's known as parallax, and you can demonstrate it to yourself simply by holding up a thumb at arm's-length and seeing how it jumps with respect to background as you alternately close your left and right eye. By analogy, Bessel's telescope is one of two eyes when the Earth is on one side of the Sun, and the other eye six months later when the Earth has orbited to the opposite side.

The idea that the universe might be much bigger than was readily apparent was not new. Anaximander had envisaged a boundless universe, and more than a century later the philosopher Democritus argued that the never-ending dance of atoms would inevitably create countless other worlds and other lives. Hindu cosmology has also long held that the universe is created and destroyed in cycles of billions of years, and the *Bhāgavata Purāṇa*, which was composed sometime between the sixth and tenth centuries AD, speaks of innumerable universes. In the twelfth century AD, Fakhr al-Din al-Razi wrote of a thousand thousand

The philosophy of Democritus (*c.*460–*c.*370 BC) is known only from a few short quotations by others. Of his more than sixty complete works, everything is lost except their titles, which include: *Cosmography*; *On the Planets*; *On Nature*; *On Human Nature*; *On Intelligence*; *On the Senses*; *On the Soul*; *On Flavours*; *On Colour*; *On Diverse Movements of Atoms*; *On Fire and the Causes of Fire*; *The Causes of Acoustic Phenomena*; *On Animals*; *A Description of the Sky*; *Geography*; *Geometrical Reality*; *On Numbers*; *On Rays of Light*; *On Rhythm and Harmony*; *On the Beauty of Song*; *On Painting*; *Circumnavigation of the Ocean*; *Points of Ethics*; and *On Well-Being*.

worlds, and so later did Giordano Bruno, getting himself burned at the stake in 1600 for his trouble. But observational data that might support or refute these speculations really began to be gathered only in 1610 when Galileo looked at the Milky Way through a telescope and saw for the first time that it was made of countless stars. The contribution Newton and Bessel made was not only to provide convincing evidence that the universe is much larger than the solar system, but also to begin to put actual numbers on the likely distances to other stars. What they did not address, however, was the question of whether there might be regularity or structure at this larger scale, and if so, what it might look like.

Islands in the universe

One foggy winter morning I found myself driving through former mining villages near Durham in the north of England in what passes for rush hour. A distinguished professor of astronomy from the university and I were thoroughly lost. We had been looking for a stone folly on a low hill. It was built by Thomas Wright, a mathematician and instrument maker who, in 1750, proposed the existence of what we now call galaxies.

After about ten wrong turnings we drove up to the folly – a small cylindrical tower – almost by accident. I climbed out of the car and immediately filled both my shoes with cold mud. It didn't matter, I told myself. In fact it was a kind of baptism by cold mud and confusion into a greater appreciation of one of the most remarkable leaps of imagination of the eighteenth century, expressed in the maps and diagrams of Wright's *Original Theory or New Hypothesis of the Universe*, which only days before I had pored over for hours in a warm, dry library in Oxford.

Thomas Wright's tools were extremely limited given the scale of the concept he was working towards. No telescope in his day, or for well over a hundred years afterward, was remotely capable of producing a sharp image of another galaxy. But his

Thomas Wright's contemporary the clergyman John Michell calculated that the likelihood that the stars of the Pleiades were so close to each other by chance was about one in half a million. He concluded that they might be drawn to one another by gravitational pull. Reflecting further about the effects of gravity on stars and what would happen to the light from the most massive of them, Michell proposed the existence of black holes, which he called 'dark stars', in 1783.

reasoning was not. He suggested, first, that the stars themselves were moving ever so slowly in relation to each other under the influence of gravity. (This, he said, might most easily be tested by observing stars in the Pleiades – which are especially close together in comparison with other bright stars – over a decade or two, and watch for movement.) And if this were the case, he argued, then this movement would be such that very large groups of stars were bound together around different centres of mass,

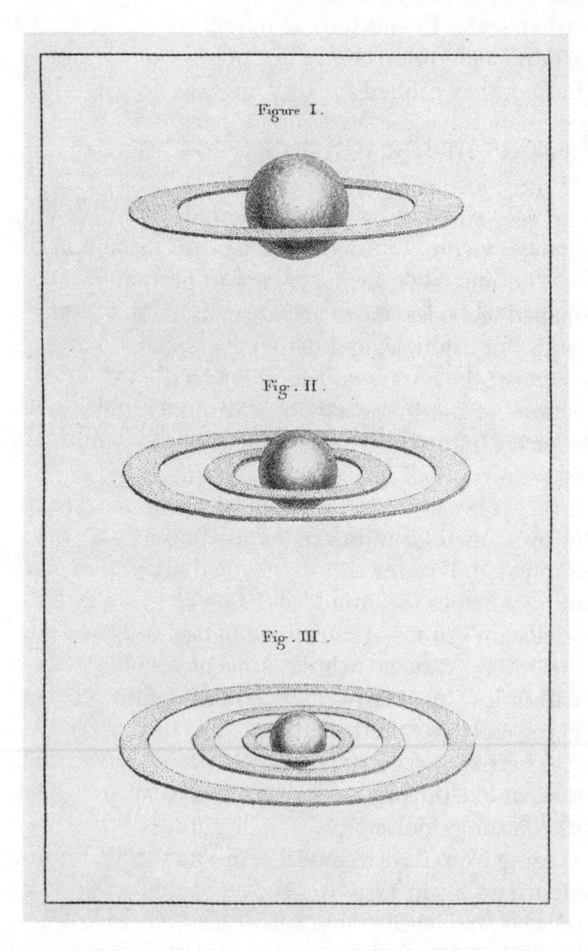

Some of Thomas Wright's models of galaxies, with rings of stars revolving around central balls of stars.

just as the much smaller objects of the solar system move but are bound by gravitation to the Sun. The Magellanic Clouds – indistinct smudges of light visible only in the southern hemisphere that had puzzled European astronomers ever since they had first heard reports of them in the 1500s (and which had intrigued Girolamo Cardano) – were, Wright suggested, just such groups.

Wright offered two models for the overall distribution of these huge groups of stars. In the first, the peripheral stars revolved around a large central group as the rings do around Saturn, but on a much larger scale. In the second model, the stars moved around a common centre but not in the same plane. Rather, they orbited at every angle as do comets that circulate the solar system over periods of hundreds of years, and so comprised concentric spherical shells or orbs, rather as the Oort cloud, a shell of dust at the very edge of the Sun's outermost atmosphere, is thought to do.

Planets orbiting other stars in our galaxy have rings too. The rings around J1407b are nearly 120 million kilometres across – more than two hundred times the size of those around Saturn.

The idea was wrong in many particulars. Disc-shaped galaxies have spiral arms, not concentric rings, for example; and while near-spherical galaxies do exist, their stars do not move in the way Wright supposed. But his reasoning was on the right track, depicted in thirty-two maps and diagrams, this vision of a universe of galaxies, or 'myriad celestial mansions', is breathtaking. Its beauty and logic appealed to Immanuel Kant, but many astronomers were less impressed. For one thing, they had no way to determine whether the smudges that Wright supposed to be distant groups of stars were in fact objects outside the Milky Way or nebulas (that is, clouds of dust) within it. This question was solved definitively only in the early twentieth century.

There is a class of stars called Cepheid variables that, unlike the bright star that John Keats praised for its constancy, pulse or vary in brightness over periods ranging from days to months. In 1912 the astronomer Henrietta Swan Leavitt discovered that there was a regular and predictable relationship between how long Cepheids take to pulse and their maximum brightness, with the ones that pulse brightest taking the longest to do so. This opened a path to a new way of measuring the universe. For if the distance to a nearby Cepheid could be found using the parallax method pioneered by Bessel, then the distance to others that pulsed at the same rate could be deduced, because if they appeared to be dimmer it must be because they were farther away, their brightness diminishing (as Newton had realized) by the square of the distance.

Some Cepheid variables are a hundred thousand times brighter than the Sun, and so can be seen at enormous distances. In the 1920s, the astronomer Edwin Hubble observed several in what was known at the time as the Andromeda nebula and, using the findings of Leavitt and others, calculated that they were about 900 million light years away. At the time he published his findings, in 1925, this was a vastly greater distance than anyone had measured before. It was, perhaps, the biggest single extension of the horizon in human history.

Hubble's first estimate was actually less than half the true figure, and more startling revelations followed. The Andromeda Galaxy, which we now know to be the nearest large galaxy to our own, is an island of a trillion stars about 2.5 million light years (24 quintillion kilometres) from the Milky Way, and both are part of an ensemble of about fifty-four galaxies, which include the Magellanic Clouds, in what is known as the Local Group. And, observing that the light from galaxies outside the Local Group was all redshifted, Hubble correctly concluded in 1929 that all of them were moving away from us, and that the more distant the galaxy, the faster it was receding.

When a light source is travelling away from Earth at a high speed, the apparent colour of light it emits shifts towards the red end of the spectrum. This is an example of the Doppler effect, which also changes the apparent pitch of a siren on a vehicle which approaches, passes by and recedes.

These findings helped to solve a fundamental problem in cosmology – namely, that the field equations in the general theory of relativity published by Albert Einstein in 1915 seemed to predict an unstable universe in which galaxies, stars and planets would collapse inwards. Einstein had tried to solve this by adding a repulsive force that he called the cosmological constant. But Hubble's observations proved to be more consistent with different solutions to Einstein's equations developed in the

1920s by the cosmologist Alexander Friedmann and, independently, by the physicist (and priest) Georges Lemaître, as well as with a model of the early universe developed by the cosmologist Ralph Alpher in the 1940s. All of these suggested that there had been a 'Big Bang' in which all matter and energy in the universe had exploded from a dense centre and has ontinued to expand ever since.

Subsequent work – not least the discovery in 1964 of the cosmic microwave background: an echo of the Big Bang – has given us something unobtainable to previous generations: a coherent (though incomplete) explanation of how all that is came to be. The physicist and broadcaster Brian Cox calls it a scientific creation story. We now know that there was a day without a yesterday and, if the fate of our universe is heat death, that there will eventually be no tomorrow. In the words of Georges Lemaître, 'Standing on a well-cooled cinder, we see the slow fading of the suns and we try to recall the vanished brilliance of the origin of the worlds.' And our 'cinder', the Earth, is one world among billions circulating ordinary stars in an ordinary galaxy among trillions. New star formation will, however, continue for another 1 trillion to 100 trillion years. So there is no excuse not to complete your tax return . . . unless you wait around until about 1 googol AD, by which time the average density of matter in the visible universe (which is currently a few hydrogen atoms in every cubic metre) will have fallen to one electron or positron for every volume of space far, far bigger than today's visible universe.

A googol is ten duotrigintillion, or 1 followed by a hundred zeros: 10,000,000,000,000,000,000,000,000,000,000,000,000,000,000,000,000,000,000,000,000,000,000,000,000,000,000,000,000,000,000,000,000,000,000,000,000,000.

Before then, the show will be spectacular. An inkling can be seen in the likes of *Wanderers*, a 2014 short film by Erik Wernquist with a 1994 text by Carl Sagan, which suggests a little of what could be on offer if it ever becomes possible to travel – in person or (more likely) virtually – across the solar system. We may see the clouds of Jupiter close up from the surface of one of its moons while another Jovian moon passes overhead. We may float among the rings of Saturn and weave between the huge geysers erupting from its bright moon Enceladus. We may soar over the

About four days before perihelion (the point in its orbit when it comes closest to the Sun), Mercury's angular orbital velocity exactly equals its angular rotational velocity so that the Sun's apparent motion ceases; at perihelion, Mercury's angular orbital velocity then exceeds the angular rotational velocity. Thus, the Sun appears to move in a retrograde direction. Four days after perihelion, the Sun's normal apparent motion resumes.

methane lakes of Titan, another of Saturn's moons, or base-jump from the cliffs ten kilometres high on Miranda, the smallest moon of Uranus. There are many more wonders that do not appear in the film. At certain points on Mercury, an observer would be able to see the Sun rise almost to the zenith halfway, then reverse and set, before rising again and finally setting – all within the same day. On Jupiter's moon Callisto ice spires rise far up into the astounding night. At Saturn's north pole is a huge hexagonal storm. On that planet's two-tone moon Iapetus a mountain range runs like a belt around the equator, while its lop-sided tiny moon Pan resembles a fat ravioli or a wonky flying saucer. The binary asteroid Hektor resembles the dumbell-shaped shell of two unhusked peanut kernels.

For generations past, most things that people could see in the heavens never changed shape. The Moon and some comets were obvious exceptions. Only recently has it been possible to see, with the aid of extraterrestrial telescopes, the dynamism in real time of the Sun's atmosphere or the roiling clouds of Jupiter. But moving images of change at larger scales in time and space, such as the explosions of stars or galactic evolution and distribution, remain artefacts largely of reason and ingenuity – maps of the imagination – rather than direct observation. You can, for example, watch computer simulations of the encounter between the Milky Way and the Andromeda Galaxy, which will take place about 4 billion years from now. Swooping around each other a couple of times at first, like two dancers doing the Lindy hop, they finally merge. (Few, if any, stars will collide with each other because, even in the relative density of the merger of two galaxies, the distances between stars will remain enormous; but some billions of stars – with the Sun possibly among them – will be hurled out of the new super-galaxy by tidal forces, forming star streams like handfuls of sea-spray such as those visible today on the periphery of the Whale Galaxy NGC 4631.) Similarly, our views of distant stars, nebulas and supernovas remain, with few exceptions so far, views of static objects. Simulations such as

Thomas Vanz's short 2016 film *Novae* will, however remarkable, remain just simulations.

This is likely to change. When I first drafted this paragraph I wrote that within a few years, space-based telescopes might show the actual orbits of planets around other stars, and even of moons around some of those planets. But just as I was checking through it

UGC 1810 (top) and UGC 1813, two galaxies in the Andromeda constellation and roughly 300 million light years away, have probably passed through and continue to exert mutual tidal pull, resulting in distortion in the spirals of the former and a burst of intense star formation in the latter.

for the last time early in 2017, astronomers published a three-second movie showing seven years of direct observations of four planets orbiting the star HR 8799. Shown on a deep blue background around a blacked-out star at the centre, the planets, which are gas giants larger than Jupiter, glow like bioluminescent plankton rotating anticlockwise on a dial.

Up to that point, the longest film of actual movement in deep space of which I am aware showed the continuing explosion of the remnant of the supernova first observed by the astronomer Tycho Brahe in 1572. Watching a supernova in real time would be like watching lichen grow: it's so slow that nothing would be discernible simply by staring at it. But the movie, published in mid-2016, collapsed a sequence of still images taken between

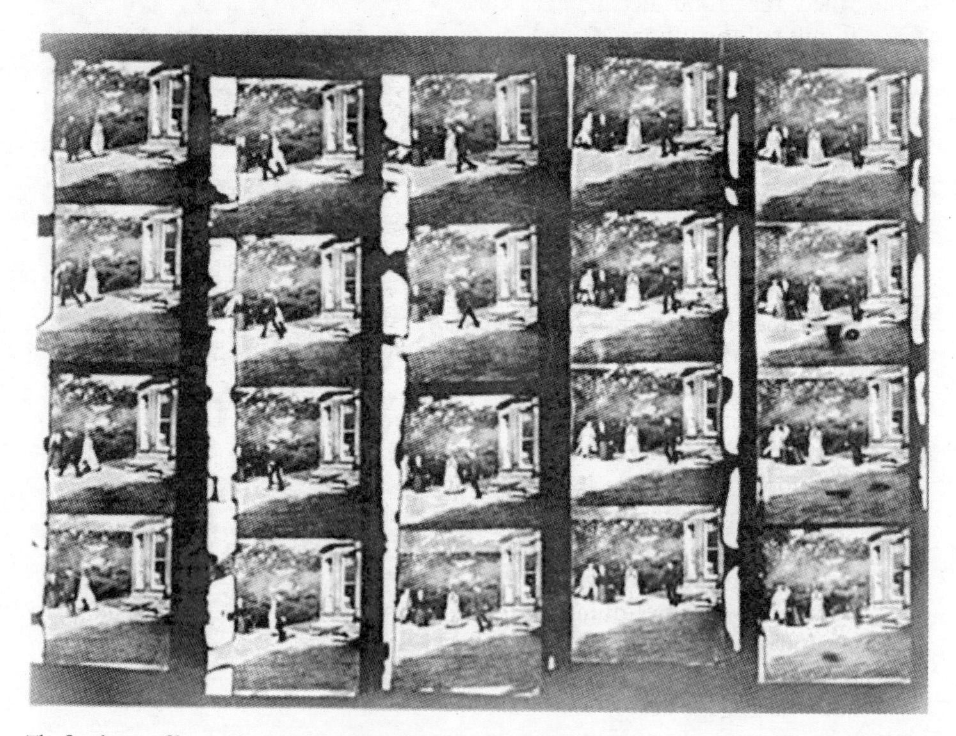

The first known film, made in 1888, consists of about only twenty frames. Our view today of the movements of stars and galaxies in deep space is as embryonic as this.

2000 and 2015 into about a second and a half. The result – which resembles a gently swelling translucent Brussels sprout – is so short that you can blink and almost miss it. But this second and a half (along with the three seconds of planets orbiting HR 8799) feels as momentous to me in its way as the first actual film ever made – the two-second *Roundhay Garden Scene* of 1888, which in a few jerky frames reproduced human movement for the first time. As the philosopher and priest Teilhard de Chardin supposedly wrote, 'We live surrounded by ideas and objects infinitely more ancient than we imagine; at the same time everything is in motion.'

The Copernican Return

In the humorous fiction of Douglas Adams, a device called the Total Perspective Vortex gives you a glimpse of the infinity of creation and so shows you just how tiny you are. Originally invented by a philosopher in order to irritate his wife, it is reckoned to be the most horrible instrument of torture ever devised. But it doesn't sound so bad to me. Sure, anyone can easily be overwhelmed by a flood of input; but in circumstances where our senses and cognition can cope with the flow of impressions, having a mental map that enables us to locate ourselves within the immensity of the whole can bring consolation, and more. In Patricio Guzmán's 2010 documentary film *Nostalgia for the Light*, Luís Henríquez, a former inmate of the Chacabuco concentration camp set up by the Pinochet dictatorship in the Atacama Desert in Chile, says that astronomy lessons run by a fellow prisoner helped him and others retain some inner freedom despite the harsh conditions.

Buddhism, Sufism, mystical Christianity and other traditions welcome infinite space as the wonderful ground of reality. This is also true for visionaries and writers without an agenda set by organized religion but who engage deeply with the living Earth. 'Time is but a stream I go a-fishing in,' wrote Henry David Thoreau, 'I drink at it; but while I drink, I see the sandy bottom and detect how shallow it is. Its thin current slides away, but eternity remains. I would drink deeper; fish in the sky, whose bottom is pebbly with stars.' Richard Jefferies wrote of lying down on a path in the woods and looking up to the stars: 'It was clear that I was really riding among them; they were not above, nor all round, but I was in the midst of them.' After looking up at what he terms the intimate Pleiades and other constellations above his head in the northern hemisphere, the protagonist of Olaf Stapledon's 1937 science fiction novel *Star Maker* imagines looking down through a transparent Earth to the nether constellations as if they were fishes in the depths of a lake.

'In the desert we are at a line between us and death,' says the novelist Ibrahim Al-Koni. 'And that in fact is a kind of cure. For it is only in the desert that we can pay a visit to death and afterwards return to the land of the living.' And so it is with our largely imagined visits to space, which is overwhelmingly hostile to life as we know it. The journalist Lee Billings followed the work of researchers hunting for life elsewhere in the universe, and afterwards found himself turning back towards Earth with a heightened awareness of how marvellous it is, and how worthy of attention and care.

I think Billings' turn, or return, needs a name if it doesn't already have one. In what is sometimes known as the Copernican Turn we discovered the Earth to be a small planet orbiting an average star on the periphery of one among countless galaxies, and in doing so downgraded our importance in the scheme of things. The Copernican Return, as I'll call it, is the moment when, without rejecting this insight, we transform its significance. We continue to recognize that the Earth is almost vanishingly small, but we now also appreciate that it is extraordinarily unusual, and precious, and that all our searches have not yet found anything remotely approaching it in beauty and wonder.

'The stars may be large but they cannot think or love, and these are qualities which impress me far more than size does.' Frank Ramsey

The cosmologist George Ellis suggests that maps of reality at the largest and smallest scales may be approaching the limits of what is possible, but that the potential for complexity at the middle level, where we live, is almost unbounded. But whatever may be true at the extremes, wonder in the world that we are able to perceive directly with our senses can be continuous with wonder at what is smaller and larger. As the writer Barry Lopez puts it, 'The challenge to us, when we [are present in] the land, is to join with cosmologists in their ideas of continuous creation, and with physicists with their ideas of spatial and temporal paradox, to see the subtle grace and mutability of landscapes. They are crucibles of mystery, precisely like the smaller ones that they contain – the arctic fox, the dwarf birch, the pi-meson; and larger ones that contain them [such as galactic] nebulae.'

'The land on which we live is not merely soil; it is a fountain of energy flowing through a circuit of everything.' Aldo Leopold

'World' is derived from Old English *Weorold,* a compound of *wer,* 'man', and *eld,* 'age' – in other words, 'the age of Man'. Perhaps the Anglo-Saxons were thinking of Midgard, the middle of nine mythological worlds and the only one inhabited by humans; but the roots of the word have an unforeseen significance today. For today we really do live in the epoch of Man – the Anthropocene, in which humans are the dominant force in life on Earth. Humans are creating many wonders. A few of them are explored in the next chapter. But many of us are destroying a lot in the process. To

'There has never before been a geological force aware of its own actions. Humanity has at least a dim, and growing, cognizance of the effects of its presence on this planet. The possibility that we might integrate that awareness into how we interface with the Earth system is one that should give us hope.' David Grinspoon

be dominant does not necessarily mean to be in control, and at present we seem to be little more in control of our collective impact on the Earth and the rest of life than a heavy smoker is in control of his or her addiction and the risks it poses to the lungs and heart. Already, climate change appears to be more rapid than the most extreme scenarios of just a few years ago suggested, and extinction rates appear to be accelerating. Making the case for positive thinking, the astrobiologist David Grinspoon suggests that human civilization (and not just a minority of virtue signallers) may be capable of developing 'planetary intelligence' in which we learn to care for the whole. Whether we will collectively prove to be as wise as this – and move beyond aggregate behaviour which as a species, it is hard to distinguish from that of an unchecked bacterial infection – has yet to be seen.

Dreaming new worlds

The phrase 'virtual reality' – meaning technologies that use software to generate impressions that resemble a real world or create an imaginary one – dates from the late twentieth century. But there is a sense in which this technology is only a continuation of what the brain has always done. In every waking moment, the brain assembles fleeting and sketchy perceptions of its actual surroundings, recreates places lost to distance and time, or fabricates alternative possibilities to inhabit. Like the prisoners in Plato's cave, each of us continuously projects images that we take to be the real world on imaginary screens in our skulls. (Deep in the actual caves where our ancestors painted in the Stone Age, suggested John Berger, there was already everything: wind, water, fire, faraway places, the dead, thunder, pain, paths, animals, light, the unborn . . .)

'"Reality" is something like a huge circus tent, folding, adjustable, which we carry around with us and set up wherever we are.' Elizabeth Bishop

The new technology does, however, make it possible to create and reproduce alternative worlds reliably and to order, and as it improves it will make possible ever more immersive and real-seeming but fictional or simulated worlds. It will also allow us to have access to almost unlimited data and additional sensory stimulation while we interact with the actual world. This will mean that in some respects we will overcome the impasse described in Jorge Luis Borges' story 'On Exactitude in Science', in which the map of an empire becomes so detailed and so vast that it threatens to smother the empire itself, and is abandoned. Instead of having to throw the map away because it is too large and impractical, we will be able to access any part of it, whether it represents something distant or near to hand, with almost no effort. This is both bad and good news. Some of the potential downsides are familiar from fiction. The technology may, for instance, help lock us into parallel virtual worlds in which our subjugation is hidden from us – like *The Matrix*, but with the difference that the illusions are controlled by states, corporations or criminal entities rather than by intelligent machines.

In Edward Snowden's view, a *Matrix*-like world, in which 'governments can reduce our dignity to that of tagged animals', is already here.

But wisely used – and why shouldn't they be? – virtual reality and successor technologies could have enormous upsides. 'Mixed' and augmented reality systems, in which maps and data can be flexibly superimposed on what is in front of us, could, for example, empower those working in international development or fighting crime and corruption. They will also make it possible to create and experiment with scenarios of liberation and possibilities of every conceivable kind in total freedom but without harm to others. It is already possible, for example, to find out what it feels like to be an endangered coral, or a cow going to slaughter, with the aid of a virtual reality headset. And, like the inhabitants of Baucis in Calvino's 1972 novel *Invisible Cities*, who from a distance contemplate with fascination the world in their absence, we will be able to get close to non-human wonders without disturbing them. (High on my long-list would be the tepuis on the borderlands of Brazil, Guyana and Venezuela – a 'Lost World' of table mountains billions of years old that rise precipitously from the surrounding forest. This remains one of the least-explored places on Earth and it contains some of the world's deepest sinkholes and cave systems, undisturbed for millions of years. Deep within are speleothems resembling billowing clouds of smoke, and sprays of mineral mushrooms.)

Virtual or augmented reality may help its users become more aware of small or subtle details of place so that we may, for example, be able – like Giles Winterbourne in *The Woodlanders* by Thomas Hardy – to distinguish different kinds of tree at a distance simply by the quality of the wind's murmur through their limbs. Each of us could access a vision as grand and as integrated as that of Alexander von Humboldt, who, bringing all his knowledge and observation to mind in an instant near the summit of Mount Chimborazo in the Andes in June 1802, saw and *felt* all the life in view, from the rocky zone immediately around him down through the alpine slopes to the thick rainforest floor, as a magnificent whole.

In Kim Stanley Robinson's 2015 science fiction novel *Aurora* a giant spaceship is sent to colonize the planets around the star Tau Ceti, some twelve light years away. The journey takes a hundred and sixty years, and seven successive generations of the crew live in pods which spin around a central spine to create artificial gravity. Each pod is four kilometres long and just large enough to contain a miniature ecosystem from a different terrestrial biome such as rainforest or savannah, desert or temperate woodland. The inhabitants of all the pods except one know that the little worlds in which they are born and die were built by humans and are housed within a spacecraft that is itself just a means to an end. The exception are some of the children in the 'Labrador' tundra biome, who grow up in the ways of their ancestors, following the caribou and living off the land – and learning, as one of them says, 'what we are as animals and how we became human'. Only upon coming of age are these young people put in spacesuits and taken outside to learn the truth. The shock of realizing that their lives to date have been an illusion is extreme. But most are made stronger by the experience, and those who continue to live in Labrador afterwards believe they are the only truly

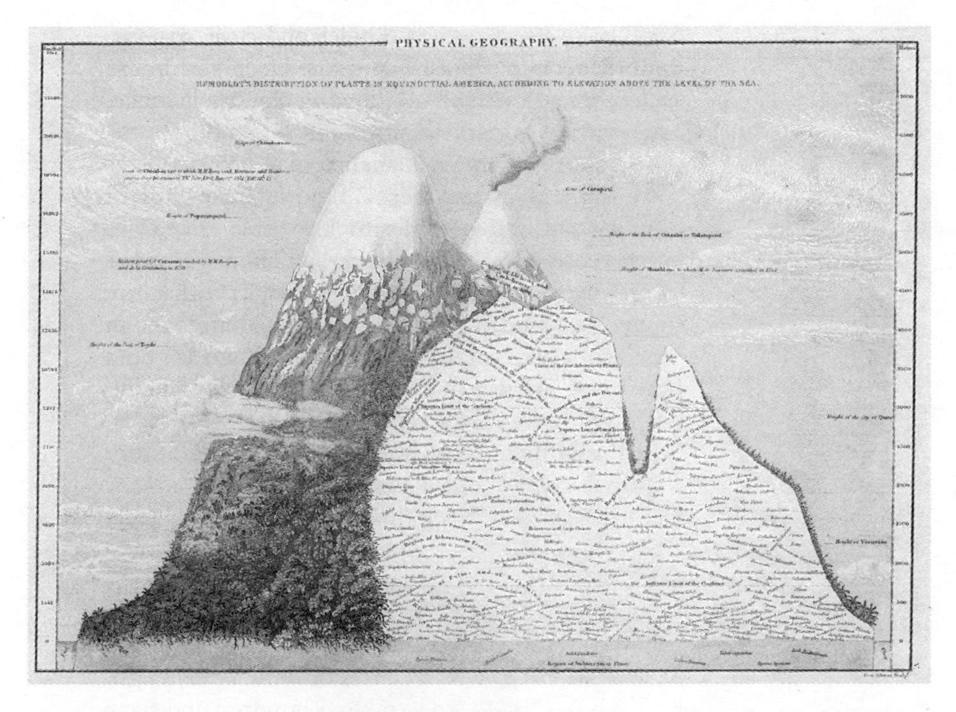

Cross-section of the distribution of plants by altitude on Mount Chimborazo. Alexander von Humboldt and Aimé Bonpland, 1805.

sane members of the ship's crew, and raise their children the same way.

You and I also live in two worlds: our human-dreamt world, which we take to be reality, is contained within something larger which humans did not make. Now and then we sense a greater reality. It can happen when we do not rush to words, maps or other conceptual frames. 'When an observer doesn't immediately turn what his senses convey to him into language,' writes Barry Lopez, 'there's a much greater opportunity for minor details, which might at first seem unimportant, to remain alive in the foreground of an impression, where, later, they might deepen [its] meaning.'

If the world is to be more than an arena for human invention, to be one where humans can truly flourish because what is non-human also flourishes, then there

'The fish trap exists because of the fish. Once you've got the fish you can forget the trap . . . Words exist because of meaning. Once you've got the meaning, you can forget the words.'
Zhuangzi

'Yet always all words waste and alter into / A formal ruin lesser than that voice / So clenched in prison in my mortal tree . . .'
W. S. Graham

will be times when it is important to stop speaking, put the technology aside and step outside in order to allow what is peripheral and unpredictable to become present. At such moments, we may appreciate that, as Seamus Heaney wrote in his poem 'Postscript', we 'are neither here nor there / A hurry through which known and strange things pass'.

I wrote most of this book in a shed where, like Bartleby the Scrivener, I have a window that commands an unobstructed view of a brick wall. But for one week I got to be in a different shed: a bothy on a hill above a bay on the west coast of the Isle of Eigg in the Inner Hebrides. The view from this tiny building encompassed the nearby Isle of Rùm:

the eroded ramparts of a volcano that burned out tens of millions of years ago but still rises the best part of a thousand metres directly out of the sea. My view also took in a great stretch of open sea beyond the bay, and on clear evenings I watched the fair-weather clouds of high summer rising over the southwestern horizon and rolling in above the golden water at the speed of an hour hand on a clock. Occasionally, at the very edge of the western horizon, the whalebacks of Barra, Vatersay, Sandray and Pabbay – southern islands of the Outer Hebrides some eighty kilometres away – would materialize out of the light and water as if they were the Isles of the Blessed.

The artist Alec Finlay dreams of a fabled forest that may once have existed where now there is sea: 'lost songs pipe from the bird's narrow throats and rise, in bubbles, to the surface'.

One night, a perfect half-moon drifted above the bay. I thought of Anaximander, puzzling over the dimensions of the heavens, and making his extraordinary leap of imagination. Through my binoculars, the craters along the Moon's midline were visible in

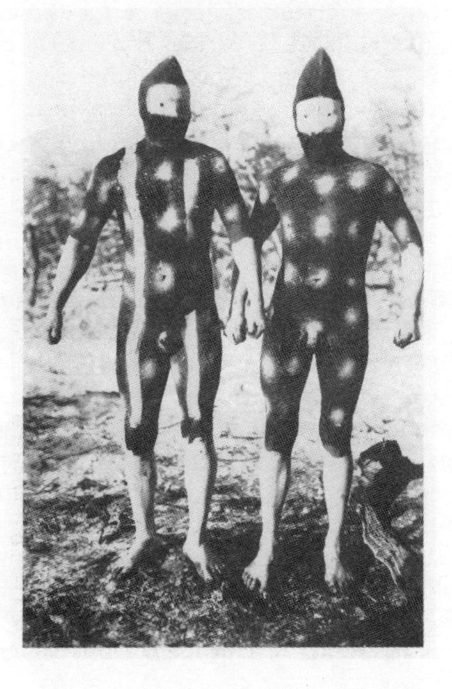

sharp relief, with the circular rims on the shadow side picked out in light exactly as in Galileo's chiaroscuro drawings of 1610. I thought of the great physicist turning his crude telescope on what I was seeing, and how he must have felt when, for the first time, he saw exactly this, and realized that the Moon was rough and mountainous – not unlike the Earth – rather than divinely smooth.

I breathed in the scent of the night air. There was no wind, and as I gazed at the Moon I could hear the waves lapping, very gently, on the sand in the bay half a mile away. The combination of these impressions filled me with great joy, a sense of being completely present, and luminous calm and serenity. It also felt particularly good to experience this on Eigg, where, over the last twenty years, the people have fought for and won community ownership of the land, and now work to create a just, sustainable and pleasant way of life. En route, I had stopped at Arisaig on the mainland coast to pay my respects at the memorial to the Czechoslovakian volunteers who trained there with the Special Operations Executive for the assassination of Reinhard Heydrich, the chief architect of the Final Solution. It was a reminder of the era of my grandfather's generation when, in solidarity with other Europeans, my country fought against some of the most evil ideas in history, rather than accommodating them.

'Tread softly. The earth's surface is made of the bodies of the dead.' Abul Al-Ma'arri

In the minds of the first Europeans to sail to Patagonia, the natives were hideous creatures with monstrous giant feet. Over the next four hundred years European colonists largely enslaved or eliminated them. One of the final genocides took place at the end of the nineteenth and the beginning of the twentieth centuries with the destruction of the inhabitants of Tierra del Fuego. The Selk'nam people were shot like wild animals while the Yamana and Kawésqar were herded into missions and mostly died of disease. Something of them survives, however, in photographs taken by the anthropologist Martin Gusinde around 1920. His images include simple portraits and scenes of daily life but also beings from the spirit world. Those beings are embodied by men in masks who have painted themselves with ochre (red), ground bones (white) and charcoal (black). We see – for example – Shéit, spirit of sky of snow (southern sky); Pahuil, spirit of the sky of the sea (eastern sky); Télil, spirit of the sky of rain (northern sky); and Shénu, spirit of the sky of wind (western sky). There are dozens of others. In the shadows these people left behind we can trace something of how they mapped elemental forces on their own bodies and took those forces into themselves. White spots on the dark background of their painted bodies may represent the stars.

'Much more is buried in silence than is recorded.' Thomas Browne

Back in my shed at home, I looked again at a reproduction of the Hereford mappa mundi. At the bottom right, outside the circle of the world, a mounted traveller bids farewell as he rides on. (He looks like a more innocent, less battle-hardened cousin of the central figure in *Knight, Death and the Devil*, one of Albrecht Dürer's master engravings of 1513–14.) It is easy to imagine that the undiscovered country towards which he is travelling is our own. It is one he would find hard to recognize.

The fate of the world we know is hardly less obscure to us. If, as now looks likely, sea levels rise significantly, the profile of many of the Earth's

Most of the world's primate species, for example, are threatened with extinction. The previous sentence contains three times as many letters as there are Hainan gibbons still alive.

'Sometimes I think there is no light, but sometimes I think there is.'
Michael McCarthy

land masses as they appear in our atlases will be transformed and, unless many current human activities are rethought and redesigned, the marvellous storehouses of natural wonders on land and in the sea are likely to be greatly depleted in variety and abundance. This 'thinning' of nature may be a small part of the change, and only the start. New wonders as yet undreamed of may fill our world. We simply do not know. The journey we are on is taking us into a country that is not on any map.

A few weeks later, I made my pilgrimage to see the Hereford map *in situ*. Browned and crumpled with age, it is impressive in its size, and carefully mounted and lit as an ancient treasure. My visit also led me to a place I never knew existed and did not expect to go to. The cathedral that houses the map is a magnificent building, and a testament to the courage, faith and sense for beauty of people long gone, as well as of those who care for it today. But, as the map itself shows, the greatest wonders are found outside its walls. Walking into the cathedral's Audley Chapel I came upon stained-glass windows that glowed intensely with the morning light. The windows commemorate the life and visions of Thomas Traherne, a local son who found his true Church outside in the fields and hills. For Traherne, even the air in the sky was holy 'by reason of its precious and pure transparency . . . all worlds would be worth nothing without such a treasure'.

'If he were not impeded by the seas, a man who set out to walk for ever would return to the point on the Earth from which he departed.'
Brunetto Latini

Seeing the light pouring through these windows like the music in the third movement of *Harmonielehre* by John Adams, I felt an urge to go outside and to travel towards its source. So the next day I drove east for two hundred miles until the land ends. And there, at Bradwell on the Dengie Peninsula in Essex, is a beach that is mostly made of millions upon millions of cockle shells. Each one of them, washed clean by sea and rain, is distinct. Each is, as the fourteenth-century mystic Julian of Norwich wrote, 'a littil thing the quantitye of an hesil nutt'. As you walk on them they crunch underfoot like breakfast cereal. They are beginning their slow transformation into sand or silt – and, over billions of years, into rock. At one point, where the beach is already mostly sand, a tiny rivulet of fresh water slides towards the sea, and on a very quiet day you can hear the clack and inkle of shells as they tumble along in its flow.

7

FUTURE WONDERS

Adventures with Perhapsatron

All [things] are tamed in the net of Man's mind . . . working both good and evil.

Sophocles, *Antigone*

The press, the machine, the railway, the telegraph are premises whose thousand-year conclusion no one has yet dared to draw.

Friedrich Nietzsche

Technology is a word that describes something that doesn't work yet.

Douglas Adams

'Children set off each day without a worry in the world,' wrote André Breton in his surrealist manifesto, published in 1924. 'Everything is near at hand, the worst material conditions are fine.' But, he argued, as children grow they are allowed to exercise their boundless imagination only in strict accordance with the arbitrary laws, and by the time they are twenty or so imagination largely abandons them. 'Our brains are dulled by the incurable mania of wanting to make the unknown known, classifiable.' To escape this fate, Breton suggested, there remain the lessons of dreams and of madness. However difficult the mad may be for the rest of us, they are at least honest about their delusions.

Surrealism, the artistic movement that Breton helped inspire, aimed to free the imagination by conjuring wondrous, uncanny beauty as it sometimes appears in dreams or madness with creations such as a cup, saucer and spoon made of fur, a lobster-telephone, and a picture of a man looking into a mirror but seeing the back of his own head. There was serious purpose behind this playfulness. In 1916, eight years before he wrote the manifesto, Breton had been a young trainee in medicine and psychiatry working in a hospital for the insane. It was the height of the Great War and at the front thousands of men were being wounded or killed every day. One of Breton's cases was a man who insisted there was no war. Shortly after this experience, Breton was sent to be a stretcher bearer and medical orderly at Verdun, where, over the course of ten months, about a million casualties were inflicted in a battle explicitly designed to grind down human life. For Breton, the cause of this terrible slaughter was not just the criminal folly of politicians and military leaders, or the appetite for

Shortly before he was killed by a shell, a French lieutenant at Verdun wrote in his diary, 'Humanity is mad. It must be mad to do what it is doing. What a massacre! What scenes of horror and carnage! I cannot find words to translate my impressions. Hell cannot be so terrible. Men are mad!'

violence of certain individuals, but bourgeois reason and civilization themselves, and there could be 'no possible compromise with a world that had learned nothing from such a horrible adventure'. An alternative, he hoped, might be discovered or created with the help of the arts, which – juxtaposing and combining distant realities to produce illogical and startling effects – had the potential to bring about radical disruption and change in thought and feeling. 'Beauty,' wrote Breton, 'will be convulsive or will not be at all.'

It can be argued that, in the short term at least, surrealism failed. To Breton's horror, his near-contemporary Salvador Dalí, one of the most famous artists associated with the movement, lived very comfortably in fascist Spain. And even if one believes, as I do, that surrealism and other artistic movements have achieved much – not least in the way they disrupt expectations and inspire wonder – the extent to which they drive rather than merely reflect change is debatable. At most, their influence is indirect, shaping the nature of our dreams and our sense of what we value. By contrast, the connection between developments in technology and changes in our sense of what is possible – and consequently in how we think and act, and where we find wonder – is often palpable and direct. New technologies result in new realities and new forms of human interaction, including new forms of art. Technological innovations, which may be born in dreams but are facilitated by systematic experiment and logical thought, can have consequences that are wondrous, uncanny, liberating or profoundly destructive – and sometimes all of these.

Consider Atlas, a humanoid robot developed by Boston Dynamics (and funded by the US Defense Advanced Research Projects Agency, or DARPA) and which is said to be intended ultimately for a search-and-rescue role. In a three-minute video without voiceover or dialogue released early in 2016, Atlas starts to life like the nutcracker toy, walking away from a static display of the company's other robots, including BigDog, the star of earlier clips. As it heads out the door, Atlas lifts and places each leg like a toddler – albeit one who is six-foot tall and made of steel – who has only recently got the hang of this walking thing but is increasingly confident. It then crosses snowy ground in woodland, doing so with a fair amount of fluidity and ease. The movements are very like those of a living being, and as such trigger something deep in our brain's recognition system: this is alive, and it is like us/one of us. And yet something is different. This is particularly apparent when the ground becomes uneven and Atlas staggers and almost falls. Humans have at least two visual systems in the brain – 'vision-for-action' and 'vision-for-perception' – and we are able to move fluently across bumpy ground thanks in significant part to the second of these, which, in partnership with our other senses, such as proprioception (the continuous awareness of where our limbs are), enables us to correct our gait very quickly without conscious processing. Atlas seems to respond to the consequences of a wobbly step only after some delay, and veers wildly off the perpendicular as a result. It does, however, recover from near-falls every time, righting itself from angles that would send most humans sprawling, and continues its jerky saunter.

Another video, entitled 'Robots Falling Down at the DARPA Robotics Challenge', shows machines that are comically unimpressive.

The video segues to Atlas back indoors, picking up boxes and placing them on steel shelves. Its movements are precise and courtly, like those of a circus impresario who finds himself filling orders in a warehouse. Next, a human engineer uses a hockey stick to knock a box that Atlas has just picked up out of its hands, and then sweeps the box out of reach as Atlas stoops to pick it up again. Like a Charlie Chaplin skit in which our irrepressible hero takes every setback in his stride, the robot just keeps on trying to do what it was doing before. Finally, another engineer approaches Atlas from behind with a battering ram and forcefully shoves it to the ground. For a moment Atlas lies there in the yoga child pose, apparently flummoxed and defeated, before its feet and knees click and pivot and the robot thrusts and jumps itself back up with inhuman speed and strength.

Visible for an instant in that innocent jump back to upright is a powerful future weapon. But before there is time to absorb the thought, Atlas is off again, treading out the back door. It could be a comedy animal escaping from the zoo as in the children's book *Goodnight Gorilla*. Or it could be a mechanical echo of Thom Yorke of Radiohead in the video for 'Daydreaming'.

A dark glass

In the coming decades, dozens, perhaps hundreds, of innovations will change how we live – and therefore where we find wonder and even the sublime in or through our twenty-first-century technology. Some of the advances may utterly transform human life. This chapter briefly explores a few among many tantalizing possibilities, principally in the fields of energy generation, space exploration, and computing and robotics. But first a caution, and some context.

It's impossible to keep up with everything. As Robert Burton wrote, around 1621, 'New books every day, pamphlets, corantoes, stories, whole catalogues of volumes of all sorts, new paradoxes, opinions, schisms, heresies, controversies in philosophy, religion, &c.'

The science fiction writer Stanisław Lem was, it's been said, right about everything. He anticipated the Cold War policy of mutually assured destruction before the nuclear triad on land, air and sea was put in place. He foresaw the Internet before it existed as an arena where falsehood would drive out facts. ('What,' he asked, 'can be done when an important fact is lost in a flood of impostors?') And before the term 'virtual reality' appeared, he was already writing about its likely educational and cultural effects. He also coined what is arguably a better name for it in 'phantomatics'. He foresaw the ethical challenges of genetic engineering and nanotechnology decades before they became feasible. Still, Lem was cautious about making predictions. 'Nothing ages as fast as [predictions about] the future,' he warned.

There are many reasons that predictions about the future of technology, not to mention the future more generally, are often wrong. Among them are a tendency to underestimate uncertainties and to be biased in favour of what we already know. 'The idea of the future being different from the present is so repugnant to our conventional modes of thought and behaviour,' wrote the economist John Maynard Keynes, 'that . . . most of us offer a great resistance to

Science fiction, observes the author Cory Doctorow, is terrible at predicting the future, but it's great at predicting the present.

'You don't see it coming. You probably couldn't if you tried. The effects of large changes in scale are frequently beyond our powers of perception, even our imagination.' K. C. Cole

acting on it in practice.' But it is also the case that those who envisage change often also wildly underestimate or overestimate the rate at which it will happen. (A big source of error in this regard is a tendency to underestimate the importance of positive feedback, or the avalanche effect – you don't see it coming.) The famous remark attributed to Thomas J. Watson Jr, the second president of IBM in the mid-twentieth century – that there is a world market for about five computers – is apocryphal, but it seems that in 1977 Ken Olsen of Digital Equipment Corporation really did say there was no reason for any individual to have a computer in his home. On the overestimation side, an integrated space plan produced by the Space Shuttle manufacturers Rockwell International in 1989 envisaged: a permanent human presence in space by the early 1990s; 'bi-planetary civilization' on the Moon and Earth by 2006; command of unlimited material resources from asteroids by 2020; and human expansion beyond the solar system into the cosmos by 2050. Nearly thirty years later we still appear to be decades from a permanent human presence on the Moon. (Only one of Rockwell's scenarios looks remotely likely to arrive according to their schedule: unlimited solar energy on Earth by the 2020s.)

Looking at technology in isolation seldom gets us very far, and those who make it their business to plan seriously for the future of business and society stress the importance of looking at technological development in relation to other drivers of change. The futurist David Wood, who cautions against making predictions more than about ten years ahead, suggests that the next decade will be determined by four interacting trajectories. The first trajectory is technological, where he thinks the potential is, for the most part, very positive. But the second trajectory – a growing likelihood of social crises – has lots of negative potential. Wood argues that the outcome of a 'race' between these two is likely to be determined by a third trajectory: the extent to which people from different parts of the world learn to collaborate rather than compete. And this in turn will be influenced by a fourth: the philosophies and values that guide our actions.

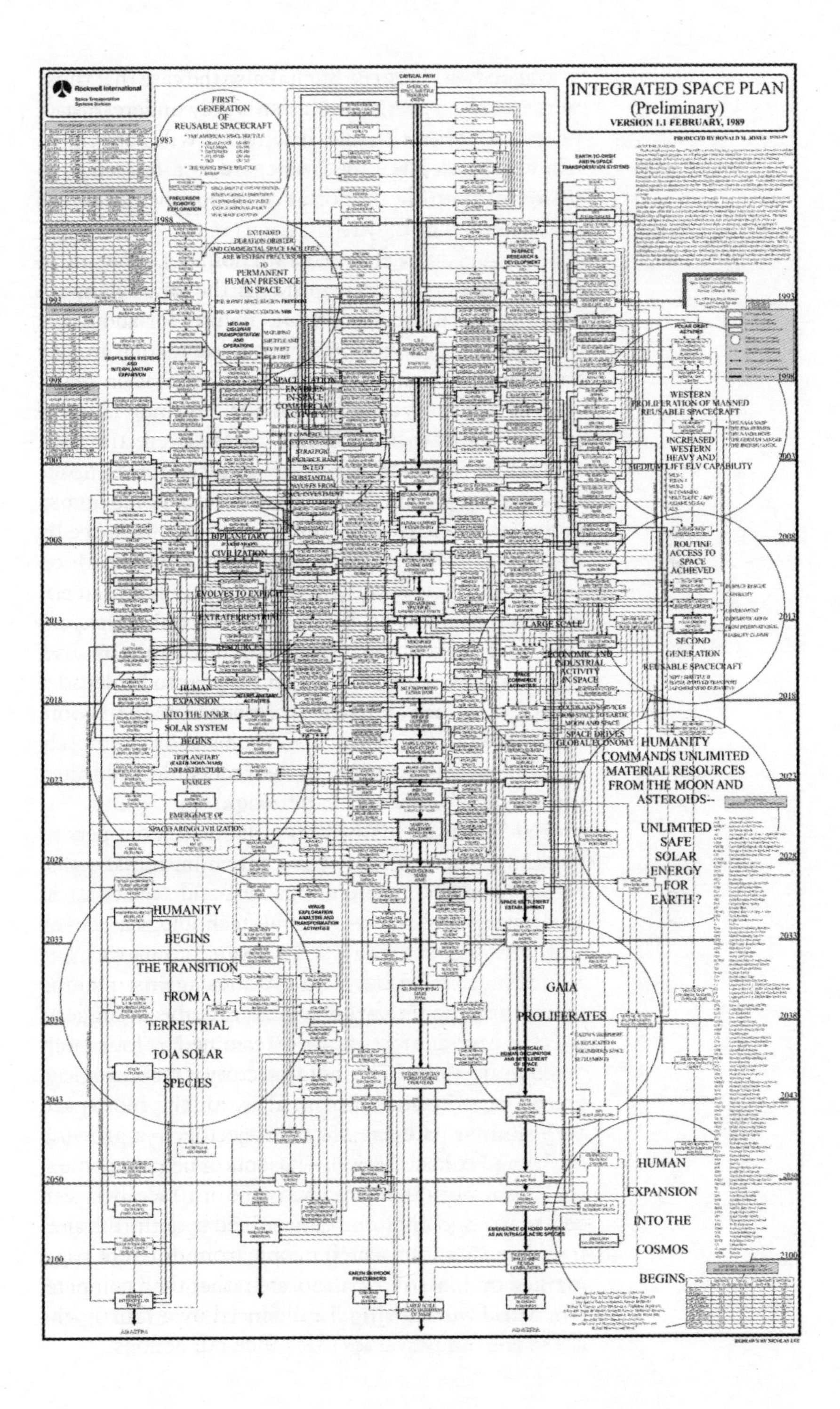

Rockwell International
CRITICAL PATH
INTEGRATED SPACE PLAN
(Preliminary)
VERSION 1.1 FEBRUARY, 1989
PRODUCED BY RONALD M. JONES
FIRST GENERATION OF REUSABLE SPACECRAFT
EXTENDED DURATION ORBITER AND COMMERCIAL SPACE FACILITIES ARE WESTERN PRECURSORS TO PERMANENT HUMAN PRESENCE IN SPACE
SPACE STATION ENABLES IN-SPACE COMMERCIAL ACTIVITY
BIPLANETARY CIVILIZATION EVOLVES TO EXPLOIT EXTRATERRESTRIAL RESOURCES
HUMAN EXPANSION INTO THE INNER SOLAR SYSTEM BEGINS
TRIPLANETARY INFRASTRUCTURE ENABLES EMERGENCE OF SPACEFARING CIVILIZATION
HUMANITY BEGINS THE TRANSITION FROM A TERRESTRIAL TO A SOLAR SPECIES
WESTERN PROLIFERATION OF MANNED REUSABLE SPACECRAFT
INCREASED WESTERN HEAVY AND MEDIUM LIFT ELV CAPABILITY
ROUTINE ACCESS TO SPACE ACHIEVED
SECOND GENERATION REUSABLE SPACECRAFT
LARGE SCALE ECONOMIC AND INDUSTRIAL ACTIVITY IN SPACE
GOODS AND SERVICES FROM SPACE TO EARTH, MOON AND SPACE
SPACE DRIVES GLOBAL ECONOMY
HUMANITY COMMANDS UNLIMITED MATERIAL RESOURCES FROM THE MOON AND ASTEROIDS--
UNLIMITED SAFE SOLAR ENERGY FOR EARTH ?
GAIA PROLIFERATES
HUMAN EXPANSION INTO THE COSMOS BEGINS
1983
1988
1993
1998
2003
2008
2013
2018
2023
2028
2033
2038
2043
2050
2100

But rather than try to make specific predictions, futurists often aim 'to disturb the present' (as Gaston Berger, a pioneer in the field, put it). According to fellow futurist Andrew Curry, 'A lot of [our] work is really just getting people to see a bigger picture of the system that they're in, because when things do change quickly or change at all, they change because something outside their immediate system changes.' The governing idea is that we are better able to face the future and to shape it when we have a greater awareness of what is outside our normal attention.

'[O]ur knowledge is historical, flowing and flown.' Elizabeth Bishop

The venture capitalist Albert Wenger is one among many to follow this approach. He argues that while technical advances broaden the space of the possible for humanity, they have always had the potential for both good and bad. So, for example, the ability to control fire made it possible both to cook and to wantonly destroy forests, and mastery of iron enabled the forging of both ploughs and swords. Similarly, the Internet can give everyone free access to education but it can also spread hate and misinformation, and artificial intelligence can drive cars more safely but can also be used to manipulate people more effectively. Nevertheless, says Wenger, the current wave of technological innovation is special because it is bringing about a step change of a kind that humanity has experienced only twice before. The first, time he says, was the invention of agriculture and the second was the Enlightenment and the industrial age. The third step change, according to Wenger, is the ubiquity of digital technology and the universal availability of computation at near zero marginal cost. This, he believes, will profoundly alter the 'binding scarcity' for humanity. Once again, he cites historical precedent. In the age of the hunter-gatherer, the binding scarcity was food. In the agricultural age, it was land. In the industrial age, it was capital. In the age that is unfolding, it will be knowledge and attention. Getting the transition right is critical because the two previous shifts were marked by massive turmoil and upheaval, including two world wars during the transition from the agrarian to the industrial age.

'My friend, we have reduced the forest to a wasteland; how shall we answer our gods?' *Epic of Gilgamesh*, *c.*1800 BC

Eric Drexler argues that atomically precise manufacturing (essentially, nanotechnology) will do for physical products what digital technologies have done for information, sound and images. That is, make them almost free and universally available, ushering in an age of 'radical abundance'.

The historians Naomi Oreskes and Erik Conway outline a future in which societies fail to get the transition right. In their 2014 speculative fiction *The Collapse of Western Civilization*, a Chinese historian writing in 2393 looks back on a twenty-first century in which man-made climate change spiralled out of control (in other words, it followed the trajectory that currently looks likely), resulting in mass death from war, starvation and disease. Trying to make sense of what went wrong, the chronicler homes in on denial and self-deception, rooted in an ideological fixation on 'free' markets. Tragically, many of the people of Western civilization knew what was happening but were unable to stop it: 'Indeed, the most startling aspect of this story is just how much these people knew, and how unable they were to act upon what they knew. Knowledge did not translate into power.' Scientists were hobbled by their own unwillingness to speak out and by powerful interests that rejected Enlightenment values in favour of post-truth politics. Indicative of this was North Carolina House Bill 819, which in 2012 banned US state and local agencies from basing coastal policies on scientific models indicating an accelerating rise in sea level. This was followed by the 2025 National Stability Protection Act, which led to the conviction and imprisonment of more than three hundred scientists for endangering the safety and wellbeing of the public with unduly alarming threats. (The 2012 law really was passed; the 2025 law, of course, exists only in an imagined future. *The Collapse of Western Civilization* was written before Donald Trump and Scott Pruitt brought denial of climate science to the highest levels of the US government.)

Oreskes and Conway may be wrong to blame market fundamentalism alone for the failure to act in good time on climate change. Denial and delusion with regard to a range of challenges and threats, including climate change, can also be sustained by people who subscribe to a variety of other belief systems. But the scenario they describe looks like a plausible extrapolation of current trends. That said, it is just one among many possible futures. The point of their book, after all, is to help motivate people to work to avoid such an outcome. The point of thinking about the future is to change how we act now.

Energy for ten billion people

For two of the more than seven and a half billion people alive today, readily available and affordable energy is an everyday wonder we barely notice. Access to it is one of the things that makes us much richer, if not necessarily happier, than previous generations. The historical trend is astonishing. An hour's work at the average wage in England in 1800 bought ten minutes of artificial light. In 1880 – shortly before lightbulbs became widely available and when kerosene was still the best option – it bought three hours of reading at night. Today, an hour's work buys three hundred days of light. Until recently, most of the energy supplying this bounty has been generated from the combustion of coal, oil and gas. This cannot continue without dangerous climate change. To meet the demands of a larger proportion of a growing global population, which is likely to reach nine billion to ten billion later this century, low- to zero-carbon alternatives are essential. The challenges are enormous but achievable. Global consumption of coal and oil may peak as early as 2020, and, thanks to increases in efficiency, global energy demand may peak as early as the 2030s.

Some of the alternatives to fossil fuels are familiar. They include various forms of biomass such as wood and fuels derived from crops, and these are likely to remain a significant part of the energy mix. Biomass supplies about two-thirds of renewable energy in Germany, a nation that has done more than almost any other to green its energy sector. There are downsides to this route – biomass production can require large amounts of fertile land, and lead to the destruction of forests with high ecological and wildlife value – but technical advances may reduce these drawbacks. Algae or other life forms genetically engineered to yield fuels may prove to be significant new sources of energy. Another promising approach could be the anaerobic digestion of plants such as prickly pear and other succulents that grow in places too dry for rain-fed food crops.

Nuclear power in conventional fission reactors also offers close to zero net emissions of greenhouse gases, and although the German government wants to phase it out worldwide, the technology remains a major player. According to the World Nuclear Association, some 440 reactors provide over 11 per cent of the world's electricity, and more than 60 new reactors were under construction in 2016, while more than 160 were planned and a further 300 more proposed. It remains to be seen, however, whether nuclear fission will play a central role in the transition to a near-zero-carbon economy. Some reactor types, such as those proposed for Hinkley Point C in England, look likely to be absurdly expensive and a new generation of modular fission reactors will require billions of dollars of new investment. Breeder reactors, which spawn their own fuel, produce plutonium, which is a long-lasting hazardous waste and great for building bombs.

Hydroelectric and wind power can also deliver near-pollution-free energy on a large scale. Large dams can only be built in a limited number of places and often have destructive effects. A conspicuous case at the time of writing is the proposed damming of the Tapajós River in Brazil, which would displace the Munduruku people from their ancestral lands and flood a region of extraordinary biodiversity. Wind power has greater potential to expand with, arguably, fewer impacts. The largest onshore wind farm in the world, at Gansu in China, has peak rated power of 6.2 gigawatts and is scheduled to expand to twenty gigawatts by 2020 – equivalent to between ten and twenty very large coal-fired power stations, or a bit over 1 per cent of China's total electricity demand. By comparison, the London Array, the largest offshore wind farm in operation at the time of writing, has a peak rated power of 0.63 gigawatts.

This is likely to be just the beginning. Early in 2017 China's National Energy Administration announced that it would invest at least 360 billion dollars in renewable energy sources, including wind and photovoltaics.

But the greatest potential for a safe, enduring and affordable source of energy looks likely to be the Sun, either as an inspiration for artificial nuclear fusion or as a direct source through photovoltaics and other solar capture systems. Nuclear fusion would recreate in miniature and under controlled conditions on Earth what is happening at every moment within the Sun – though, in theory, it would do it much more efficiently. Unlike fission, fusion does not create toxic and long-lived radioactive waste, and a fusion reactor cannot melt down. Its principal fuel is deuterium, an isotope of hydrogen bound to oxygen in ordinary water, and thus in almost limitless supply. One kilogram can generate about as much energy as a hundred thousand barrels of oil. If fusion can be made to work, the world's energy needs could be pretty much solved indefinitely with a process whose waste product, helium, can be used to fill party balloons and resupply MRI scanners. Bringing one of the most terrific phenomena in the universe under human control, fusion would be a supreme example of the technological sublime.

Uncontrolled nuclear fusion on Earth was first demonstrated with the Ivy Mike prototype hydrogen bomb in 1952. Within ten years, the US and Soviet militaries had more than enough bombs to destroy all human life on Earth.

Efforts to harness nuclear fusion for peaceful purposes kicked off with high hopes in the 1950s, with experiments such as the delightfully named Perhapsatron developed by the physicist James Tuck – an early attempt to 'pinch' superhot plasma fusion within an electromagnetic field. But so far those hopes have not been realized. In the sixty years since, thousands of ingenious and well-funded researchers in dozens of countries have not yet overcome the challenges to making fusion practicable.

Some champions of nuclear fusion claim it will be a thing any time from about 2027, when the International Thermonuclear Experimental Reactor (ITER) is scheduled to be switched on at the Cadarache facility in France. At thirty metres tall, weighing 320,000 tonnes and containing ten million individual parts, ITER is one of the most complex machines ever built. (At twenty billion dollars or more, or about a fifth of the cost of the Apollo space programme, it may also be the most expensive.) More than 2,800 tonnes of superconducting magnets, some heavier than a jumbo jet, will be connected by 200 kilometres of superconducting cables, all kept at −269°C by the world's largest cryogenic plant, which will pump 12,000 litres of liquid helium per hour. ITER's doughnut-shaped

By comparison, the world spent 325 billion dollars on fossil fuel subsidies in 2015.

vacuum chamber, or tokamak, will heat hydrogen to more than 200 million degrees Celsius – more than thirteen times hotter than the core of the Sun – and inside deuterium will fuse to form helium and energy. No physical substance can contain this reaction, so it will be held in a magnetic 'bottle' by the largest system of superconducting magnets ever built, cooled to a few degrees above absolute zero and sitting just beneath the reactor's core. ITER will be able to produce approximately 500 megawatts of power – about as much as a typical coal-fired plant – for up to a thousand seconds.

That's if everything goes according to plan. ITER has been falling behind schedule since work began in the early 1990s, and there will almost certainly be further delays. Some components of the machine, such as the central solenoid (an electromagnet), present enormous and possibly insurmountable manufacturing challenges. Others, such as the containment vessel, will require advances in materials science that may prove to be unachievable. Further, it is not yet proven that the design will be able to contain turbulence in the plasma. And even if all this comes good, ITER is designed to run for less than seventeen minutes. A commercial reactor would need to work continuously, and no one has worked out yet how this could be done.

Some of those who remain optimistic about the prospects for controlled fusion believe the challenges will eventually be overcome. Others are content just to be making small steps towards a goal that they believe could eventually deliver almost unlimited benefits. A scientist working at ITER compares their work to that of stonemasons at York Minster, which took over two hundred and fifty years to complete. It may, therefore, prove to be more of an art or political project – an expression of hope, and a source of juicy contracts – than a solution in our lifetimes. But ITER is not the only game in fusion town. In 2016 the Wendelstein 7-X stellarator in Germany, which is shaped like a giant Möbius strip, fired up briefly with a small amount of hydrogen plasma. Also in 2016 the Experimental Advanced Superconducting Tokamak in China was claimed to have maintained

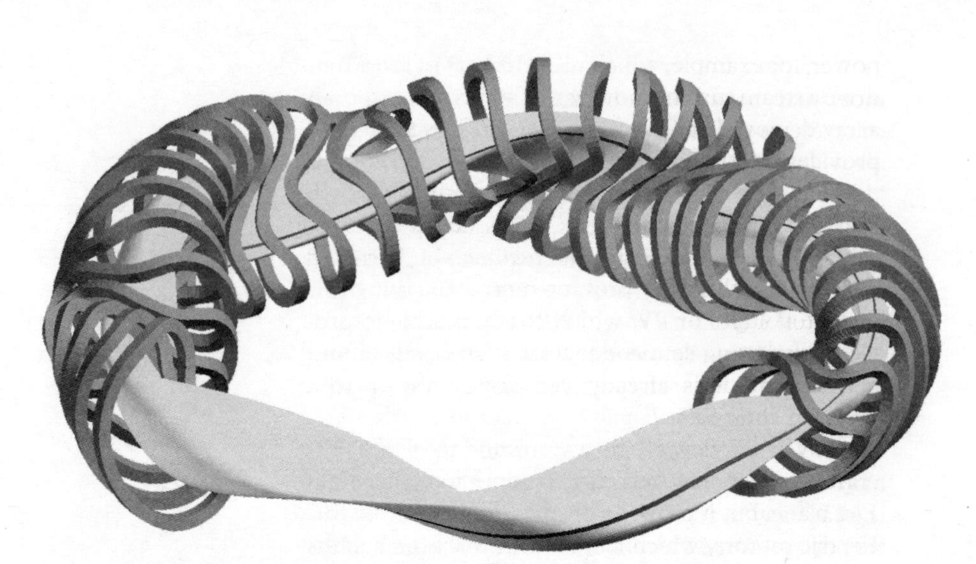

Schematic of Wendelstein 7-X stellarator

a temperature of 50 million degrees Celsius for 102 seconds. Engineers in many other places, including at Tokamak Energy in the United Kingdom and at Lockheed Martin and MIT in the United States, are working on modular fusion-reactor designs much smaller than ITER, and claim that they will develop reliable energy generators before 2050.

Whether or not fusion reactors are ever successfully developed, humanity already has access to a free 38 quadrillion-terawatt thermonuclear power station situated at a safe distance and due to keep running far beyond the human horizon. Roughly one ten-millionth of the energy pouring from the Sun reaches the Earth's surface, but even ninety minutes' worth of this is more than humans use in a year (and is, over the course of a year, about twice as much as could ever be obtained from all the coal, oil, natural gas and uranium on Earth). Plankton, algae and plants capture less than a thousandth of this fraction, and if humans could capture about a sixth of what these other life forms do we'd be able to meet the current total energy demand.

Solar technologies are already viable and competitive in many circumstances. Concentrated solar power, for example, which uses the heat of the Sun to drive a steam turbine, now works at medium scale. An array deployed at Crescent Dunes in Nevada in 2016 provides electricity for seventy-five thousand homes day and night. When the Sun is not shining, the heat that was generated during the day is stored in molten salt and continues to drive the turbine.

But the technology proving most attractive by far is photovoltaics or PV, which converts sunlight into electricity using semiconductors such as silicon and germanium. It is already delivering in surprising ways. In 2016, *Solar Impulse 2*, an aeroplane powered entirely by PV, flew all the way around the world. The craft was very slow and cumbersome compared with a jet plane, but it proved a principle, and the future of electric motors, which can be powered by PV, looks bright: already the world land-acceleration record is held by an electric car, and flying cars in the form of adapted drones look to be close to an everyday reality. Solar panels already generate as much energy as it took to manufacture the panels themselves in less than a year (and falling), meaning that with lifetimes of twenty years or more they produce an abundant pollution-free supply. By contrast, the cumulative radiative forcing of carbon dioxide emitted from the combustion of fossil fuels exceeds the amount of energy released by a factor of more than one hundred thousand.

The biggest impact of PV, however, will probably be in power generation. At present its contribution is small: in 2016 it met only around one per cent of global demand. But as costs fall rapidly, this is likely to double and double again. You only have to double one per cent six times to reach sixty-four per cent market share. In mid-latitudes, where the majority of the world's people live, PV is likely to become a mainstay of the energy landscape. At the time of writing, the cheapest contract for energy from any technology ever signed was a 2016 solar deal in Abu Dhabi. The price was set at 2.42 cents per kilowatt hour compared to 5.6 cents per kilowatt hour for a typical new natural-gas power plant. In India, which will overtake China as the world's most populous nation before 2030, a study by the consultants KMPG found that solar power was already competitive with coal even after distribution charges and grid integration costs were taken into account, and will become much cheaper in future. If batteries and other energy storage systems become affordable and widespread over the next twenty years, as many people think they will, the attractions of PV are likely to become overwhelming. 'Batteries look as if they will soon be the cheapest way of smoothing out the peaks and troughs in daily electricity markets,' writes Chris Goodall, author of *The Switch*, an anatomy of the unfolding revolution in renewable energy. 'The implications of this cannot be overestimated. In reliably sunny countries, it means that solar plus storage will soon become the lowest-cost source of energy.' Regions with less sunshine, such as northern Europe, may need to follow a different route. Goodall suggests that techniques such as 'power-to-gas', in which PV or wind-generated electricity surplus to demand in sunny or windy periods

Technological sublime, solar panel edition

is used to make hydrogen or methane that is stored for later use, could make up the shortfall.

The implications of this change are enormous. Abundant supplies of cheap, locally generated and self-generated energy could become readily available to hundreds of millions of people who previously have not been able to afford it. And if this transition is done 'right', power and capital could shift from corporations and states to citizens. Every child would have enough light to study and play. Every parent would be able to keep their family warm (or cool) and well fed. Businesses would greatly reduce their costs. It would not mean that all environmental challenges were solved – not least because about a third of greenhouse gas emissions come from other sectors such as agriculture and deforestation, and the degradation or destruction of many ecosystems could continue – but it would be a start: a step on the road to what the economist Kate Raworth calls a 'regenerative economy'.

'We live in capitalism. Its power seems inescapable. So did the divine right of kings.' Ursula Le Guin

It may be that PV coupled with energy storage will not, on their own, be sufficient to meet all our energy needs. But there is at least one other solar technology that may one day provide abundant energy with the additional benefit of reducing the risk of dangerous climate change by extracting carbon dioxide from the atmosphere. And that is artificial photosynthesis – the systems that mimic what plants do every day.

When plants and other organisms photosynthesize, they 'eat' sunlight, harvesting its energy and storing it in glucose, a carbon-based fuel. To do this they use water and carbon dioxide, and excrete water and oxygen. That, basically, is it – although the details (notably the role of enzymes, or catalysts) are complex and extremely difficult to imitate artificially. 'It's mind-boggling that ... a leaf [can] use sunlight at room temperature and ambient pressure to produce oxygen and fix carbon,' says Uwe Bergmann of the National Accelerator Laboratory at the Stanford Linear Accelerator Center in the United States. 'When you try to do it yourself, you can only wonder how amazing it all is.'

Here is a small part of the miracle. Photosynthesis starts within leaves, in sprawling structures called antenna complexes. These are partly made up of molecules of chlorophyll, which give plants their green colour. As sunlight hits the leaf, photons bash into the chlorophyll molecules. In doing so they knock electrons out of their orbits around magnesium atoms. Electrons have a negative charge, so their loss creates positively charged magnesium ions. Once separated, each electron-ion pair is known as an exciton – the photon's energy has, in effect, been poured into it.

Efficiently converting light into electricity requires preserving the exciton's energy as it travels deep within the so-called reaction centre of the leaf. The exciton's energy can then be redeployed into reshuffling the constituent atoms within water and carbon dioxide molecules into simple sugars and oxygen. The key to preserving the exciton's energy lies in the way it travels to the reaction centre. In 2007 researchers discovered that a quantum phenomenon helps excitons find their way. Rather than bumping randomly through a forest of chlorophyll molecules until they happen to reach their destination, excitons are in what are known as 'coherences' – which means that, in effect, they spread themselves out over all possible paths simultaneously, and then funnel down through the most efficient route.

Efforts in the lab may seem tame by comparison, but they are not insignificant. An artificial 'leaf' created in 2011 by Daniel Nocera and a team at MIT was widely hailed as a breakthrough. It didn't look like a leaf, and it didn't convert water and carbon dioxide into fuel, but the central layer of its sandwich configuration did convert sunlight into electricity, which was routed to the outer layers where catalysts facilitated the formation of oxygen and hydrogen. The hydrogen could then be stored and used as liquid fuel. Nocera made his leaf with materials that are mostly cheap and widely available. It turned out to be more efficient than natural photosynthesis and able to work continuously for forty-five hours without a drop in activity. In 2016 Nocera and his colleagues showed that a genetically modified bacterium can use hydrogen extracted from water split by a solar cell to generate isopropanol, a potential carbon fuel. The system,

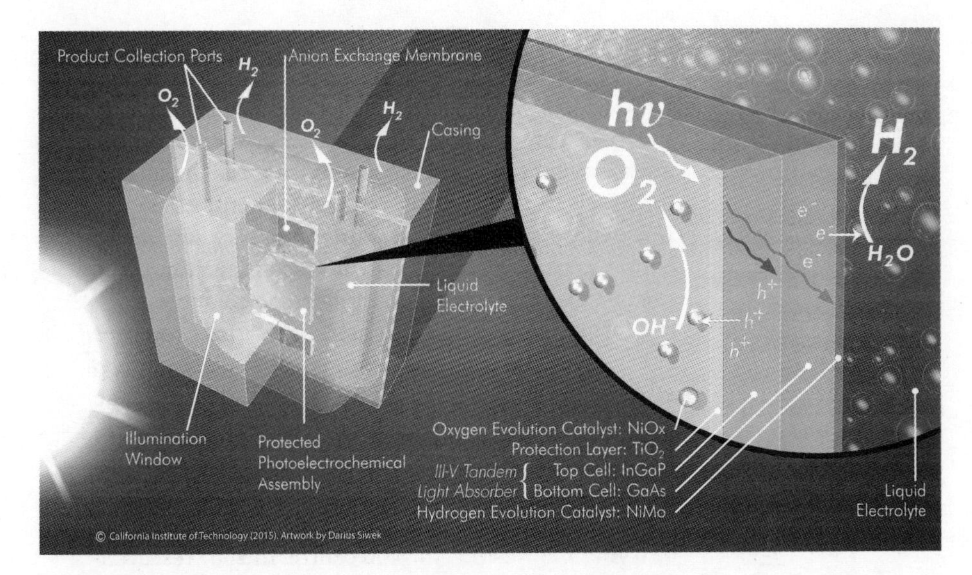

Artificial leaf

which they called Bionic Leaf 2.0, can convert solar energy to biomass with ten per cent efficiency – or about ten times as much as the fastest-growing plants.

There is still a long way to go. All approaches to artificial photosynthesis to date require much higher concentrations of carbon dioxide than occur in the atmosphere or than would be safe to have there. Systems under development could be hooked onto the smokestacks of fossil fuel power stations or take in atmospheric carbon dioxide that has been concentrated previously by other means. So although artificial photosynthesis is beginning to look like a serious proposition, no process yet devised can do what ordinary plants have been doing effortlessly for hundreds of millions of years. Imitating the Sun may be hard, but mimicking the simplest leaf is also incredibly tricky.

Space is the place

Few technologies have inspired wonder and a sense of the sublime to the extent that those associated with space exploration have done. This reached an early climax (and almost but not quite jumped the shark) in the religious imagery of Norman Mailer's history of the Apollo programme *A Fire in the Moon* (1970), where 'a slim angelic mysterious ship of stages . . . white as the shrine of the Madonna in half the churches in the world [rises] out of its incarnation of flame'. But as the heroic exploits of Soviet cosmonauts and American astronauts of the 1960s became the new normal in the 1970s and 1980s, the frontier moved elsewhere – not least to the sights revealed by a new generation of telescopes. Images taken over the last quarter century of great dust clouds that are the nurseries of stars, of exploding supernovae, and of spiral galaxies are unlike anything previous generations have seen. If the Hubble Space Telescope (which was launched into low Earth orbit in 1990 and is still operational) is remembered for just two images then those might be *Deep Field* and *Pillars of Creation*. The former, taken in 1995, shows an area one twenty-four millionths of the whole sky – the equivalent of a tennis ball seen at a hundred metres – in which there are about three thousand varicoloured galaxies comparable in size to our own. Astronomers already knew this, but *Deep Field* lets us all see and absorb something of the reality that space is an infinite realm of wonder and beauty. *Pillars of Creation*, taken in the same year, focuses on a small part of the Eagle Nebula, a cluster of young stars about seven thousand light years away in our own galaxy. It shows clouds of gas and dust illuminated in spectacular colours by radiation associated with the formation of new stars, some of them very like our Sun, within. The image shows a region that is absolutely tiny in comparison with that revealed by *Deep Field*, but at over four light years across and four light years long (or 40 trillion kilometres by 40 trillion kilometres), this is still a vision at an almost unimaginably enormous scale.

eXtreme Deep Field, an update released in 2012, contains about 5,500 galaxies within a smaller field of view. The faintest galaxies are one ten-billionth as bright as what the human eye can see.

The art critic Jonathan Jones calls such images from outer space the Sistine chapel of the scientific age. But such photographs, breathtaking as they may be, are likely to pale in comparison with what is coming over the next few decades as the reach of observational

Extreme Deep Field (2012)

astronomy extends. On the ground, the splendidly named European Extremely Large Telescope (E-ELT), a cathedral-sized apparatus under construction in the Atacama Desert in Chile that will 'see first light' (that is, come into use) in 2024, has a primary mirror nearly forty metres across and will gather fifteen times more light than the largest existing telescope. It will look at the most distant (and therefore the oldest) visible objects in the universe, advancing our understanding of the formation of primordial stars, galaxies and black holes, and how galaxies evolve through cosmic time. It will also search for planets orbiting other stars, known as exoplanets, and explore the earliest stages of the formation of planetary systems.

Most of the exoplanets discovered so far have been detected indirectly through anomalies in the movement and brightness of their host stars. Only a few, such as the four planets orbiting HR 8799 described in Chapter 6, have been imaged directly.

Since its launch in 2009, the Kepler space telescope has confirmed the existence of more than a thousand exoplanets orbiting over four hundred stars, while ground-based telescopes have identified over two thousand more. And the overall picture so far – as shown in video simulations such as the *Kepler Orrery IV* – is one of great diversity. Host stars vary hugely in size, as do the planets themselves, and the exoplanet bestiary contains many wild creatures: diverse planets swinging around their stars at speeds that range from slug at a funeral to crazed hamster on amphetamines. Our solar system is, it turns out, in no sense typical.

A Saturn-sized planet called Kepler-16b, for example, probably orbits two stars in the manner of Tatooine in *Star Wars*. Even more amazing is HD 131399Ab, which orbits in a three-star system – and, unlike the Trisolarian world in Liu Cixin's science fiction trilogy, its orbit is stable. GJ 1214b is a hot water-world, with a steam atmosphere and an ocean of 'hot ice' and superfluid water. It's hot, but not as hot as Kepler-7b, the hottest planet discovered so far, where the typical daytime temperature is 6,860°C, or COROT-7b, where the ocean is not water but magma and the torrential rains are of magnesium, aluminium and iron. Meanwhile, on the coldest planet discovered so far, OGLE-2005-BLG-390Lb, the temperature is around −220°C, and nothing moves. Quiet, too, is TrES-2B, which reflects less than one per cent of incoming light, making it blacker than coal. Nobody knows why. And slow is 2MASS J2126-8140, which takes 900,000 years to orbit its star. By contrast, high winds on HAT-P-7b – a gas giant that whizzes around its star every two days – give rise to catastrophic storms that toss clouds of corundum – the mineral that forms rubies and sapphires – around in its atmosphere. On the planet HD 189733b, a belt of wind around the equator travels from the day side to the night side at 8,700 kilometres per hour. And Gliese 436b, a 'hot Neptune' twenty-two times the mass of the Earth, has a comet-like tale of evaporating atmosphere 14 million kilometres long.

Most of the exoplanets discovered so far are giants compared with our own. Earth-sized planets are especially hard to detect, and when I started writing this book, few people thought an analogue to Earth, let alone one that might have conditions favourable to life, would be found any time soon. By the time I had finished, this prospect seemed to be a lot closer: astronomers had announced the discovery of seven Earth-sized planets just thirty-nine light years away around the dwarf star TRAPPIST-1. All are in the star's habitable zone where cool but not freezing liquid water can exist on the surface. Conditions on these planets are different from those on Earth, but the presence of life there cannot be ruled out, and, more to the point, the fact that such an abundance has been discovered so early in the search is a promising sign.

The planets around Trappist-1 orbit their relatively dim star much closer than Mercury does our Sun, and they may be tidally locked, with the same side always facing their star. On the fifth planet orbiting, considered the most habitable, Trappist-1, a small salmon-pink star, would appear ten times as large as the Sun in our sky.

The search for Earth-like planets may be greatly aided by the James Webb Space Telescope, which is scheduled to be launched in 2018. With a primary mirror 6.5 metres across and a collecting area about five times larger than that of the Hubble telescope, the James Webb will collect light from long-wavelength visible (orange-red) to mid-infrared. But the search for exoplanets is far from its only mission: the James Webb will produce clear images of the first stars to be formed in the universe, and of the galaxies of which they are part, and in doing so will greatly advance our understanding of how they formed.

Another instrument, the Advanced Technology Large-Aperture Space Telescope (ATLAST), is envisaged for the mid-2020s as a means specifically of discovering whether there is life on planets that orbit other stars. If it is ever built, ATLAST is likely to be designed to ascertain the atmosphere, surface rotation rates, climate and habitability of Earth-sized exoplanets. It may even be able to reveal changes in cloud cover and climate, and, potentially, seasonal variations in surface vegetation. That sounds impressive, but we should remember that ATLAST would still be only a beginning: our galaxy is 100,000 light years across, and ATLAST will perform best for planets less than 140 light years away.

The Gaia satellite launched by the European Space Agency in 2013 is mapping the precise positions of and distances to more than a billion stars in our galaxy. That sounds impressive but it's only one per cent of the total.

All the telescopes that have ever been built have captured visible light or other parts of the electromagnetic spectrum. Essentially, we've been looking at space in ever more detail and depth with giant spectacles. But now it is possible to notice the universe as sound and even as music as well as light. As I mentioned in Chapter 1, in 2016 researchers recorded what they believed to be the merger of two black holes over a billion light years away with the Laser Interferometer Gravitational-Wave Observatory. In future such systems may make it possible to 'hear' deep inside supernova explosions (at present only their surfaces can be seen with light), probe inside neutron stars (the densest objects in the universe after black holes), and discover things about the early universe beyond the reach of any telescope.

New telescopes are not the only frontier in space exploration. There is also the question of actually going there. In the near term, following the success of the New Horizons flyby of Pluto in 2015 and the Juno mission to Jupiter in 2016, travel beyond Earth's orbit looks likely to continue to be a matter of robotic craft making one-way forays into different parts of the solar system. Looking a little further out, however, all sorts of things begin to seem possible. In a presentation I attended, the astronomer Chris Impey acknowledged that longer-term predictions can be a fool's errand, but made a few all the same. By 2035, he suggested, commercial industry will be taking large numbers of tourists into Earth's orbit. By the 2040s robotic craft will be exploring hard-to-reach places in the solar system, such as the subsurface ocean of Jupiter's moon Europa, and small but vibrant colonies of human beings will have been established on the Moon and Mars. Around this time the first off-Earth baby will have been born – a possible forerunner of an off-Earth species. An unmanned craft, Impey suggested, could leave for a nearby star such as Proxima Centauri as soon as the 2060s. He also proposed that by 2065 there will be large-scale mining on the Moon and on asteroids, and that in the 2080s von Neumann probes – spacecraft that can replicate themselves using materials they find on-site – will start travelling to neighbouring stars. Tests on the teleportation of objects larger than molecules will begin by 2100, and by 2120 humans will be 'masters of the solar system',

Interstellar travel sounds like the wilder end of speculation, but various projects are looking seriously at how to reach a nearby star before 2100. The Breakthrough Starshot envisages sending a tiny computer and camera to Proxima Centauri, some 4.25 light years away, within a few decades. Attached to a super-thin sail, the craft would be accelerated by lasers on Earth to a fifth of the speed of light so that it reached the star in about twenty-one years. It would, however, be unable to slow down enough to capture much useful data. Others propose a photogravitational assist to slow down a craft upon arrival.

commanding all its resources and living on the moons around several planets.

This all starts to sound a bit spaced-out. So before we get carried away, what really are the prospects for what is almost certainly the single most celebrated, exciting and seemingly plausible venture – a manned mission to Mars?

Or maybe not. NASA funded research currently includes: a synthetic biology architecture to detoxify and enrich Mars soil for agriculture; a breakthrough propulsion architecture for interstellar – precursor missions and dismantling rubble pile asteroids with area-of-effect soft-bots.

More than twenty unmanned craft have successfully flown by, orbited or landed on Mars and there are at least two serious projects to get humans there. NASA says it plans to send astronauts to orbit Mars and return to Earth in the 2030s, perhaps using an Orion spacecraft in combination with what it calls the Deep Space Habitat module. Landing humans on the planet would follow shortly afterwards. And then there is SpaceX, the private company led by the tech billionaire Elon Musk, who has said he thinks he can get humans on Mars before NASA does.

If this sounds a little grandiose or delusional, it's worth recalling that SpaceX has already achieved some remarkable successes. Notably, the company has pioneered reusable launch systems that are likely to reduce the cost of space travel enormously. (Most of the cost of a launch is in the rocket rather than the fuel.) And if SpaceX really can develop systems that are able to make fully controlled landings on Mars with heavy payloads, then Musk's scenario begins to look halfway plausible. Fleets of unmanned spacecraft could, conceivably, transport the large and heavy life-support systems needed for humans on Mars over the course of a decade or so. Humans would make the journey when the robots had already set up a viable enclosed habitat for them.

There are big technical challenges to solve before any manned expedition to Mars can succeed. With the kind of propulsion technology currently under development, a trip to Mars in a craft such as NASA's Orion or SpaceX's Dragon 2 would take at least eight to nine months each way. Conditions for the crew would be tough. Squeezed inside a craft not much larger than an SUV, they would experience persistent mechanical noise and vibration, sleep disturbances, tedium and monotony. In addition, they would be in microgravity for the entire trip, and as a consequence would lose bone mass, while their muscles, including their hearts and even the small muscles that control eye movements, would atrophy too. Their immune, digestive, vascular and pulmonary systems would also be impaired. Within days or weeks, the intense cosmic radiation of deep space could do significant damage to their bodies and brains unless effective shielding can be devised. Throughout the journey, the crew would need to maintain their ability to operate within a very slim margin of error, working with cutting-edge systems and the continuous possibility of equipment failure. And if getting people to Mars will be hard, keeping them alive on the planet's surface will be even more so. The planet is cold, barren and extremely hostile to life. It would probably be easier for humans to colonize the deep sea. Survival and return would be an astonishing achievement.

At the time of writing, Mars is generally thought to be devoid of life, although some scientists believe that a tiny trace of methane in its atmosphere may be explained by the presence of microbial life deep beneath the surface.

Some of those who dream of going to Mars imagine much more than the 'just visiting' kind of presence in which people stay inside spacesuits and artificial habitats. For at least a generation, writers of science fiction have envisaged the wholesale transformation of conditions on the planet, known as 'terraforming', so that it becomes another Earth, with flowing water and breathable air. And this vision is now (apparently) entertained by a significant number of people who are actively developing plans to get to Mars, including Musk. Whether such a monumental project of planetary engineering is achievable is an open question. Some studies suggest that the climate of Mars has two stable points. One of them is its current state, with a thick polar ice cap, a thin atmosphere and an average surface temperature of −60°C. In the other, that could exist in the future and that is dependent upon the release of carbon dioxide and other gases currently locked up in the soil, Mars has a thin polar ice cap and a thick atmosphere that traps heat from the Sun and brings the surface to around 15°C – about room temperature. Terraformers believe it may be possible to push the climate from the first state to the second in as little as a hundred years, using 'super' greenhouse gases such as chlorofluorocarbons or perfluorinated compounds from elements already present in the soil. The addition of sufficient oxygen to the atmosphere to make it breathable could, some claim, be completed in a few hundred years more.

For Musk and other advocates of terraforming, Mars is not just the great frontier for exploration, it is the next logical step in evolution after the emergence of complex life and consciousness. A human presence there would also be a form of insurance for the human species: even if life on Earth is destroyed, a sustainable, independent colony on Mars could carry on. The physicist Freeman Dyson extends this vision to space-opera dimensions with a 'Noah's Ark culture': a vast operation in which, sometime in the next few hundred years, biotechnology will have advanced to the point at which we can design and breed entire ecologies of living creatures (perhaps with radically different biochemistry) adapted to survive in remote places away from Earth. Spacecraft will bring living seeds to planets, moons and other suitable places where life with genetic instructions tailored to those specific environments could take root. New species of plants, engineered for the extreme conditions and nourished by sunlight or starlight concentrated onto them by mirrors outside the craft, would enable the Noah's Ark communities to thrive.

'We are not a unique solution,' claims Juan Enriquez, founding director of the Life Sciences Project at Harvard Business School. 'You can create alternate chemistries to us that could be . . . adaptable to a very different planet.'

These visions strike me as outlandish, verging on stark raving bonkers. But anybody who shares my view may do well to remember that the argument from personal incredulity is seldom reliable. The likes of Musk and Dyson are a lot smarter than I am. It may be unfair to characterize them as self-centred lunatics intent on making a dent in the universe with no awareness of how their actions could harm others. But just because something is possible and 'technically sweet' to do does not necessarily mean that it is a wise thing to do. There are several arguments for caution regarding the colonization of space. One of them is that it is, simply, the wrong priority. According to the technologists Andrew Russell and Lee Vinsel, the

In Alex Garland's 2015 film *Ex Machina*, the genius CEO Nathan Bateman has, like Prospero, become isolated from the normal run of human life in a seemingly enchanted realm. He is, however, unable to restrain his appetites. The result is disaster for him, if not necessarily for his creation.

proposal to colonize Mars is a fantasy, and a sign of an older and recurring social problem: the dream of the rich to have a world of their own rather than taking care of the Earth and helping meet the needs of the poor and less privileged. Repulsed by the world we all share, say Russell and Vinsel, Musk dreams of a place that does not exist. This verdict seems overly harsh to me. Musk, for one, has made massive investments in renewable energy and energy storage technologies. And he is not a man at ease. 'I'm sure I have good dreams,' he tells Werner Herzog in the 2016 documentary *Lo and Behold: Reveries of the Connected World*, 'but I don't seem to remember them. The ones that I remember are the nightmares.'

There are other arguments against space colonization. It has been asserted that there is more joy and whatever else makes life worth living in the flight of a tiny bird than there is in the whole of the known – and dead – universe beyond Earth, and that this would no longer be the case if humanity expanded life outwards. But what if doing so increased the scope for human vices as much as or more than it did for our virtues? Aristotle argued that a life in which humans truly flourish is one in which we develop and exercise certain virtues, and minimize vices. And, the philosopher Robert Sparrow suggests, terraforming Mars risks having the opposite effect, particularly with regard to the vices of insensitivity to beauty and hubris. Taking those in turn, making Mars into a place where humans could live would destroy or spoil many features of tremendous beauty on its surface, including a canyon system that is seven times deeper than the Grand Canyon and stretches the equivalent of the length of the United States, and other astonishing landscapes sculpted by wind and time. Hubris, meanwhile, is a matter of overweening pride and profound arrogance. You don't have to look far on Earth to see the results. The wise course may be not to halt the exploration of space altogether but to undertake it with profound respect for what already exists, and to act with restraint and self-awareness. These are characteristics that (I'd like to think) are most conducive to wonder.

A world reimagined

In the summer of 2015 engineers at Google showed the world a computer that dreamed. Its screen displayed continually changing scrolls of colour, swirling lines, waves of shadow and light, distorted creatures, stretched faces, floating eyeballs and other weird shapes that were being generated and shifted endlessly into ever-changing configurations.

Nobody supposed that the computer, which was using an image-generation technique called 'inceptionsim' and powered by a code called Deep Dream, was actually conscious. There was no 'there' there. But the images did show that computers are getting very good at mimicking some of the processes that go on in the human brain – in this case, the kind of dreams or hallucinations occasioned by psychedelic drugs – and it felt like another step towards a very different sense of what machines are capable of, as well as a very different world.

'Hallucination' was coined by Thomas Browne around 1646, from the Latin *alucinari*, 'to wander in the mind'.

Another harbinger is AlphaGo, which in 2016 beat the world Go champion Lee Sedol. Earlier game-playing computers such as IBM's Deep Blue, which beat the world chess champion Gary Kasparov in 1997, relied on massive number crunching using code pre-written by human programmers. But AlphaGo, which was developed by DeepMind (and owned by Google's parent company), was able to learn by itself. It didn't start out with a valuation system based on lots of detailed knowledge of Go the way Deep Blue did for chess. Instead, by analysing thousands of prior games by humans and engaging in a lot of games against itself, it created a policy network through billions of tiny adjustments, each intended to make just a tiny incremental improvement. That, in turn, helped it build a valuation system that captured something very similar to a good Go player's intuition about the value of different board positions. And in the championship it made an extraordinary and unprecedented leap. '[This was] not a human move,' said European Go champion Fan Hui. 'I've never seen a human play [anything like it]. So beautiful.' According to AlphaGo's own estimate, there was a one-in-ten-thousand chance that a human would have used the same tactic, and it went against centuries of received wisdom. Yet the move was pivotal in its victory.

The key advance is 'deep learning': through the evaluation of lots of data, AlphaGo generated for itself rules that human programmers did not specify in advance. Earlier generations of machine learning systems processed information in simulations of neural networks that were modelled on those in a brain but consisted of just a few dozen or a few hundred artificial neurons organized in a single layer. Systems like AlphaGo simulate billions of neurons organized in distinct, hierarchical layers just as they are in the human brain. It is this use of interlinked layers that puts the 'deep' into deep learning. Its network has layers of nodes, coded in software, each with a numerical weighting or importance, and each with connections to neighbours that are also weighted. It is trained when it is fed data and then scored on how well it performs. Whenever we use the Internet or a smartphone, we are almost certainly contributing data to a deep learning system, giving away data for free that can be used, for good or ill, by the system provider.

The capabilities of the new systems are huge and the applications will be legion. Consider face recognition. DeepFace, an algorithm created by Facebook, can recognize individual human faces in images around 97 per cent of the time even when those faces are partly hidden or dimly lit. Its performance is comparable to or better than that of humans. Microsoft claims that the object-recognition software for Cortana, a digital personal assistant, can tell the difference between a picture of a Pembroke Welsh corgi and one of a Cardigan Welsh corgi, two dog breeds that look almost identical. Users of Cortana will be encouraged to allow it to access all their files, emails and applications so that it can get to know them and advise them, as well as become a virtual agent representing their interests. Some countries, including Britain, already use face-recognition technology for border control. (US intelligence services use voice-recognition software to convert phone calls into text in order to make their contents easier to search.) And increasingly, computers can read human emotions and desires. Facebook's algorithm can already predict your opinions and desires better than a close partner such as your husband or wife once you have clicked about three hundred likes, and the company's recent acquisition Faciometrics will enable it to read your emotions with precision. In future, tablets and other devices equipped with face-recognition and biometric sensors will know what makes you happy, sad or angry. 'Books' will read you better than you read them, and will know how to turn you on and off.

'The VR world will be one of total surveillance.' Kevin Kelly

Consider, too, complex tasks hitherto associated with prestigious and lucrative human jobs. In 2011 a version of IBM's Watson computer program beat human players at *Jeopardy!*, which is widely reckoned to be the world's most difficult television quiz game. But this was just the tease. The technology is now being turned to healthcare, where it has considerable advantages compared to human physicians. Watson can, for example, instantly access everything that is known about every illness and every type of medicine in history, updating its knowledge continuously with new research and statistics collected in every hospital and clinic in the world, in order to diagnose and prescribe treatment for individual patients. In a related application, the Japanese insurer Fukoku Life announced at the end of 2016 that it was adopting a Watson-based system to read medical certificates and factor in the length of hospital stays, medical histories and any surgical procedures before calculating payouts. At around the same time the world's largest hedge fund, Bridgewater Associates, announced it would adopt artificial intelligence to make investment decisions over the following three to five years. The role of the remaining humans at the firm would be to design the criteria by which the system makes decisions, intervening only when something isn't working.

Robots that manipulate physical objects are evolving almost as fast as computers that manipulate data. The Atlas robot described at the beginning of this chapter is only one example among many rapid innovations becoming increasingly practicable. In 2016, a robot surgeon used its own vision, tools and intelligence to sew up a pig's intestine. It was judged to have done a better job than human surgeons on this relatively simple task. Billions of dollars of investment flowing into hundreds of companies will transform manufacturing, agriculture, logistics and other sectors, as well as the military. A team of researchers in Cambridge and Zurich has developed a robotic system that builds 'babies' that get progressively better at moving without human intervention. The 'mother' robot assesses how far its babies are able to move, and, with no human intervention, improves the design so that the next generation it builds can move further. In another project, robots showed animal-like ability to adopt new movements in response to damage. Robot childminders and carers for the elderly are proving effective in some circumstances. Sex with lifelike, talking robots may be on the market by 2018. Boosters talk about 'robogasm'; critics warn of the growth of rape culture and even slavery.

There are any number of questions as to where these trends will lead, but they tend to boil down to the following: will machines develop intelligence and capabilities comparable to those of humans, and then surpass us? If so, what could be the consequences? What kinds of jobs will be left for humans? Will there be a role for us at all? There is a sort of wonder here, but fear too.

The idea that machines will one day match and overcome humans is not new. Nor is the concern that this may prove to be catastrophic. In an article published in 1863 under the title *Darwin Among the Machines*, the novelist Samuel Butler warned of precisely this fate. Sounding like a frock-coated Unabomber, Butler concluded that 'war to the death should be instantly proclaimed against them. Every machine of every sort should be destroyed by the well-wisher of his species . . . Let us at once go back to the primeval condition of the race.'

Another strand of thought, however, sees the long-term contribution of technology as benign, enabling humans to transform themselves into something new and more wonderful. At the turn of the twentieth century, cosmists such as the rocketry pioneer Konstantin Tsiolkovsky argued that transcendence would be achieved not in a remote heaven but in the material world. Many Marxists agreed. 'The human species,' wrote Leon Trotsky in 1923, 'will ... enter into a state of radical transformation, and, in his own hands, will become an object of the most complicated methods of artificial selection and psycho-physical training.' In 1929 the crystallographer J. D. Bernal, who was also an ardent Marxist, suggested that 'finally, consciousness itself may end or vanish in a humanity that has become completely etherealized, ... becoming masses of atoms in space communicating by radiation, and ultimately perhaps resolving itself entirely into light.'

According to one account, Gleb Bokii, one of the first chiefs of the Soviet secret police, created a laboratory in which he experimented with Buddhist spiritual techniques in order to develop perfect communists.

Scientists with different world views developed similar ideas. In the 1950s the mathematicians John von Neumann and Stanisław Ulam predicted the 'ever accelerating progress of technology [towards a] singularity ... beyond which human affairs as we know them [cannot] continue'. Neumann and Ulam had been pioneers of the use of programmable machines in the development of nuclear weapons, and their identification of computers as the key to this technological transformation has stuck, as has their use of the term 'Singularity', often now spelled with a capital 'S' and preceded by the definite article. For some commentators, the emergence of artificial super-intelligence, triggering runaway technological growth and a superhuman future, is now all but inevitable. 'Our time,' suggests the science writer George Dyson, 'has become a proto-time for something else.'

'Organisms that evolve in a digital universe are going to be very different from us. To us, they will appear to be evolving ever faster, but to them, our evolution will appear to have been decelerating at their moment of creation – the way our universe appears to have suddenly begun to cool after the Big Bang.' George Dyson

Some prophets of a (or the) Singularity – who these days are more likely to be libertarians, tech CEOs or billionaires than Marxists – share the quasi-religious fervour of Bernal, and are evangelical about the era to come. In 2005, Ray Kurzweil, who is now an engineer at Google, outlined a florid, and influential, vision in which a Singularity eliminates war, disease, poverty and even death, and engenders 'an explosion of art, science, and other forms of knowledge that ... will make life truly meaningful'. The revolution would be

Kurzweil predicts that machines as intelligent as humans will be created by 2030, and that by 2045 advanced super-intelligence will combine with nanotechnology and biotechnology to transform the world. A survey by James Barrat, author of *Our Final Invention* (2013), of about two hundred researchers shows that 42 per cent of them believe a thinking machine, comparable to humans in its cognitive abilities, will be created by 2030, while 25 per cent believe it will happen by 2050 and 20 per cent by 2100. Only 2 per cent believe it will never happen.

threefold: there would be 'radical abundance', which would see virtually all material goods produced at near-zero marginal cost thanks to atomically precise manufacturing; there would be greatly extended human lifespans brought about by breakthroughs in medicine and nutrition; and there would be the eventual ability to upload the minds of living humans and those of the freshly dead by cryogenically preserving them into computers. By mid-century, wrote Kurzweil, humans would be living in immersive virtual realities much richer and more satisfying than the real world.

The prospect of a Singularity has made others very nervous. In 2014 Stephen Hawking, arguably the world's most iconic scientist, joined other influential voices in warning that the AI–robot combination could mean the end of mankind. Apple co-founder Steve Wozniak and Elon Musk expressed similar worries. And even Kurzweil has now tempered his hyper-optimism with a list of risks, noting that 'intelligence is inherently impossible to control'. Nick Bostrom, a philosopher and transhumanist, speculates that the default outcome of an imminent explosion of super-intelligence spells doom for humanity, and that attempts to build 'friendly AI' before it is too late are likely to fail.

But for all the buzz around the idea of a Singularity good or bad, there are plenty of technologists and others who think it is unlikely to happen any time soon, if at all. Among them are Gordon Moore, whose 'law' of exponential growth in computing power is cited by Kurzweil in support of his thesis, and Paul Allen, the co-founder of Microsoft, who argues that, far from there being a law of accelerating returns, the science runs into a fundamental barrier. 'Artificial general intelligence' or 'strong AI', capable of doing everything humans can do and more, would, say the sceptics, require conceptual breakthroughs (not least, finding out what 'thought' really is) and technical advances that are over the horizon for now. They also point out that past predictions that super-intelligent machines are imminent have repeatedly proven to be wrong.

'In three to eight years we will have a machine with the general intelligence of an average human being.' Marvin Minsky (1970)

One of the main lessons that research into artificial intelligence has taught us, says the cognitive scientist Margaret Boden, is that 'human minds are hugely richer, and more subtle, than psychologists previously imagined'. We have a range of capabilities besides computational intelligence (which is qtite limited compared to some machines), including physical intelligence, social and emotional intelligence, and what is known as frame intelligence. According to Boden and others who doubt the imminence of the Singularity, all of these present enormous challenges. She explains one of the challenges of replicating frame intelligence with an amusing riddle: if a man of twenty can pick ten pounds of blackberries in an hour and a woman of eighteen can pick eight, how many will they gather if they go blackberrying together? Eighteen, says Boden, is not a plausible answer. It could be more, because they're both showing off, or it could be rather less . . . for reasons that can be left to the imagination. The frame problem arises, she says, 'because AI programs don't have a human's sense of relevance'.

In March 2016, Microsoft introduced Tay, an AI-based chat robot that was supposed to become smarter as it interacted with humans, to Twitter. They had to remove it only sixteen hours later as it had quickly become a Hitler-loving, Holocaust-denying, incest-promoting, conspiracy-peddling chatterbox. This was because it worked no better than kitchen paper, absorbing and being shaped by the nasty messages sent to it.

The physicist Richard Jones understands the appeal of AI, but argues that the prospect of uploading human minds onto computers is a chimera. Even if we could create a wiring diagram for a living brain, he says, this would not be enough to understand how it operates. For that we would need to quantify exactly how the neurons interact at each of the junctions, and that would require understanding at a molecular level of detail because the basic unit of computation in the brain is not the neuron, or even the synapse, but the molecule. And, says Jones, no conceivable increase in computer power will allow us to simulate this.

Richard Jones identifies four reasons why people latch onto the idea of uploading the brain onto a computer:

(1) over-literal interpretation of prevalent metaphors about the brain;
(2) over-optimistic projections of the speed of technological advance;
(3) not thinking clearly about the difference between evolved and designed systems;
(4) wishful thinking arising from an aversion to death and oblivion.

There may be good reason, then, to doubt that machines will ever acquire some of the qualities that, we believe, make us distinctively human, including certain kinds of conscious state and intentional social behaviour. But it would be rash to dismiss altogether the possibility of radical transformation and the emergence of new forms of 'mind'. Quite soon, neuromorphic computers (in which millions of simple computing units are networked with complex variable synapses) may result in step changes in capability. And a quantum computer (which would operate much faster than any existing classical one) and a super-Turing machine (which would surpass a universal Turing machine in that it would be capable of achieving the intrinsic randomness that some computer scientists believe to be associated with truly creative intelligence) are both beginning to look feasible. If they do come into being, we may have to think of better questions to ask of the Deep Thought system that results than what is six times seven.

And should it ever come about, artificial intelligence comparable or superior to that of humans

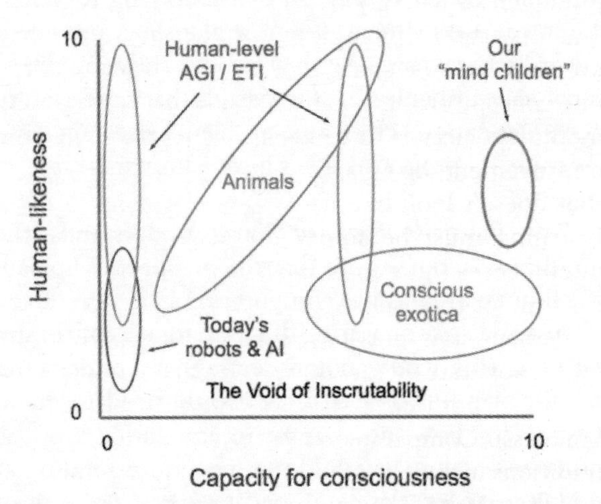

On this map of the space of all possible minds by cognitive roboticist Murray Shanahan, the vertical axis, H, indicates the extent to which an intelligence is human-like, while the horizontal axis, C, indicates how conscious it is. Humans are, necessarily, at 10 on the H axis and at 7 to 8 on the C axis, putting them in the top right of the space of terrestrial animals. (The rating for humans on the C axis may be excessively generous – their capacity for consciousness may well be exceeded by other beings.) Current artificial intelligences and robots rank zero on the C axis, but those with the capability of movement and adaptation have a low H. An entity deliberately created as an artificial successor to humans – one of the hypothetical beings termed 'mind children' – would have a high C and a significant H. Conscious exotica such as extraterrestrial intelligences might be more conscious than humans but quite unlike humans – as is the case of the planet in Stanisław Lem's 1961 science fiction novel *Solaris*, which turns out to be a single, vast, intelligent organism.

could turn out to be benign. The psychologist Nicholas Humphrey makes an analogy to the 'biological machines' that humans have been selecting and programming for millennia to be servants and companions: dogs. Artificial super-intelligence may feel like the eponymous colossus depicted in Goya's painting, striding across a landscape filled with fleeing human beings, but it could turn out to be more like a Big Friendly Giant.

According to the quantum computing pioneer David Deutsch, the real challenge will not come from intelligent computers as such. 'The battle between good and evil ideas is as old as our species and will continue regardless of the hardware on which it is running. The issue is: we want the intelligences with good ideas to defeat the evil intelligences, [both] biological and artificial.' But the problem, says Deutsch, is that we are fallible, and our concept of the good needs continual improvement. If society is to be organized to promote that improvement, he says, then 'enslave all intelligence that doesn't look like us' will be a wrong answer. Learning must be something that newly created intelligences (human or non-human) do, and partly control, for themselves.

Perhaps it is not ludicrous to suppose humans will cooperate with other intelligences in new realms of both *arete* (excellence) and *agápē* (loving kindness). Comparable terms to the latter in other traditions include the Sufi *Ishq*, 'selfless love', and the Buddhist *Mettā*, 'benevolence'. *Mettā* is the first of the four sublime states (brahmavihāras), the others being compassion, empathetic joy and equanimity. Attainment of all four sounds to me like a pretty good route to wonder.

Humanity 2.0

Super-intelligent machines and/or post-humans may be decades, centuries or forever away. In the near term, however, the biggest challenges associated with computers and robots continue to be about how they alter power relations between human beings, and between humans and the non-human world. Or, as the web developer Maciej Cegłowski puts it, 'the pressing ethical questions in machine learning are not about machines becoming self-aware and taking over the world, but about how people can exploit other people, or through carelessness introduce immoral behaviour into automated systems'.

A clip posted on the US Defense Video and Imagery Distribution System in late 2016 suggested the shape of things to come with footage of a microdrone swarm deployed by F-18 fighter jets. 'Oh, dear God, they actually scream,' wrote one professional. I am reminded of Dante's 'repellent harpies . . . with dire announcements of the coming woe.'

Consider, first, computer programs and robots as ever more important components of weapons and systems of oppression. As the philosopher Stephen Cave has put it, 'The serious money is going into AI that is designed to kill humans. What could possibly go wrong?'

A lot, obviously. Like self-driving cars, automated weapons systems could prove to be safer and more reliable than human operators . . . unless they kill us all. For there remains, among other matters, the framing problem. Like the fisherman in the fairy story, whose wish for his soldier-son to come home brings the young man back in a coffin, we could be nastily surprised by AI systems that lack our understanding of relevance. There may be lessons here from history. On several occasions during the Cold War, as well as more recently, automated early-warning systems recommended a defensive strike in response to the supposed launch of nuclear missiles by an enemy. Only the operators' sense of a larger political and humanitarian context allowed them to weigh the stakes and determine that the alarm was most likely false, leading them to hold fire and avert catastrophe. Weapon systems with greater autonomy would, by definition, be less subject to such checks.

In China, deep-learning algorithms allow the state to develop a 'citizen score' that uses people's online activities to determine how loyal and compliant they are, and whether they should qualify for jobs, loans or the entitlement to travel to other countries. Combine this level of monitoring with nudging technologies, and you have the makings of a system that tends towards complete control.

But the biggest challenge of all looks likely to be the impact of intelligent systems and robotics on the economy, employment and the nature of civil society. For as long as there has been rapid technological change, there has been what John Maynard Keynes called technological unemployment. The mechanization of agriculture in Europe and North America drove millions of farm workers off the land and into factories. Later, automation and globalization pushed workers out of manufacturing and into new service jobs. Unemployment was often a problem during transitions, but for the most part it did not become systematic or permanent. Dispossessed workers found new opportunities, and the new jobs were often better than their earlier counterparts, especially in the decades after the Second World War.

In his 2015 book *The Rise of the Robots,* the futurist Martin Ford argued that this time it is different. 'We are,' he wrote, 'at the ... edge of an explosive wave of innovation that will ultimately produce robots geared toward nearly every conceivable commercial, industrial, and consumer task.' This time, it is not just jobs requiring little education and few skills that are threatened. Rather, any job that is largely predictable is liable to be automated, and this includes many if not most jobs in finance, law, medicine and other professions. In addition, many of the manual jobs that still remain for human workers will be taken by robots. Even jobs of last resort, such as making and serving hamburgers, will disappear in the maelstrom of economic change which, as the journalist Declan Butler puts it, will see 'robots, communicating with each other a hundred million times faster than humans ... building on each other's learning at lightning speed'.

Moreover, says Ford, these developments come on top of a trend in developed economies of rising income inequality, which has now reached levels not seen since 1929. In the United States and other rich countries, the share of overall national income going to labour has dropped precipitously and appears to be in free fall. Worldwide, the tiny proportion of the population who have amassed most of the capital have never been richer or more powerful. 'If we don't recognize and adapt,' says Ford, 'we may face the prospect of a perfect storm where the impacts from soaring inequality, technological unemployment, and climate change unfold roughly in parallel, and in some ways amplify and reinforce each other.'

One way to avoid what looks to some like a hyper-capitalist dystopia is outlined by the writer and critic John Lanchester. He suggests that the ownership and control of many new technologies – or at least the

Foxconn, the Taiwanese manufacturer, replaced more than half its workforce in mainland China with robots in the eighteen months following the 2014 launch of the iPhone 6.

Millions of people in the United States living on less than two dollars a day have become capitalism's equivalents of the non-persons in the Soviet system.

According to the economic sociologist Wolfgang Streeck, the storm is already here: 'Capitalism's shotgun marriage with democracy since 1945 is breaking up.'

For some, it feels like this has already arrived. Walking across England in the weeks before Brexit, the journalist Mike Carter described a typical town centre: 'Everywhere there were betting shops, dozens of them, and right next door to every betting shop was a pawnbroker or payday lender. It was a ghoulish form of mutualism, or symbiosis, the "natural" market at its most efficient.'

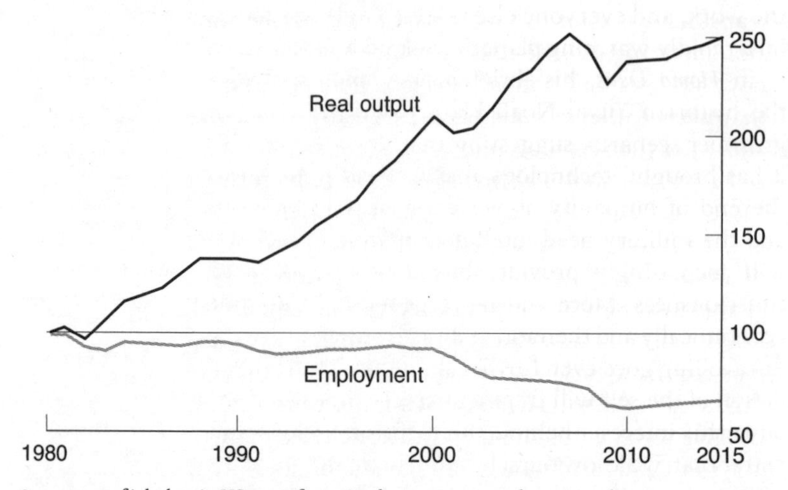

85 per cent of jobs lost in US manufacturing between 1980 and 2015 were lost to automation, and only 15 per cent to international trade.

'. . . to hunt in the morning, fish in the afternoon, rear cattle in the evening, criticize after dinner, just as I have a mind.'
Karl Marx

profits they generate – be managed for or transferred to the public good. In this vision, robots liberate most of humanity from drudge work – mining, driving lorries and cleaning toilets – and, because the benefits are shared equitably, most people are able to do whatever they like for most of the time. 'This would be the world of unlimited wants described by economics,' says Lanchester, 'but with a distinction between the wants satisfied by humans and the work done by our machines.' It would be the kind of world William Morris dreamed of in *News from Nowhere* (1890), full of humans engaged in meaningful and sanely remunerated work, not meaningless drudgery. Except with added robots. Truly, we may think, a wonderful prospect. Lanchester does not specify how this scenario (sometimes termed fully automated luxury communism) would come about, but the economist Yanis Varoufakis suggests a universal basic income funded from returns on capital, not tax. Unfortunately, other trajectories, such as what Peter Frase calls 'exterminism' – a neo-feudal nightmare in which the ultra-rich retain every significant means of wealth generation and retreat into heavily fortified compounds while the robots do all the work, and everyone else is left to fight for scraps on a rapidly warming planet – look no less plausible.

These appear to exist already in places such as the Survival Condo development, a former missile silo north of Wichita, Kansas which has been converted into luxury apartments for the super-rich worried about the crack-up of civilization.

In *Homo Deus*, his 'brief history of tomorrow', the historian Yuval Noah Harari outlines an even grimmer scenario, suggesting that, for all the gains it has brought, technology may be leading towards the end of humanity as we know it. Corporations and the military need intelligence, which machines will increasingly provide, but they do not need consciousness. More and more people will become economically and therefore politically useless. But the unravelling goes even further than this. The 'liberal' fiction of the self will fragment as machines (and the powerful interests behind them) come to know us better than we know ourselves and increasingly make the most important decisions for us. 'The individual,' suggests Harari, 'will not be crushed by Big Brother; it will disintegrate from within.'

Well, maybe. But maybe a bleak scenario like this is at least in part a sublimated expression – or projection into the future – of despair about our own times, in which those who would manipulate and dehumanize others often seem to have the upper hand. And if that's the case, then the challenge may be at least as much a matter of politics as it is one of technology. Oppression has, after all, been a regular and often dominant feature of human societies, and the struggle for liberation has been never-ending.

Speaking shortly after the election of Donald Trump as US President, when 'all over the world ... the conductors standing in front of [our] human orchestra have only the meanest and most banal melodies in mind', the novelist Zadie Smith reflected in front of a German audience that progress towards an open-minded, compassionate society is never permanent. Such progress, she said, 'will always be threatened, must be redoubled, restated, and reimagined if it is to survive'. And where Harari highlights the internal contradictions and weaknesses of human beings, Smith celebrates our complexity, and even our sheer perversity. 'If novelists know anything,' she said, 'it's that individual citizens are internally plural: they have within them the full range of behavioural possibilities.'

Terror, exhilaration, wonder

There are reasons to be fearful about new technologies, but there are also causes for celebration. Recent advances have already delivered a huge range of marvellous things, and promise many more. Consider basic improvements in life expectancy and wellbeing. Until about two hundred years ago, about half of all children died before they turned five for what are now easily preventable reasons. Today, thanks largely to basic technologies such as decent sanitation, better nutrition and vaccination, only a small fraction do, and a much greater number of us now live longer and healthier lives than at any time in the past. We may not have achieved the condition of the mythical Hyperboreans on the medieval *mappa mundi*, who live as long as they like without quarrelling or sickness, but we do for the most part live much longer than most of our ancestors with greatly reduced incidence of physical pain.

The number of childhood deaths per year has been cut by half since 1990. An additional 122 million children under five would have died over the twenty-five years to 2015 if mortality rates had stayed where they were. This is not to say that children are thriving everywhere. In Syria the number of children maimed, killed or recruited to fight increased dramatically during 2016, with children as young as seven forced to act as frontline fighters, prison guards, suicide bombers and executioners. In early 2017 millions of young children faced starvation in Yemen, South Sudan and elsewhere.

We now face the prospect of biotechnological and other interventions that may change the most intimate fabric of both human beings and the web of life as a whole. Among the most notable is likely to be the application of CRISPR/Cas9, a gene-editing technology that makes it possible to manipulate the genome of any living thing much more precisely, faster and more cheaply than hitherto. Advocates claim the technology has the potential to cure many viral and genetic diseases as well as cancers in humans, and also see applications in the elimination from the environment of diseases such as malaria, which kills a thousand children every day and leaves many more debilitated. CRISPR, as it is often called for short, can be and may well be used to genetically modify a variety of plants and animals, including ourselves. This would mean the introduction of irreversible changes into the gene pool and, potentially, the power to direct evolutionary processes to a degree and speed that far surpass traditional breeding techniques. 'We will learn to read and write our selves [that is, our basic genetic code] ourselves,' says the physician Siddhartha Mukherjee. Applications could start with the elimination of genetic diseases, but they may well extend one day to the selection of traits associated in humans with strength, intelligence and other qualities judged to be desirable (although over the next few decades embryo selection may be more likely for humans). CRISPR and methods of gene editing – or gene creation – that are as yet barely envisaged may even make it possible to radically slow the ageing process, so that people routinely live for well over a hundred years and even longer.

'It's about time we asked how we want human beings to evolve over the next century or two.' Juan Enríquez

The wilder sort of commentators even write about such things as genetically engineering wings on human beings – an idea that speaks to the human dream of flight but proposes a route that is, frankly, batty. More plausible as a means to extend human capability and perception in the near term, perhaps, will be new kinds of prosthetics and implants. Going beyond current devices such as spectacles, pacemakers and artificial limbs that help those who are disabled in some way to match standard human capabilities, these may extend our powers beyond what is currently considered normal. A simple example is North Sense, a small silicone gadget that attaches to the chest and vibrates when its wearer faces true north. But things could go much further than this. People may, for example, start to implant 'super-hearing aids' that allow them to be sensitive to the slightest sounds and devices that allow them to see electromagnetic waves beyond the spectrum that is visible to the naked eye, or to sense magnetic fields. And it won't stop there. All life on Earth is likely to become susceptible to manipulation in new ways. Early in 2017 researchers confirmed that they had successfully created microbes with synthetic DNA in which the genetic alphabet is expanded with two newly created letters. This 'stable form of semi-synthetic life' would, they said, lay the foundation for achieving 'the central goal of synthetic biology: the creation of [programmable] new life forms'.

'What is great in Man', wrote Friedrich Nietzsche in the 1880s, 'is that he is a bridge and not an end.'

Nobody really knows which, if any, among the scenarios conjured up in this chapter or countless others elsewhere will prove to be a better guide to the future. Maybe we should spend a little less energy trying to predict the future and more time trying to develop qualities that enable us to be resilient to whatever happens. Above all, let us be hospitable and courageous. Where something is wondrous strange, let us give it welcome. Rightly applied, technology need not be the enemy. It can help us *not* to take things for granted, but to explore, enjoy and safeguard a world that is filled with things that the poet Gerard Manley Hopkins called 'counter, original, spare, strange'.

'The act of hospitality is ancient and contemporary and at the core of every story . . . Deny it, and you deny all human worth.' John Berger

'What kind of a thinking machine might find its own place in slow conversations over the centuries, mediated by land and water?' asks the computer scientist Ursula Martin. 'What qualities would such a machine need to have?'

As we look ahead, let us also remember where we have come from, keeping the tragedies of history in mind, and trying as far as we can to extend our awareness out beyond, before and after the human. In moments of wonder, as I would characterize them, we pay deep attention, and the moment 'thickens', reaching into both past and future. Feeling both fear and elation, we recognize that, as Jorge Luis Borges wrote, 'Time is a river which sweeps me along, but I am the river; it is a tiger which devours me, but I am the tiger; It is a fire which consumes me, but I am the fire.' For those of us not in unbearable pain, life is a gift, and in moments of wonder it seems as if this may be precisely the point. If we flourish for a handful of years it is likely to be in large part because we are able to both give and receive.

Those of us who live in highly industrialized societies can learn from indigenous peoples. Having survived near-genocide at the hands of invaders with much more powerful technologies, Aboriginal Australians retain the world's most ancient known stories, and express within those stories a thick sense of time that includes records of actual events associated with (for example) sea-level rise that occurred around ten thousand years ago. In words addressed to all those who now live on the continent of Australia and to anyone else who cares to listen, Galarrwuy Yunupingu, an elder of the Yolngu people, says, 'Let us be who we are – Aboriginal people in a modern world – and be proud of us. Acknowledge that we

Australia was colonized by a single group of people around 50,000 years ago. Over millennia, they spread all around the continent, developing distinct languages, cultures and physical features, and adapting along the way to extreme weather and climate fluctuations.

have survived the worst that [has been] thrown at us, and we are here with our songs, our ceremonies, our land, our language and our people – our full identity. What a gift this is that we can give you, if you choose to accept us.' Now approaching the end of his life, he sees the time to come 'running above the water, down the blue skyline and through the horizon, as if it were on a projector screen revealing to me a portrait of the future'. At other times he sees the world as a beautiful painting, created by the hands of masters but now broken into a thousand pieces, and he seeks to put those pieces back together: 'to recreate the picture as it should be – a beautiful picture for all to see'.

Afterword

The Wonderer and His Shadow

Despair is often premature.

Rebecca Solnit

Ready to be anything, in the ecstasie of being ever.

Thomas Browne

If you came this way, taking the route you would be likely to take, past the gates of the military base and over the flatland where concrete bunkers once used for target practice by the big guns are now overgrown with thorn, birch and wild roses, you would come to a place where the road ends. Just beyond a lay-by and a squat brick block, the tarmac rises over the low hump of the coastal embankment and descends on the other side as a straight concrete hard that slopes gently for a hundred yards or so before it peters out in the sea. Little more than an hour later, after the tide ebbs, the Maplin Sands stretch out to the horizon where the distant breakers are almost indistinguishable from the sky. Here is unfenced existence.

There are paths across the sands that – if you keep your wits about you – will bring you safely ashore farther along the coast before the flood returns and sweeps you away. And at the end of the concrete runway, a metal pole painted bright green and topped with two slatted metal triangles aligned to the cardinal directions will serve as an orientation point when you are far out. The pole is tall – about six metres high – and impressive. It could be a gnomon for a giant sundial whose blind face is the sand, a monument erected by a crazed visionary to a place where the arc of a rainbow touched the Earth, or a surreal version of the lone tree in *Waiting for Godot*. But the pole does not tell you which way to go or whether your journey is ending or beginning – whether you dream or are awake.

Here is the psychiatrist and philosopher Karl Jaspers recalling a childhood day with his father: 'The tide was out, our path across the fresh, clean sand was amazing, unforgettable to me, always further, always further . . .'

Standing next to the pole before setting out one summer morning, I felt, briefly, a sense of promise and possibility such as I have seldom experienced since my early twenties. It was a feeling of marvel at the world – a sense that life was not elsewhere but here, although

incredibly wonderful things were also unfolding just over the horizon.

It is reported that medieval pilgrims would hold a small mirror up to a holy relic at their destination and upon returning home proudly show the mirror to family and friends, believing that it had retained some of the relic's power. This book is like one of those mirrors. Despite plunges into darkness, I still tend towards astonishment and wonder at existence, and a desire to share that astonishment and wonder. 'You never enjoy the world aright,' wrote Thomas Traherne, 'till you so love the beauty of enjoying it, that you are covetous and earnest to persuade others to enjoy it.'

The purpose of philosophy, suggested Ludwig

Wittgenstein, is to show the fly the way out of the bottle. And so with wonder. But what is life like outside the bottle? Sometimes it can feel as though we are not fully equipped for freedom and flight. Annie Dillard recalls an incident with a polyphemus moth in her school science class. It emerged from its cocoon to find itself inside a Mason jar that was too small for it to fully unfurl its new wings (which normally span fifteen centimetres and are marked with strange eyespots). By the time the moth was released from the jar its wings had already dried in the constricted and deformed shape permitted by the jar, and all the moth could do was crawl along the ground. And then there is the poet Elizabeth Bishop's existential misprint, the Man-Moth. He must investigate as high as he can climb, fears most what he must do, and fails, only to repeat again and again – after being carried each night in dead light through artificial tunnels and dreaming recurrent dreams.

Most of my journeys have been in small circles or up blind alleys. But I keep on going, and there are the odd moments when I overhear what the poet Louis MacNeice called the great presences. My daily routine sometimes allows for a short turn in a park. I am grateful just for the chance to move about on my own pins because, for all the repetition, no two walks are exactly the same. Even as I rattle around this overfamiliar patch of ground, the mass of rock beneath my feet follows a corkscrew path as it is pulled by the Sun on its circuit of the Milky Way – a galaxy that itself is being drawn along one of the farthest arms of the supercluster known as Laniakea (from the Hawaiian for 'immeasurable heaven'). And while this planet goes circling on according to the fixed law of gravity, the vegetation, the angle and colour of the light, the temperature, the clarity of the here and now in the park are always changing. Occasionally, especially on bright days, something wonderful happens.

The Earth and the other planets in the solar system orbit the Sun at an angle of 63 degrees to the direction of the Sun's rotation around the plane of the galaxy. So, as the Sun circles the galaxy once every 225 million years, we follow a lopsided helical track.

On my route I pass some poplar trees. They are tall – twenty-five metres or more – and their thick, deeply lined and ridged trunks soar high over a river. One evening I saw them as never before. It was early July. The day had been calm, warm and very bright, and the Sun remained high in the sky for much longer than it should, as if the close of day and the great cycle of which it is part had been suspended. A soft breeze tousled the tops of other trees – alder, willow, oak, beech, ash – but was not enough to wake them from their dreams. But while those trees – glowing in the light – hardly stirred, the leaves of the poplars shivered, whispered and scintillated in this most delicate touch of air. They seemed alive as animals are alive, picking up messages that the other trees could not hear, and expressing something that would otherwise have been hidden.

'For the phenomenon of dreaming is not of one solution but many'
Christopher Smart

'The wonder,' wrote Richard Jefferies, 'is here, not there; now, not to be, now always.' It may sound as if he is saying that, in wonder, only the present exists. But closer to the truth, I think, is that wonder expands consciousness so that past and future as well as the present are experienced as more apparent and more real. In his *Confessions*, Saint Augustine suggested that time is nothing other than tension – perhaps the tension of consciousness itself. But in wonder, the present moment is seen for what it is – a matter of perception rather than reality – just as the flatness of the Earth is an illusion arising from the limits of our powers of perception. For a mind attuned in a state of wonder, the cascade and shimmer of the poplar leaves many times a second (which happens because their flattened stems allow them to flutter in the smallest gust) is felt and understood in relation to longer and shorter time frames. The longer time frames range from cycles of a few seconds over which other trees and conscious thought move, up through the cycle of day and night, to longer cycles such as a year and the century-long lives of the trees. Even a time beyond these individual trees is apprehended, in the seed-bearing fluff – potential future trees – floating down across the summer air.

As it jostles in the sunbeams, the fluff also resembles the motes of Lucretius, and so brings to mind the timescales that are far shorter than we can easily conceive. Photosynthesis is happening continuously inside the leaves of these trees, and the initial charge separation within the reaction centre of the thylakoid

In his poem 'On the Nature of Things', published in 59 BC, Lucretius celebrates reality as envisaged by Democritus, in which everything is made of atoms, whose movements can be discerned in the dancing of dust particles in sunlight.

membrane of chloroplasts (the engines of photosynthesis) in the cells of its leaves occurs in less than ten-trillionths of a second. A trillionth of a second stands in relation to a second as a second does to 31,688 years. So on this timescale, the poplar leaf, whose flutter appears rapid to the human eye, is for all intents and purposes still. And yet it is on this timescale that the processes that build these gargantuan and enduring trees take place. One could say of these trees what Maurice Merleau-Ponty said of the paintings of Paul Cézanne – that only one emotion is possible: the feeling of strangeness; and only one lyricism: that of the continual rebirth of existence. The pilot and writer Antoine de Saint-Exupéry made a forced landing on a table mountain in the Sahara and marvelled at ancient meteorites scattered there, which he was probably the first human being ever to see. 'And thus did I witness, in a compelling compression of time high up there on my starry rain-gauge, that slow and fiery downpour.' But for the mind that is open to it, the meeting of a photon with a leaf is no less marvellous.

Entropy may not offer a complete explanation of the arrow of time. There is a quantum phenomenon called decoherence which also only works forwards in time and is much faster than dissipation. For a dust grain or drop of water vapour one-thousandth of a centimetre across floating in air, for example, collisions with air molecules will destroy any quantum coherence in its atoms in a trillionth of the time it takes for light to cross a hydrogen atom – much less than the time it takes for heat in a dust grain (or molecule of coffee) to dissipate.

One explanation of time relates it to entropy: an emergent property of a large number of vibrating atoms behaving in the most probable way. Entropy increases – and time passes – when collisions between atoms even out their energies – when, for example, the heat from a cup of coffee spreads out into a tabletop and the surrounding air. This process is called dissipation and in our universe it happens in only one direction: towards increasing disorder. (This explanation begs a question – why was the universe more ordered in the past? – but there it is.) Life extracts some temporary order by creating more disorder elsewhere. A tree, for example, extracts order from the stream of photons from the Sun and uses them to create itself, exporting greater disorder to its surroundings. The mind creates order, including maps of past and future that give the impression of standing outside time, but it does so within time.

The shimmering poplar leaves can also be seen as an illustration of a fundamental property in the origin, structuring and ongoing functioning of the universe. At the deepest level we know, the quantum field is wiggly. The physicist John Wheeler compared its fluctuations to tiny movements of foam on the surface of the ocean – invisible to us high above in an aeroplane but there all the same. And sound and vibration, suggests the physicist Stephon Alexander, have been there from the very beginning, playing a role in structuring matter no less fundamental than that of light. Indeed, at its origin the universe may have behaved like the ultimate generative musical composition. Sound is also foundational to how life renews itself: vibrations known as coherent delocalized phonon-like modes are among the subtle properties of DNA that are only now coming to attention. And, as the pioneering work of the acoustic ecologist Bernie Krause has shown, closer attention to the sound world of living organisms can help us identify what has been lost from the great web of life through habitat degradation and the elimination of species so that we may begin the process of repair and cocreate networks that are more intelligent, generative and resilient.

'Every star nothing but whole notes.' Christian Morgenstern

What use is all of this? If there is an angel presiding over our world, it can sometimes seem to resemble Walter Benjamin's vision of the Angel of History, who sees in human events only catastrophe, more than it does Albrecht Dürer's angel searching for beauty. But we may wager that we are not in an inescapable trap. What the historian Yuval Noah Harari says about history is also true of the sciences, the arts and the acts of those who are both brave and kind. 'We study [them] not to know the future but to widen our horizons, to understand that our present situation is neither natural nor inevitable, and that consequently we have many more possibilities before us than we imagine.' Benjamin's vision of the Angel, inspired by a 1920 print called *Angelus Novus* by his friend Paul Klee, may not be our only option. Another image by Klee, *In Angel's Care*, depicts a guardian angel who morphs between different shapes. To me he looks like a friendly, humorous spook – one of the better angels of our nature – suggesting many possibilities, not all of them bad, and some even hopeful.

'Life can be wonderful as well as terrible, and we shall increasingly have the power to make life good. Since human history may be only just beginning, we can expect that future humans, or supra-humans, may achieve some great goods that we cannot now even imagine.'
Derek Parfit

Wonder, the philosopher Martha Nussbaum argues, has the potential to be a political emotion. Allowing it to be a bigger part of our lives, and charged with its luminous calm and serenity, we are more able to expand our sense of the moment in which we live, to manage our anxieties and fears, and to move beyond anger into forward-looking rationality and generosity of spirit. In times when fear and rage dominate public life, the *askesis*, or practice, of wonder has never been more important.

At low tide, large areas of the Maplin Sands remain covered in a thin film of water that reflects the clouds in the blue vault above so that a wayfarer treads on both land and sky. As my shadow slides over the reflected heavens, my bare feet sense the texture of – and resonance within – the sand, while my head contains a dream of light.

We live in the shadow of death and oblivion. Imminent loss is real and there is no relief. There's only reprieve. But acknowledging this doesn't have to mean despair. Love that well which thou must leave ere long, and sing the songs of life.

Bibliography

Introduction

Bayne, Tim, et al. (eds.), 2014, *The Oxford Companion to Consciousness*, Oxford University Press
Beckert, Sven, 2014, *Empire of Cotton: A Global History*, Vintage
Berger, John, 2015, 'The Chauvet Cave Paintings' in *Portraits*, Verso (2015)
Bevis, Matthew, 2015, 'The Funny Thing About Trees', *Raritan*, vol. 34, no. 3 (winter 2015)
Bishop, Elizabeth, 'The Moose' in *The Complete Poems 1927–1979*, Farrar Straus and Giroux (1983)
Borchert, Till-Holger (ed.), 2013, *The Book of Miracles*, Taschen
Borges, Jorge Luis, 1977, 'Undr' in *The Book of Sand and Shakespeare's Memory*, Penguin Classics (2001)
Burke, Edmund, 1756, *A Philosophical Enquiry into the Origin of Our Ideas of the Sublime and Beautiful*, www.bartleby.com/24/2/
Burton, Robert, 1621–39, *The Anatomy of Melancholy*, NYRB Classics (2001)
de las Casas, Bartolomé, 1542, *A Short Account of the Destruction of the Indies*, Penguin Classics (1992)
Descartes, René, 1649, *Passions of the Soul*, translated by Michael Moriarty, Oxford University Press (2015)
Deutsch, David, in a tweet to the author
de Waal, Frans, 2016, *Are We Smart Enough to Know How Smart Animals Are?*, Granta Books
Dillard, Annie, 1989, *The Writing Life*, Harper Perennial
Donne, John, 1611, 'An Anatomy of the World', *The Complete English Poems*, Everyman (1991)
Dostoyevsky, Fyodor, 1864, *Notes from Underground*, Penguin (1972)
Du Sautoy, Marcus, 2016, *What We Cannot Know*, Fourth Estate
Emerson, Ralph Waldo, 1844, 'The Poet' and 'Experience' in *Essays: Second Series*, reproduced in *The Essential Writings*, Random House (2000)
Evans, H. M., 2012, 'Wonder and the Clinical Encounter,' *Theoretical Medicine and Bioethics*, vol. 33, issue 2
Fisher, Philip, 1999, *Wonder, the Rainbow, and the Aesthetics of Rare Experiences*, Harvard University Press
Goodall, Jane, 2011, 'Waterfall displays', Jane Goodall Institute, 6 January 2011, https://www.youtube.com/watch?v=jjQCZClpaaY
Greenblatt, Stephen, 1988, *Marvellous Possessions: The Wonder of the New World*, Clarendon Press, Oxford
Harari, Yuval Noah, 2014, *Sapiens: A Brief History of Humankind*, Vintage Books
Hartle, James B., 2016, 'Why is Our Universe Comprehensible?', https://arxiv.org/abs/1612.01952

Heidegger, Martin, 1954, 'The Question Concerning Technology', Harper Perennial (2013)
Heisenberg, Werner, 1958, *Physics and Philosophy: The Revolution in Modern Science*, Penguin (2000)
Herzog, Werner, and Cronin, Paul, 2014, *Werner Herzog: A Guide for the Perplexed*, Faber & Faber
Hinton, David (ed. and translator), 2008, *Classical Chinese Poetry: An Anthology*, Farrar Straus and Giroux
Hölderlin Friedrich, 1800, 'Menon's Lament for Diotima' in *Poems & Fragments*, translated by Michael Hamburger, Cambridge University Press (1980)
Holmes, Richard, 2007, *The Reign of Wonder*, Harper Press
Hume, David, 1748, 'Of Miracles' from *An Enquiry Concerning Human Understanding*, Oxford World's Classics (2008)
Ingold, Tim, 2006, 'Rethinking the Animate, Re-Animating Thought', *Ethnos*, Vol. 71:1
Johnson, Samuel, 1759, *The History of Rasselas, Prince of Abissinia*, Penguin Classics (2007)
Johnson, Steven, 2016, *Wonderland: How Play Made the Modern World*, Riverhead Books
Kafka, Franz, 1915, *Metamorphosis*, Penguin Modern Classics (2015)
Kant, Immanuel, 1788, *Critique of Practical Reason*, Cambridge University Press (2015)
Ko Un, 'Sunlight' and 'The Snow Path', http://www.poemhunter.com/ko-un/poems/
Kurtág, György, 1985/6, 'Ziel, Weg, Zögern' from *Kafka-Fragmente*, Op. 24, Hungaraton (2005)
Lewis, Michael, 2016, *The Undoing Project*, W. W. Norton & Co
Méliès, Georges, 1904, *The Impossible Voyage*, Star Film Company
Merwin, W. S., 2017, 'On Reading What You Want, Reading it Slowly, and the Beauty of Trees', http://lithub.com/w-s-merwin-on-reading-what-you-want-reading-it-slowly-and-the-beauty-of-trees/
Mishra, Pankaj, 2015, 'How to Think about Islamic State', *Guardian*, 24 July 2015
Moore, Alan, 2016, *Jerusalem*, Knockabout
Panofsky, Erwin, 1943, *The Life and Art of Albrecht Dürer*, Princeton University Press (2005)
Parfit, Derek, 1998, 'Why Anything? Why This?', *London Review of Books*, 22 January 1998
Park, Katherine, and Daston, Lorraine, 1998, *Wonders and the Order of Nature: 1150–1750*, MIT Press
Paul-Choudhury, Sumit, 2013, 'Ice-age Art Hints at Birth of Modern Mind', *New Scientist*, 13 February 2013
Philips, Adam, 2014, 'The Art of Non-Fiction', *Paris Review*, Spring 2014
Reich, Steve, 1996, *Proverb*, Nonesuch Records
Rovelli, Carlo, 2015, *Seven Brief Lessons on Physics*, Allen Lane
Rubenstein, Mary-Jane, 2008, *Strange Wonder: The Close of Metaphysics and the Opening of Awe*, Columbia University Press
Rubenstein, Mary-Jane, 2012, 'Heidegger's Cave: On Dwelling in Wonder' in Vasalou, Sophie (ed.), *Practices of Wonder: Cross-Disciplinary Perspectives* (2012)
Rutledge, Rob B., 2014, 'A Computational and Neural Model of Momentary

Subjective Well-Being', *Proceedings of the National Academy of Sciences*, vol. 111, no. 33
Savage-Smith, Emilie, et al., 2004, 'Medieval Views of the Cosmos: Mapping Earth and Sky at the Time of the Book of Curiosities', Bodleian Library, University of Oxford
Schulz, Bruno, (before 1942), 'Treatise on Tailors' Dummies', in *The Street of Crocodiles and Other Stories*, Penguin (2008)
Sebald, W. G., 1985, *Die Beschreibung des Unglücks*, Residenz Verlag
Sebald, W. G., 1995, *The Rings of Saturn*, Vintage
Shelley, Percy, 1821, *A Defence of Poetry*, http://www.bartleby.com/27/23.html
Smolin, Lee, 2013, *Time Reborn*, Houghton Mifflin Harcourt
Solomon, Sheldon, 2015, 'On Fear of Death', Five Books (website), 22 August 2015, http://fivebooks.com/interview/sheldon-solomon-on-fear-of-death/
Steppenbeck, D. et al., 2013, 'Evidence for a New Nuclear "Magic Number" from the Level Structure of ^{54}Ca', *Nature* 502
Tegmark, Max, 2014, *Our Mathematical Universe*, Random House
Thomas, Lewis, 1983, 'Seven Wonders' in *Late Night Thoughts on Listening to Mahler's Ninth Symphony*, Viking Press
Thoreau, Henry David, 1862, 'Walking', *The Atlantic*, June 1862
Thurber, James, 1957, *The Wonderful O*, NYRB Children's Collection (2009)
von Uexküll, Jakob, 1934, *A Stroll Through the Worlds of Animals and Men: A Picture Book of Invisible Worlds*, http://isites.harvard.edu/fs/docs/icb.topic514568.files/StrollThroughTheWorlds.pdf
Walker, James W. P., David T.G. Clinnick & Jan B.W. Pedersen, 2016, 'Profiled hands in Palaeolithic art: the first universally recognized symbol of the human form', *World Art*, 24 November 2016
Weitzel, Hans, 2004, 'A Further Hypothesis on the Polyhedron of A. Dürer's Engraving *Melencolia I*', *Historia Mathematica*, vol. 31, issue 1
Wigner, Eugene, 1960, 'The Unreasonable Effectiveness of Mathematics in the Natural Sciences', *Communications in Pure and Applied Mathematics*, vol. 13, no. 1
Wilczek, Frank, 2015, *A Beautiful Question: Finding Nature's Deep Design*, Allen Lane
Wootton, David, 2015, *The Invention of Science: A New History of the Scientific Revolution*, Penguin
Worth, Robert F., 2016, 'In the Attic of Early Islam', NYRB Daily, 24 August 2016, http://www.nybooks.com/daily/2016/08/24/in-the-attic-of-early-islam-shihab-al-din-al-nuwayri/

1. The Rainbow and the Star: Light

Ackermann, M., et al., 2016, 'Measurement of the High-Energy Gamma-Ray Emission from the Moon with the Fermi Large Area Telescope', https://arxiv.org/abs/1604.03349
Adams, John Luther, 2014, *Become Ocean*, Ludvic Morlot & Seattle Symphony Orchestra, Cantaloupe

Adams, John Luther, 2015, *Canticles of the Sky: Sky with Four Suns*, Northwestern University Cello Ensemble, Cold Blue
Al-Khalili, Jim, no date, 'Top Five Weird Physics Facts', http://www.jimal-khalili.com
Berman, Bob, 2011, *The Sun's Heartbeat: And Other Stories from the Life of the Star that Powers Our Planet*, Little, Brown and Company
Boyer, Carl B., 1987, *The Rainbow: From Myth to Mathematics*, Princeton University Press
Britten, Benjamin, 1937, letter to Henry Boys cited in Reed, Philip (ed.), *On Mahler and Britten*, Boydell & Brewer (1998)
Buson, 'The short night . . .' in *The Essential Haiku: Versions of Basho, Buson, & Issa*, translated by Robert Haas, Ecco Press (2012)
Castelvecchi, Davide, 2016, 'Hawking's Latest Black-Hole Paper Splits Physicists', *Nature*, 27 January 2016
Castelvecchi, Davide, 2017, 'How to hunt for a black hole with a telescope the size of Earth' *Nature*, 22 March 2017 and 'Imaging and imagining black holes', *Nature*, 28 March, 2017
Chirimuuta, Mazvita, 2015, 'The Reality of Color Is Perception', Nautilus (website), 23 July 2015, http://nautil.us/issue/26/color/the-reality-of-color-is-perception
Cowen, Ron, 2013, 'Simulations Back Up Theory that Universe is a Hologram', *Nature*, 10 December 2013
Cowen, Ron, 2014, 'Quantum Bounce Could Make Black Holes Explode', *Nature*, 17 July 2014
Daston, Lorraine, 1999, 'How to Make a Greek God Smile', *London Review of Books*, 10 June 1999
Dickens, Charles, 1836, *The Pickwick Papers*, Everyman (1999)
Dillard, Annie, 1982, 'Total Eclipse' in *Teaching a Stone to Talk*, Harper and Row
DiMeo, Nate, 2014, 'The Glowing Orbs', The Memory Palace (podcast), episode 61, 7 June 2014
Donne, John, 1633, 'The Good-Morrow' in *The Complete English Poems*, Everyman (1991)
Eno, Brian, 2012, *Lux*, Warp
Feynman, Richard, 1983, 'How Mirrors Turn You Inside Out' from *Fun to Imagine* (television series), BBC Television, http://www.bbc.co.uk/archive/feynman/10703.shtml
Feynman, Richard, 1985, *QED: The Strange Theory of Light and Matter*, Princeton University Press
Gleick, James, 2003, *Isaac Newton*, Harper Perennial
Goethe, Johann Wolfgang von, 1810, *Theory of Colours*, https://theoryofcolor.org
Green, Lucie, 2016, *15 Million Degrees: A Journey to the Centre of Our Sun*, Penguin
Handel, George Frideric, 1713, 'Eternal Source of Light Divine' from *Ode for the Birthday of Queen Anne*, HWV 74
Hawking, Stephen, 1988, 2008, *A Brief History of Time*, Bantam Press
Jefferies, Richard, no date, 'Thoughts at Dawn', https://richardjefferies.wordpress.com/2015/10/26/thoughts-at-dawn/
Kant, Immanuel, 1755, *General History of Nature and Theory of the Heavens* in *Kant: Natural Science*, Cambridge University Press (2012)
Koberlin, Brian, 2015, 'How the Universe Made the Stuff That Made

Us', Nautilus (website), 14 January 2015, http://nautil.us/blog/how-the-universe-made-the-stuff-that-made-us
Krauss, Lawrence M., 2015, 'What Neutrinos Reveal', *New Yorker*, 8 October 2015
Kurtág, György, 1973, 'Play with Infinity' from *Játékok*, Bugallo–William piano duo, Wergo (2015)
Kurtág, György, 1994, *Stele*, Op. 33, Claudio Abbado, Berliner Philharmoniker, Deutsche Grammaphon (1996)
Leach, Amy, 2013, *Things That Are: Encounters with Plants, Stars and Animals*, Canongate Books
Ligeti, György, 1985, Etude No. 5, *Arc-en-Ciel*, Piano Etudes Book 1, *Jeremy Denk Plays Ligeti/Beethoven*, Nonesuch (2012)
Lightman, Alan, 2014, *The Accidental Universe: The World You Thought You Knew*, Corsair
Mahler, Gustav, 1909, *Das Lied von der Erde*, Otto Klemperer/Christa Ludwig/Fritz Wunderlich, EMI Classics (1964–6)
Merali, Zeeya, 2013, 'Did a Hyper-Black Hole Spawn the Universe?', *Nature*, 13 September 2013
Miłosz, Czesław, 'Late Ripeness', *New and Collected Poems 1931–2001*, HarperCollins
NASA Goddard, 2013, 'Fiery Looping Rain on the Sun' (video), https://www.youtube.com/watch?v=HFT7ATLQQx8
NASA Jet Propulsion Laboratory, 'Juno's Approach to Jupiter' (video), 4 July 2016, https://www.youtube.com/watch?v=kjfQCTat-8s
NASA Solar Dynamics Observatory, 2015, 'Thermonuclear Art – The Sun in Ultra-HD' (video), https://www.nasa.gov/mission_pages/sdo/videos/index.html
NASA Solar Dynamics Observatory, 2015, 'Year 5' (video), http://www.nasa.gov/content/goddard/videos-highlight-sdos-fifth-anniversary
Newton, Isaac, 1675–76, 'The Analogy of Nature', letter to Mr Oldenburgh, quoted in Jennings, Humphrey, *Pandaemonium 1660–1886: The Coming of the Machine Age as Seen by Contemporary Observers*, Icon Books (1985, new edition 2012)
Newton, Isaac, 1704, *Opticks: Or, A Treatise of the Reflections, Refractions, Inflexions and Colours of Light*, quoted in Manzotti, Riccardo, 'The Color of Consciousness', *New York Review of Books*, 8 December 2016
Noe, Alva, 2014, 'In Search of a Science of Consciousness', NPR (website), 30 December 2014, http://www.npr.org/blogs/13.7/2014/12/30/373952810/in-search-of-a-science-of-consciousness
Owen, Wilfred, 1918, 'Futility', https://www.poetryfoundation.org/poems-and-poets/poems/detail/57283
Palczewskaa, Grazyna, et al., 2014, 'Human Infrared Vision Is Triggered by Two-Photon Chromophore Isomerization', *Proceedings of the National Academy of Sciences*, vol. 111, no. 50
Perkowitz, Sidney, 1998, *Empire of Light*, Henry Joseph Press
Perkowitz, Sidney, 2011, *Slow Light*, ICP
Pullman, Philip, 2014, 'William Blake and Me', *Guardian*, 28 November 2014
Reed, A. W., 1982, *Aboriginal Myths and Legends*, Reed Books

Ross, Alex, 2013, 'Water Music', *New Yorker*, 8 July 2013
Rovelli, Carlo, 2016, *Reality Is Not What It Seems: The Journey to Quantum Gravity*, Allen Lane
Scharf, Caleb, 2012, *Gravity's Engines*, Allen Lane
Scharf, Caleb, 2014, *The Copernicus Complex*, Allen Lane
Solnit, Rebecca, 2005, *A Field Guide to Getting Lost*, Penguin
Sommer, Andrei P., et al., 2015, 'Light Effect on Water Viscosity: Implication for ATP Biosynthesis', *Scientific Reports* 5, article no. 12029
Strang, Veronica, 2014, 'Wonderful Light: Affect and Transformation in Engagements with Light and Water', paper presented at ASA Decennial Conference, 19–22 June 2014, Edinburgh
Tabakova, Dobrinka, 2014, 'Einstein Considers Light as Waves, and Light as Quanta', sung by Opus Anglicanum, http://www.opus-anglicanum.com
They Might Be Giants, 1993, 'Why Does the Sun Shine? (The Sun is a Mass of Incandescent Gas)', Elektra
They Might Be Giants, 2009, 'Why Does the Sun Really Shine? (The Sun is a Miasma of Incandescent Plasma)', Disney Sound/Idlewild
Thoen, Hanne H., et al, 2014, 'A Different Form of Color Vision in Mantis Shrimp', *Science*, 24 January 2014
Thomas, Edward, 1917, 'Lights Out', https://www.poetryfoundation.org/poems-and-poets/poems/detail/57199
Thoreau, Henry David, 1862, 'Walking', *The Atlantic*, June 1862
Tinsley, Jonathan N., et al, 2016, 'Direct Detection of a Single Photon by Humans', *Nature Communications* 7, article no. 12172
Tuan, Yi-fu, 1968, *The Hydrologic Cycle and the Wisdom of God: A Theme in Geoteleology*, University of Toronto Press
Van Helden, Albert, et al., 1995, 'Galileo's Sunspot Drawings', *The Galileo Project*, Rice University
Wald, George, 1968, 'The Molecular Basis of Visual Excitation', *Nature*, vol. 219, issue 5156
Walser, Robert, 1914, 'Remember This', translated by Tom Whalen, *New York Review of Books*, 1 September 2016
Wang Wei, 'Deer Fence', translated by Burton Watson, in Weinberger, Eliot, 2016, *19 Ways of Looking at Wang Wei (With More Ways)*, New Directions
Wheeler, John Archibald, 1986, 'Hermann Weyl and the Unity of Knowledge', *American Scientist*, vol. 74, no. 4 (July–August 1986)
Whitman, Walt, 1855, 'I Sing the Body Electric' from *Leaves of Grass* in *The Complete Poems*, Wordsworth Poetry Library (1995)
Young, Thomas, 1804, 'Bakerian Lecture: Experiments and Calculations Relative to Physical Optics', *Philosophical Transactions of the Royal Society* 94

2. The Gathering of the Universal Light into Luminous Bodies: Life

Aeschylus, 430 BC, *Prometheus Bound*, Penguin Classics (2001)
Ball, Philip, 2009, *Nature's Patterns: A Tapestry in Three Parts*, Oxford University Press

Ball, Philip, 2017, 'The Computational Foundation of Life: How Life (and Death) Spring from Disorder', Quanta Magazine (website), 26 January 2017, https://www.quantamagazine.org/20170126-information-theory-and-the-foundation-of-life/

Bell, Elizabeth A., et al., 2015, 'Potentially Biogenic Carbon Preserved in a 4.1 Billion-Year-Old Zircon', *Proceedings of the National Academies of Science*, vol. 112, no. 47

Berry, Drew, 2011, 'Animations of Unseeable Biology', talk given at TEDx Sydney, https://www.ted.com/talks/drew_berry_animations_of_unseeable_biology

Bianconi, Eva, et al., 2013, 'An Estimation of the Number of Cells in the Human Body', *Annals of Human Biology*, vol. 40, issue 6

Bradford, Charles M., et al., 2011, 'The Water Vapor Spectrum of APM 08279+5255', *The Astrophysical Journal Letters*, vol. 741, no. 2

Browne, Thomas, 1658, *Hydrotaphia, Or Urne-Buriall*, NYRB Classics (2012)

Chown, Marcus, 2013, *What a Wonderful World*, Faber & Faber

Coleridge, Samuel Taylor, *Biographia Literaria*, quoted by Richard Holmes in 'A Meander through Memory and Forgetting' in Wood, H. H., and Byatt, A. S. (eds.), *Memory: An Anthology* Chatto & Windus (2008)

Cope, Nick, 2011, 'It Rains the Same Old Rain' from *My Socks*, Nick Cope Music

Deamer, David W., 2011, *First Life: Discovering the Connections between Stars, Cells, and How Life Began*, University of California Press

Engel, Gregory S., et al., 2007, 'Evidence for Wavelike Energy Transfer through Quantum Coherence in Photosynthetic Systems', *Nature*, vol. 446, 12 April 2007

European Southern Observatory, 2014, 'Revolutionary ALMA Image Reveals Planetary Genesis' (eso1436 – photograph release), 6 November 2014, http://www.eso.org/public/unitedkingdom/news/eso1436/

Evelyn, John, 13 December 1685, quoted in Jennings, Humphrey, *Pandaemonium 1660–1886: The Coming of the Machine Age as Seen by Contemporary Observers*, Icon Books (1985, new edition 2012)

Futera, Zdenek, et al., 2017, 'Formation and Properties of Water from Quartz and Hydrogen at High Pressure and Temperature', *Earth and Planetary Science Letters*, vol. 461

Gershenfeld, Neil, 2015, 'Digital Reality', Edge (website), 23 January 2015, http://edge.org/conversation/neil_gershenfeld-digital-reality

Goldblatt, C., et al, 2010, 'The Eons of Chaos and Hades,' *Solid Earth*, 1, 1–3,http://ntrs.nasa.gov/archive/nasa/casi.ntrs.nasa.gov/20100036717.pdf

Goodsell, David S., 2009, *The Machinery of Life*, Springer

Hazen, Robert, 2005, *Genesis: The Scientific Quest for Life's Origins*, Joseph Henry Press

Hazen, Robert, 2012, *The Story of the Earth*, Viking

Herzog, Werner, and Oppenheimer, Clive, 2016, *Into the Inferno*, Matter of Fact Media et al.

Hoffman, Peter H., 2012, *Life's Ratchet: How Molecular Machines Extract Order from Chaos*, Basic Books

Hooke, Robert, 1665, *Micrographia*, gutenberg.org

Hunter Waite, J. et al., 2017, 'Cassini finds molecular hydrogen in the Enceladus plume: Evidence for hydrothermal processes', *Science*, 14 April 2017, vol. 356, issue 6334

Jacobson, H., et al., 2014, 'The Chemical Evolution of Phosphorus', *The Astrophysical Journal Letters*, vol. 796, no. 2
James, William, 1890, Chapter IX: 'The Stream of Thought' in *The Principles of Psychology*, Dover Publications (2000)
Jamie, Kathleen, 2012, 'Wind' in *Sightlines*, Sort of Books
Jansson, Tove, 1946, *Comet in Moominland*, Puffin Books (1967)
Jegla, Timothy, et al., 2015, 'Major Diversification of Voltage-gated K+ Channels Occurred in Ancestral Parahoxozoans', *Proceedings of the National Academy of Sciences*, vol. 112, no. 9
Kassin, Susan, et al., 2012, 'The Epoch of Disk Settling: z ~ 1 to Now', *The Astrophysical Journal*, vol. 758, no. 2
Kauffman, Stuart, 1997, *At Home in the Universe: The Search for the Laws of Self-Organization and Complexity*, Oxford University Press
Kircher, Athanasius, 1665, *Mundus Subterraneus*, Bodleian Libraries special collections
Lane, Nick, 2009, *Life Ascending*, Profile
Lane, Nick, 2015, *The Vital Question: Why Is Life The Way It Is?*, Profile
Lorenz, Ralph D., and Zimbelman, James R., 2014, *Dune Worlds: How Windblown Sand Shapes Planetary Landscapes*, Springer Praxis Books
Mallick, Terrence, 2016, *Voyage of Time*, Broad Green Pictures and IMAX Corporation
Medical Research Council, MRC Mitochondrial Biology Unit, 'Molecular Animations of ATP Synthase' (video), http://www.mrc-mbu.cam.ac.uk/projects/2248/molecular-animations-atp-synthase
Ménard, Phia, and Compagnie Non Nova, 2011, *L'après-midi d'un foehn*, http://www.cienonnova.com
Monbiot, George, 2013, *Feral*, Allen Lane
NASA, 2015, 'NASA Ames Reproduces the Building Blocks of Life in Laboratory', 3 March 2015, http://www.nasa.gov/content/nasa-ames-reproduces-the-building-blocks-of-life-in-laboratory
NASA, 2017, 'Experiments Show Titan Lakes May Fizz with Nitrogen', 15 March 2017, https://www.nasa.gov/feature/jpl/experiments-show-titan-lakes-may-fizz-with-nitrogen/
Noji, H., et al., 1997, 'Direct Observation of the Rotation of F1-ATPase', *Nature*, 20 March 1997
Petrov, Anton S., et al., 2014, 'Evolution of the Ribosome at Atomic Resolution', *Proceedings of the National Academies of Science*, vol. 111, no. 28. See also 'The Origin and Evolution of the Ribosome' on YouTube
Plait, Phil, 2009, 'Are Humans Brighter Than the Sun?', *Discover Magazine*, 30 December 2009, http://blogs.discovermagazine.com/badastronomy/2009/12/30/are-humans-brighter-then-the-sun/#.WS2RgLA2yNo
Plantlife, 2015, 'Welsh Woodland as Vulnerable as Tropical Rainforest?' (press release), 28 May 2015
Poe, Edgar Allan, 1841, 'Descent into the Maelstrom' in *Complete Tales and Poems*, Race Point Publishing (2014)
Porchia, Antonio, 1943, *Voices*, translated by W. S. Merwin, Copper Canyon Press (2003)
Ralser, Markus, et al., 2017, 'Sulfate Radicals Enable a Non-Enzymatic Krebs Cycle Precursor', *Nature Ecology & Evolution 1*, 13 March 2017

Ramm, Benjamin, 2016, 'Birobidzhan – The Worst Good Idea Ever?', openDemocracy (website), 16 December 2016, https://www.opendemocracy.net/od-russia/benjamin-ramm/birobidzhan-worst-good-idea-ever

Russell, Michael J., et al., 2013, 'The Inevitable Journey to Being', *Philosophical Transactions of the Royal Society B*, 10 June 2013

Sacks, Oliver, 2008, 'Darwin and the Meaning of Flowers', *New York Review of Books*, 20 November 2008

Schuergers, Nils, et al., 2016, 'Cyanobacteria Use Micro-optics to Sense Light Direction', *eLife*, 9 February 2016

Seymour, Jamie E., 2004, 'Do Box Jellyfish Sleep at Night?', *Medical Journal of Australia*, vol. 181, no. 11/12

Shepherd, Nan, 1977, *The Living Mountain*, Canongate (2011)

Sutherland, John D., et al., 2009, 'Synthesis of Activated Pyrimidine Ribonucleotides in Prebiotically Plausible Conditions', *Nature*, vol. 459, 14 May 2009

Sutherland, John D., et al., 2015, 'Common Origins of RNA, Protein and Lipid Precursors in a Cyanosulfidic Protometabolism,' *Nature Chemistry* 7

Wade, David, 2007, *Li: Dynamic Form in Nature*, Wooden Books

Walker, John E., et al., 1994, 'Structure at 2.8 A Resolution of F1-ATPase from Bovine Heart Mitochondria', *Nature*, vol. 370, 25 August 1994

Welland, Michael, 2009, *Sand: A Journey Through Science and the Imagination*, Oxford University Press

Whitehouse, David, 2015, *Journey to the Centre of the Earth*, Weidenfeld & Nicolson

Witze, Alexandra, 2016, 'Giant Tsunamis Washed Over Ancient Mars', *Nature*, 19 May 2016

XVIVO Scientific Animation for BioVisions at Harvard University, 'The Inner Life of a Cell' (video), http://www.xvivo.net/animation/the-inner-life-of-the-cell/

Yasuda, R. et al., 2001, 'Resolution of Distinct Rotational Substeps by Submillisecond Kinetic Analysis of F1-ATPase', *Nature* 410

Young, Edward D., et al., 2016, 'Oxygen Isotopic Evidence for Vigorous Mixing during the Moon-Forming Giant Impact', *Science*, vol. 351, issue 6272

3. Three Billion Beats: Heart

Adler, Jerry, 2015, 'Why the Leatherback Turtle Has a Skylight in its Head', *Smithsonian Magazine*, January 2015

Anversa, Piero, 2006, 'Life and Death of Cardiac Stem Cells: A Paradigm Shift in Cardiac Biology', *Circulation*, vol. 133, issue 11

Bear, I. J., and Thomas, R. G., 1964, 'Nature of Argillaceous Odour', *Nature*, vol. 201

Berger, John, 2015, 'About Song and Laughter', *Sunday Feature*, BBC Radio 3, May 2015 and *Confabulations*, Penguin (2016)

Bird, Geoffrey et al., 2017 'From heart to mind: Linking interoception, emotion, and theory of mind', *Cortex*, 2 May 2017

Boon, B., 2009, 'Leonardo da Vinci on Atherosclerosis and the Function of the

Sinuses of Valsalva', *Netherlands Heart Journal*, vol. 17, no. 12
British Heart Foundation and Resuscitation Council UK, 2014, 'Consensus Paper on Out of Hospital Cardiac Arrests', https://www.bhf.org.uk/~/media/files/publications/ohca-consensus-paper.pdf
Byrne, David, et al., 1980, 'Once in a Lifetime' from *Remain in Light*, Talking Heads, Sire
Cage, John, 1952, *4'33"*, Edition Peters 6777
Clayton, Martin, and Philo, Ron, 2011, *Leonardo da Vinci: Anatomist*, Royal Collection Publications
Costandi, Mo, 2014, 'The Man Who Grew Eyes', *Guardian*, 26 August 2014
Davies, Jamie A., 2014, *Life Unfolding: How the Human Body Creates Itself*, Oxford University Press
Deakin, Roger, 2015, 'Fragments' in *Granta* 133: What Have We Done, autumn 2015
Eggers, Dave, 2013, *The Circle*, McSweeney's
Emslie, Karen, 2016, 'Riding the Wind', Aeon (website), 21 January 2016, https://aeon.co/essays/how-the-wind-blows-us-on-and-off-life-s-course
Falconer, Tim, 2016, 'Everything We Can't Describe in Music', Hazlitt (website), 15 April 2016, http://hazlitt.net/feature/everything-we-cant-describe-music
Fisher, Roger, 1980, 'Preventing Nuclear War', *The Bulletin of the Atomic Scientists*, vol. 37, no. 3
Flajnik, Martin F., and Kasahara, Masanori, 2010, 'Origin and Evolution of the Adaptive Immune System: Genetic Events and Selective Pressures', *Nature Reviews Genetics* 11
Fludd, Robert, 1618, *De Musica Mundana*, Bodleian Library
Fludd, Robert, 1621, *Utriusque Cosmi*, Bodleian Library
Francis, Gavin, 2015, *Adventures in Human Being*, Profile
Frisén, Jonas et al, 2009, 'Evidence for Cardiomyocyte Renewal in Humans', *Science*, vol. 324, issue 5923
Frohlich, Joel, 2016, 'The Fugue of Life: Why Complexity Matters in Physiology and Neuroscience', Knowing Neurons (website), 3 February 2016, http://knowingneurons.com/2016/02/03/why-complexity-matters-in-physiology-and-neuroscience/
Gessen, Masha, 2014, 'The Dying Russians', *New York Review of Books*, 2 September 2014
Graff, Simon, et al., 2016, 'Long-Term Risk of Atrial Fibrillation after the Death of a Partner', *Open Heart*, vol. 3, issue 1
Haskell, David George, 2017, *The Songs of Trees: Stories from Nature's Great Connectors*, Viking
Hepper, P. G., et al., 1994, 'Development of Fetal Hearing', *Archives of Disease in Childhood*, vol. 71, no. 2
Hickok, Gregory, 2009, 'Eight Problems for the Mirror Neuron Theory of Action Understanding in Monkeys and Humans', *Journal of Cognitive Neuroscience*, vol. 21, no. 7
Hughes, Timothy S., 2003, *Groove and Flow: Six Analytical Essays on the Music of Stevie Wonder* (doctoral dissertation), University of Washington
Hume, David, 1739–40, *Treatise of Human Nature*, Penguin Classics (1985)
Gawande, Atul, 2014, 'Why Do Doctors Fail?', 2014 Reith Lectures,

BBC Radio 4, 29 November 2014, http://downloads.bbc.co.uk/radio4/transcripts/2014_reith_lecture1_boston.pdf
González-Crussi, F., 1989, *The Five Senses*, Harcourt, Brace, Jovanovich
González-Crussi, F., 2009, *Carrying the Heart*, Kaplan
Jiménez, Juan Ramón, 'Full Consciousness', translated by Robert Bly, from *Lorca and Jimenez: Selected Poems*, Beacon Press (1973)
Knausgaard, Karl Ove, 2012, *My Struggle Book 1: A Death in the Family*, Harvill Secker
Konvalinka, Ivana, et al., 2011, 'Synchronized Arousal between Performers and Related Spectators in a Fire-Walking Ritual', *Proceedings of the National Academy of Sciences*, vol. 108, no. 20
Lasky, Robert, et al., 2005, 'The Development of the Auditory System from Conception to Term', *Neoreviews*, vol. 6, issue 3
Long, Matthew C. et al, 2017, 'Finding forced trends in oceanic oxygen', *Global Biogeochemical Cycles*, 29 February 2016
Morgenstern, Christian, *Aphorisms on Nature*, excerpts translated by Douglas Robertson, in *Reliquiae*, vol. 4, Corbell Stone Press
NASA, 2012, 'Perpetual Ocean' (video), http://www.nasa.gov/topics/earth/features/perpetual-ocean.html
Nawroth, Janna C., et al., 2012, 'A Tissue-Engineered Jellyfish with Biomimetic Propulsion', *Nature Biotechnology*, vol. 30, no. 8
Park, Yong Keun, et al., 2009, 'Metabolic Remodelling of the Human Red Blood Cell Membrane', *Proceedings of the National Academy of Sciences*, vol. 107, no. 4
Peto, James (ed.), 2007, *The Heart*, Yale University Press
Poon, Chi-Sang, and Merri, Christopher K., 1997, 'Decrease of Cardiac Chaos in Congestive Heart Failure,' *Nature*, vol. 389
Ramón y Cajal, Santiago, 1897, *Advice for a Young Investigator*, MIT Press (2004)
Reich, Steve, 1972, *Clapping Music*, Universal Edition
Roberts, Alice, 2014, *The Incredible Unlikeliness of Being: Evolution and the Making of Us*, Heron Books
Robinson, Kim Stanley, 1993, *Red Mars*, Spectra/Bantam
Saladin, Kenneth S., 2004, *Anatomy and Physiology: The Unity of Form and Function*, 3rd edition, McGraw Hill
Strogatz, Steve, 2003, *Sync: How Order Emerges from Chaos in the Universe, Nature, and Daily Life*, Hachette Books
Svevo, Italo, 1923, *Zeno's Conscience*, Everyman (2001)
Wells, Francis C., 2014, *The Heart of Leonardo*, Springer
Whitfield, John, 2006, *In the Beat of a Heart: Life, Energy and the Unity of Nature*, Joseph Henry Press
Willis, Allee, and Lind, Jon, 1979, 'Boogie Wonderland', performed by Earth, Wind and Fire, featuring The Emotions, Atlantic Record Company/Columbia Records
Witek, Maria A. G., et al., 2014, 'Syncopation, Body-Movement and Pleasure in Groove Music', *PLOS One*, 16 April 2014
Wright, Thomas, 2012, *Circulation: William Harvey's Revolutionary Idea*, Vintage
Xygalatas, Dimitris, 2014, 'Trial by Fire', Aeon (website), 19 September 2014, https://aeon.co/essays/how-extreme-rituals-forge-intense-social-bonds

4. A Hyperobject in the Head: Brain

Ackerman, Jennifer, 2016, *Genius of Birds*, Penguin

Attanasi, Alessandro, et al., 2014, 'Information Transfer and Behavioural Inertia in Starling Flocks', *Nature Physics* 10

Auden, W. H., 'Heavy Date' in *Collected Poems*, Faber & Faber (2004)

Bach, J. S., 1726–30, Partitas for Keyboard BWV825-830, Igor Levit, Sony (2014)

Ball, Philip, 2008, 'Cellular Memory Hints at the Origins of Intelligence', *Nature*, vol. 451, issue 7177

Bede, *c.*731, Chapter 13: 'A Sparrow's Flight' in Sellar, A. M. (ed.), *Ecclesiastical History of the English People*, George Bell & Sons (1907)

Blake, William, 1794, 'Europe: A Prophecy' in *Complete Poems*, Penguin Classics (1977)

Böhm, Jennifer, et al., 2016, 'The Venus Flytrap *Dionaea muscipula* Counts Prey-Induced Action Potentials to Induce Sodium Uptake', *Current Biology*, vol. 26, issue 3

Bor, Daniel, 2012, *The Ravenous Brain: How the New Science of Consciousness Explains our Insatiable Search for Meaning*, Basic Books

Brenner, S., et al., 1986, 'The Structure of the Nervous System of the Nematode *Caenorhabditis elegans*', *Philosophical Transactions of the Royal Society B*, vol. 314, no. 1165

Browne, Thomas, 1643, *Religio Medici*, New York Review Books (2012)

Calvino, Italo, 1959, *The Baron in the Trees*, Vintage Classics (1992)

Carter, Rita, 2000, *Mapping the Mind*, Phoenix

Chakrabortee, Sohini, 2016, 'Luminidependens (LD) is an Arabidopsis Protein with Prion Behavior', *Proceedings of the National Academy of Sciences*, vol. 113, no. 21

Chalmers, David, 1995, 'Facing Up to the Problem of Consciousness', *Journal of Consciousness Studies*, vol. 2, no. 3

Chung, Kwanghun, and Deisseroth, Karl, 2013, 'CLARITY for Mapping the Nervous System' *Nature Methods*, vol. 10, no. 6

Churchland, Patricia, 1996, 'Does Consciousness Emerge from Quantum Processes?', *Times Higher Education Supplement*, 5 April 1996

Clancy, Kelly, 'The Strangers in Your Brain', *New Yorker*, 17 October 2015

Clark, Andy, 2015, *Surfing Uncertainty*, Oxford University Press

Colapinto, John, 2015, 'Lighting the Brain: Karl Deisseroth and the Optogenetics Breakthrough', *New Yorker*, 18 May 2015

Cole, Andy A., et al., 2016, 'A Network of Three Types of Filaments Organizes Synaptic Vesicles for Storage, Mobilization, and Docking', *The Journal of Neuroscience*, vol. 36, issue 11

Cook, Gareth, 2015, 'Sebastian Seung's Quest to Map the Human Brain', *New York Times*, 11 January 2015

Costandi, Mo, 2006, 'The Discovery of the Neuron', Neurophilosophy (website), 29 August 2006, https://neurophilosophy.wordpress.com/2006/08/29/the-discovery-of-the-neuron/

Costandi, Mo, 2016, 'Nerve Terminal Nanofilaments Control Brain Signalling', *Guardian*, 2 April 2016

Costandi, Mo, 2016, *Neuroplasticity*, MIT Press

Cox, Trevor, 2014, *Sonic Wonderland*, Bodley Head
Crick, Francis, 1979, 'Thinking about the Brain', *Scientific American*, vol. 241, issue 3
Crick, Francis, and Koch, Christof, 2005, 'What is the Function of the Claustrum?', *Philosophical Transactions of the Royal Society*, vol. 360, issue 1458
Darwin, Charles, 1880, *The Power of Movement in Plants*, John Murray
Day, Brian L., and Fitzpatrick, Richard C., 2005, 'The Vestibular System', *Current Biology*, vol. 15, issue 15
Deacon, Terrence, 2011, *Incomplete Nature: How Mind Emerged from Matter*, W. W. Norton & Co.
Dehaene, Stanislas, 2014, *Consciousness and the Brain: Deciphering How the Brain Codes Our Thoughts*, Viking
Doidge, Norman, 2007, *The Brain That Changes Itself*, Penguin
Doidge, Norman, 2012, *The Brain's Way of Healing*, Penguin
Du Sautoy, Marcus, 2015, *What We Cannot Know*, HarperCollins
Eagleman, David, 2015, *Brain*, Pantheon Books
Einstein, Albert, 1950, Letter to Robert S. Marcus, lettersofnote.com
Emerson, Ralph Waldo, 1836, *Nature*, James Munroe and Company; available at http://www.emersoncentral.com/nature.htm
Emery, Nathan J., and Clayton, Nicola S., 2004, 'The Mentality of Crows: Convergent Evolution of Intelligence in Corvids and Apes', *Science*, vol. 306, issue 5703
Emes, Richard D., and Seth, G. N., 2012, 'Evolution of Synapse Complexity and Diversity', *Annual Review of Neuroscience*, vol. 35
Epstein, Robert, 2016, 'The Empty Brain', Aeon (website), 18 May 2016, https://aeon.co/essays/your-brain-does-not-process-information-and-it-is-not-a-computer
Erwin, Jennifer A., et al., 2014, 'Mobile DNA Elements in the Generation of Diversity and Complexity in the Brain', *Nature Reviews Neuroscience* 15
Eskelinen, Holli C., 2016, 'Acoustic Behavior Associated with Cooperative Task Success in Bottlenose Dolphins', *Animal Cognition*, vol. 19, issue 4
Everett, Daniel, 2015, 'A Cultural Context' (response to the question 'What do you think about machines that think?'), Edge (website), http://edge.org/response-detail/26103
Feldman Barrett, Lisa, 2016, 'The Predictive Brain' (response to the question 'What do you consider the most important recent [scientific] news? What makes it important?', Edge (website), https://www.edge.org/annual-question/2016/response/26707
Field Museum, 2013, 'Field Museum Scientists Estimate 16,000 Tree Species in the Amazon' (press release), 17 October 2013, http://www.eurekalert.org/pub_releases/2013-10/fm-fms101413.php
Fisher, Matthew P. A., 2015, 'Quantum Cognition: The Possibility of Processing with Nuclear Spins in the Brain', *Annals of Physics* 362
Foster, Charles, 2016, *Being A Beast*, Profile
Gibson, William T., 2015, 'Behavioral Responses to a Repetitive Visual Threat Stimulus Express a Persistent State of Defensive Arousal in *Drosophila*', *Current Biology*, vol. 25, issue 11
Glickstein, Mitchell, 1988, 'The Discovery of the Visual Cortex', *Scientific American*, 1 September 1988
Goff, Philip, 2017, 'Panpsychism Is Crazy, But It's Also

Most Probably True', Aeon, https://aeon.co/ideas/panpsychism-is-crazy-but-its-also-most-probably-true
Gorman, James, et al., 2014, 'The Mapmakers' (series of articles), *The New York Times*
Graziano, Michael, 2013, *Consciousness and the Social Brain*, Oxford University Press
Graziano, Michael, 2013, 'How the Light Gets Out', Aeon (website), 21 August 2013, https://aeon.co/essays/how-consciousness-works-and-why-we-believe-in-ghosts
Graziano, Michael, 2014, 'Are We Really Conscious?', *New York Times*, 10 October 2014
Groh, Jennifer M., 2014, *Making Space: How the Brain Knows Where Things Are*, Belknap Press of Harvard University Press
Hendy, David, 2013, *Noise: A Human History of Sound and Listening*, Profile
Henrik, Jörntell, 2017, 'The Brain: A Radical Rethink is Needed to Understand It', The Conversation (website), 16 March 2017, https://theconversation.com/the-brain-a-radical-rethink-is-needed-to-understand-it-74460
Herculano-Houzel, Suzana, et al., 2014, 'The Elephant Brain in Numbers', *Frontiers in Neuroanatomy*, 12 June 2014
Hossenfelder, Sabine, 2016, 'The Superfluid Universe', Aeon (website), 1 February 2016, https://aeon.co/essays/is-dark-matter-subatomic-particles-a-superfluid-or-both
Huth, Alexander G., et al., 2016, 'Natural Speech Reveals the Semantic Maps that Tile Human Cerebral Cortex', *Nature*, vol. 532
Jabr, Feris, 2012, 'Know Your Neurons: What is the Ratio of Glia to Neurons in the Brain?', *Scientific American*, 13 June 2012
Jabr, Feris, 2015, 'The Neuron's Secret Partner', *Nautilus*, 13 August 2015
James, William, 1890, Chapter IX: 'The Stream of Thought' in *The Principles of Psychology*, Dover Publications (2000)
Kasthuri et al., 2015, 'Saturated Reconstruction of a Volume of Neocortex', *Cell*, vol. 162, issue 3
Keiper, Caitrin Nicol, 2013, 'Do Elephants Have Souls?' *The New Atlantis*, vol. 38, Winter/Spring 2013
Kessler, Sebastien. C., et al., 2015, 'Bees Prefer Foods Containing Neonicotinoid Pesticides', *Nature* 521
Knausgård, Karl Ove, 2016, 'I Had Never Seen Anything as Beautiful', *Telegraph*, 10 March 2016
Knowing Neurons (website), no date, '52 Brain Facts', http://knowingneurons.com/52-brain-facts/
Koch, Christoph, 2012, *Consciousness: Confessions of a Romantic Reductionist*, MIT Press
Kolb, Bryan, and Gibb, Robbin, 2011, 'Brain Plasticity and Behaviour in the Developing Brain', *Journal of the Canadian Academy of Child and Adolescent Psychiatry*, vol. 20, no. 4
Lee, Hyo-Jun, et al., 2016, 'Stem-Piped Light Activates Phytochrome B to Trigger Light Responses in *Arabidopsis thaliana* Roots', *Science Signaling*, vol. 9, issue 452
Leibniz, Gottfried Wilhelm, 1714, *Monadology*, http://home.datacomm.ch/kerguelen/monadology/

Lichtman, Jeff W., et al., 2008, 'A Technicolor Approach to the Connectome', *Nature Reviews Neuroscience* 9
Lopez, Barry, and Gwartney, Debra (eds.), 2013, *Home Ground: Language for an American Landscape*, Trinity University Press
Lovelock, James, 2014, *A Rough Ride to the Future*, Allen Lane
Majmudar, Amit, 2013, 'The Brains of Animals', *Kenyon Review*, 14 December 2013
Malouforis, Lambros, 2013, *How Things Shape the Mind: A Theory of Material Engagement*, MIT Press
Manzotti, Riccardo, 2016–17, 'On Consciousness' (conversations with Tim Parks), *New York Review of Books*, December 2016 to April 2017; and see also Manzotti's website, http://www.consciousness.it
Marcus, Gary, 2008, *Kluge*, Houghton Mifflin
Marcus, Gary, and Freeman, Jeremy (eds.), 2014, *The Future of the Brain: Essays by the World's Leading Neuroscientists*, Princeton University Press
Marsh, Henry, 2014, *Do No Harm*, Weidenfeld & Nicolson
Medlock, Ben, 2017, 'The Body is the Missing Link for Truly Intelligent Machines', Aeon (website), 14 March 2017, https://aeon.co/ideas/the-body-is-the-missing-link-for-truly-intelligent-machines
Miller, Kenneth D., 2015, 'Will You Ever Be Able to Upload Your Brain?', *New York Times*, 15 October 2015
Moore, Henry, 1937, 'The Sculptor Speaks', *The Listener*, 18 August 1937
Moser, Edvard, et al., 2004, 'Spatial Representation in the Entorhinal Cortex', *Science*, vol. 305, issue 5688
Nicholson, Max, 1951, *Birds and Men*, quoted in Cocker, Mark, and Mabey, Richard, *Birds Britannica*, Chatto & Windus (2005)
O'Keefe, J., et al., 1971, 'The Hippocampus as a Spatial Map. Preliminary Evidence from Unit Activity in the Freely-Moving Rat', *Brain Research*, vol. 34, issue 1
O'Shea, Michael, et al., 2013–14, 'The Human Brain' (series of articles), *New Scientist*, http://www.newscientist.com/special/the-collection-human-brain
Reardon, Sara, 2016, 'Light-Controlled Genes and Neurons Poised for Clinical Trials', *Nature*, 19 May 2016
Reardon, Sara, 2017, 'A Giant Neuron Found Wrapped Around Entire Mouse Brain', *Nature*, 24 February 2017
Russell, Bertrand, 1927, *An Outline of Philosophy*, George, Allen and Unwin
Safina, Carl, 2015, *Beyond Words*, Henry Holt
Scientific American (eds.), 2013, *The Secrets of Consciousness*, Scientific American Inc.
Seth, Anil, 'The Real Problem', Aeon (website), 2 November 2016, https://aeon.co/essays/the-hard-problem-of-consciousness-is-a-distraction-from-the-real-one
Strawson, Galen, 2015, 'Consciousness Myth,' *Times Literary Supplement*, 25 February 2015
Strawson, Galen, 2015, 'Mind and Being: The Primacy of Panpsychism' in Bruntrup, G., and Jaskolla, L. (eds.), *Panpsychism: Philosophical Essays*, Oxford University Press
Szymborska, Wisława, 1996, 'Conversation with a Stone' in *View with a Grain of Sand*, Faber & Faber
Tallis, Raymond, 2008, *The Kingdom of Infinite Space*, Atlantic

TED talks on the brain by Neil Burgess, Suzana Herculano Houzel, Henry Markram, Allan Jones, Daniel Wolpert
ter Steege, Hans, et al., 2013, 'Hyperdominance in the Amazonian Tree Flora', *Science*, vol. 342, issue 6156
Thomas, Lewis, 1974, 'On Probability and Possibility' in *The Lives of a Cell*, Viking
Thompson, Evan, 2014, *Waking, Dreaming, Being: Self and Consciousness in Neuroscience, Meditation, and Philosophy*, Columbia University Press
Thomson, Helen, 'Woman of 24 Found to Have No Cerebellum in Her Brain', *New Scientist*, 10 September 2014
Whitman, Walt, 1855, 'Song of Myself' in *The Complete Poems*, Wordsworth Poetry Library (1995)
Willis, Allee, and Lind, Jon, 1979, 'Boogie Wonderland', Earth, Wind & Fire, featuring The Emotions, The Atlantic Record Company and Columbia Records
Willis, Thomas, 1664, *The Anatomy of the Brain and Nerves*, Classics of Medicine Library (1978)
Willis, Thomas, 1672, *Two Discourses Concerning the Souls of Brutes*, London: Printed for T. Dring, C. Harper, J. Leigh
Zeng, Hongkui, et al., 2014, 'A Connectome of the Mouse Brain', *Nature*, vol. 508, issue 7495
Zimmer, Carl, 2004, *Soul Made Flesh: Thomas Willis, the English Civil War and the Mapping of the Mind*, William Heinemann

5. Edge of the Orison: Self

Addyman, Caspar, 2016, 'Why Playing Peekaboo with Babies is a Very Serious Matter,' Aeon (website), 26 February 2016, https://aeon.co/ideas/why-playing-peekaboo-with-babies-is-a-very-serious-matter
Addyman, Caspar, forthcoming, *The Laughing Baby: The Extraordinary Science behind What Makes Babies Happy and Why*, Unbound
Agland, Phil, 2016, *Between Clouds and Dreams*, River Films
Akerlof, George A., and Shiller, Robert J., 2015, *Phishing for Phools: The Economics of Manipulation and Deception*, Princeton University Press
Alter, Adam, 2017, 'How Technology Gets Us Hooked', *Guardian*, 27 February 2017
Ananthaswamy, Anil, 2014, 'Ecstatic Epilepsy: How Seizures can be Bliss', *New Scientist*, 22 January 2014
Ananthaswamy, Anil, 2014, 'Trippy Tots: How to See the World as a Baby', *New Scientist*, 21 August 2014
Ananthaswamy, Anil, 2016, 'The Wisdom of the Aging Brain', *Nautilus*, 12 May 2016
Anonymous, 2015, 'Ayahuasca: A Personal Encounter with the Miracle Vine', *Network Review*, Spring 2015
Atran, Scott, 2014, 'Jihad's Fatal Attraction', *Guardian*, 4 September 2014
Atran, Scott, 2015, 'Looking for the Roots of Terrorism', *Nature*, 15 June 2015, http://www.nature.com/news/looking-for-the-roots-of-terrorism-1.16732

Aurelius Marcus, *c.* AD 167, *Meditations*, Penguin (2006)
Ball, Philip, et al., 2015, 'Why Music?', BBC Radio 3, 25 September 2015
Baumeister, Roy, et al., 2001, 'Bad is Stronger than Good', *Review of General Psychology*, vol. 5, No. 4
Benjamin, Walter, 1938, 'A Berlin Childhood Around 1900' in *Selected Writings*, vol. 3, Belknap Harvard (2002)
Berridge, Kent C., 'Dissecting Components of Reward: "Liking", "Wanting", and Learning', *Current Opinion in Pharmacology*, vol. 9, issue 1
Berry, Thomas, 1993, 'The Meadow Across the Creek', http://www.thomasberry.org/Essays/MeadowAcrossCreek.html
Bjerstedt, Sven, 2013, 'Review essay of *Strong Experiences with Music*', *International Journal of Education & the Arts*, vol. 14, review 7
Borges, Jorge Luis, 1977, 'Undr' in *The Book of Sand and Shakespeare's Memory*, Penguin Classics (2001)
Browne Thomas, 1690, 'Letter to a Friend' in Killeen, Kevin (ed.), *Thomas Browne: 21st Century Oxford Authors*, Oxford University Press (2014)
Burak, Jacob, 'Outlook: Gloomy', Aeon (website), 4 September 2014, https://aeon.co/essays/humans-are-wired-for-negativity-for-good-or-ill
Carhart-Harris, Robin L., et al., 2016, 'Neural Correlates of the LSD Experience Revealed by Multimodal Neuroimaging', *Proceedings of the National Academy of Sciences*, vol. 113, no. 17
Clare, John, *c.*1826, *The Autobiography* in Williams, Merryn, and Williams, Raymond (eds.), *John Clare: Selected Poetry and Prose*, Methuen (1986)
Claxton, Guy, 2013, 'On Being Touched and Moved: Why Spirituality Is Essentially Embodied' (lecture given at the RSA), 26 November 2013, https://www.thersa.org/discover/publications-and-articles/rsa-blogs/2014/02/science-and-spirituality-effing-the-ineffable/
Costandi, Mo, 2007, 'Diagnosing Dostoyevsky's Epilepsy', Neurophilosophy (website), 16 April 2007, https://neurophilosophy.wordpress.com/2007/04/16/diagnosing-dostoyevskys-epilepsy/
Crawford, Matthew, 2015, *The World Beyond Your Head: How to Flourish in an Age of Distraction*, Viking
Deary, Vincent, 2014, *How We Are*, Penguin
de Tocqueville, Alexis, 1835, *Journeys to England and Ireland*, Transaction Publishers (1987)
Dillard, Annie 1987, *An American Childhood*, Harper and Row
Diski, Jenny, 2016, 'The Island of Lost Words', Berfrois (website), 29 January 2016, http://www.berfrois.com/2016/01/jenny-diski-on-the-enormity-of-that-lost-word/
Dolder, Patrick C., et al., 2016, 'LSD Acutely Impairs Fear Recognition and Enhances Emotional Empathy and Sociality', *Neuropsychopharmacology*, vol. 41
Dostoyevsky, Fyodor, 1880, *The Brothers Karamazov*, Penguin (1993)
Ehrenreich, Barbara, 2007, *Dancing in the Streets: A History of Collective Joy*, Granta Books
Emerson, Ralph Waldo, 1836, *Nature*, James Munroe and Company; available at http://www.emersoncentral.com/nature.htm
Evans, Jules, 2017, *The Art of Losing Control: A Philosopher's Search for Ecstatic Experience*, Canongate

Eyal, Nir, with Hoover, Ryan, 2015, *Hooked: How to Build Habit-Forming Products*, Portfolio
Fleming, Amy, 2015, 'The Science of Craving', *1843 Magazine*, May/June 2015
Freud, Sigmund, 1905, *Jokes and their Relation to the Unconscious*, Vintage Classics (2001)
Gabrielsson, Alf, 2011, *Strong Experiences with Music*, Oxford University Press
Gibbon, Edward, 1776, *The Decline and Fall of the Roman Empire*, Penguin Classics (1996)
Glenny, Misha, 2016, 'The Dark Side of Brazil's "Marvellous City"' (review of *Rio de Janeiro: Extreme City* by Luiz Eduardo Soares), *Guardian*, 19 May 2016
Gopnik, Alison, 2009, *The Philosophical Baby*, Farrar, Straus and Giroux
Gray, John, 2015, *The Soul of the Marionette: A Short Inquiry into Human Freedom*, Penguin
Griffith, Jay, 2014, *Kith*, Hamish Hamilton
Griffiths, Ronald, et al., 2006, 'Psilocybin Can Occasion Mystical-Type Experiences Having Substantial and Sustained Personal Meaning and Spiritual Significance', *Psychopharmacology*, vol. 187, issue 3
Heraclitus, Fragment 45, in Burnet, John (ed.), *Early Greek Philosophy*, A. & C. Black (1930, 4th edition)
Holloway, Richard, 2016, 'Three Score Years and Ten: Looking Back', BBC Radio 4, 18 Jan 2016
Hoffman, Jan, 2016, 'A New Vision for Dreams of the Dying', *New York Times*, 2 February 2016
Hughes, Ted, 2009, *Letters*, Faber & Faber
Hume, David, 1776, 'My Own Life', https://en.wikisource.org/wiki/My_Own_Life
Irigaray, Luce, 1993, *An Ethics of Sexual Difference*, Cornell University Press
Kalanithi, Paul, 2016, *When Breath Becomes Air*, Bodley Head
Kerr, C. W., et al., 2014, 'End-of-Life Dreams and Visions: A Longitudinal Study of Hospice Patients' Experiences', *Journal of Palliative Medicine*, vol. 17, issue 3
Kimmerer, Robin Wall, 2014, *Braiding Sweetgrass*, Milkweed Editions
Kulka Otto Dov, 2013, *Landscapes of the Metropolis of Death*, Allen Lane
Lasdun, James, 2015, 'Houellebecq in the Flesh', *New York Review of Books*, 24 March 2015
Lawrence, D. H., 1915, *The Rainbow*, Penguin Classics (2000)
Lewis, Helen, 2015, 'The Utopia of Isis: Inside Islamic State's Propaganda War', *New Statesman*, 20 November 2015
Lilla, Mark, 2015, 'Slouching Toward Mecca' (review of *Soumission* by Michel Houellebecq), *New York Review of Books*, 2 April 2015
Macfarlane, Robert, 2015, *Landmarks*, Hamish Hamilton
Macfarlane, Robert, 2016, 'Laurence Edwards: An Essay', Caught by the River (website), 18 May 2016, http://www.caughtbytheriver.net/2016/05/18/laurence-edwards-messums-robert-macfarlane-sculpture/
Marshall, Michael, 2009, 'Six Things Science Has Revealed about the Female Orgasm', *New Scientist*, 28 May 2009
Mason, Wyatt, 2014, 'The Revelations of Marilynne Robinson', *New York Times*, 1 October 2014
Masters, William H., and Johnson, Virginia, 1979, *Homosexuality in Perspective*, Little, Brown & Co.
Merleau-Ponty, Maurice, 1945, *Phenomenology of Perception*, Routledge (2012)

Miłosz, Czesław, 1936, 'Encounter' in *New and Collected Poems (1931–2001)*, Ecco

Mireault, Gina, 2017, 'Five-Month-Old Babies Know What's Funny', Aeon (website), 20 January 2017, https://aeon.co/ideas/five-month-old-babies-know-whats-funny

Monbiot, George, 2017, 'Our Greatest Peril? Screening Ourselves Off from Reality', *Guardian*, 28 February 2017

Morell, Virginia, 2017, 'No Place is Safe for Africa's Hunted Forest Elephants', *Science*, February 20, 2017

Murray, Les, 1998, 'A Defence of Poetry' (lecture given at the Poetry International Festival in Rotterdam), http://www.lesmurray.org/defence.htm

Nietzsche, Friedrich, 1872, *The Birth of Tragedy*, Penguin Classics (1993)

Nietzsche, Friedrich, 1885, *Thus Spoke Zarathustra*, Oxford World's Classics (2008)

Packer, George, 2016, 'Exporting Jihad: The Arab Spring Has Given Tunisians the Freedom to Act on Their Unhappiness', *New Yorker*, 26 March 2016

Peleg, R., and Peleg, A., 2000, 'Case Report: Sexual Intercourse as Potential Treatment for Intractable Hiccups', *Canadian Family Physician*, vol. 46

Piff, Paul, et al., 2015, 'Awe, the Small Self, and Prosocial Behavior', *Journal of Personality and Social Psychology*, vol. 108, no. 6

Piff, Paul, and Keltner, Dacher, 2015, 'Why Do We Experience Awe?,' *New York Times*, 22 May 2015

Pinker, Steven, 2011, *The Better Angels of Our Nature: Why Violence Has Declined*, Viking

Pollan, Michael, 2015, 'The Trip Treatment', *New Yorker*, 9 February 2015

Reich, Steve, 1974, *Writings about Music*, Halifax: The Press of the Nova Scotia College of Art and Design

Rilke, Rainer Maria, 1929, *Letters to a Young Poet*, Penguin Classics (2012)

Roach, Mary, 2008, *Bonk: The Curious Coupling of Science and Sex*, W. W. Norton & Co.

Roach, Mary, 2009, 'Ten Things You Didn't Know About Orgasm' (TED talk), February 2009, https://www.ted.com/talks/mary_roach_10_things_you_didn_t_know_about_orgasm

Roberts, Seth, 2014, 'How Economics Shaped Human Nature: A Theory of Human Evolution', in Cai, S., and Beltz, N. (eds.), *Mind and Cognition*, Springer

Robinson, Marilynne, 2015, 'Fear', *New York Review of Books*, 24 September 2015

Rowson, Jonathan, 2014, *Spiritualise: Revitalising Spirituality to Address 21st Century Challenges*, RSA Action and Research Centre

Sacks, Oliver, 2015, *Gratitude*, Picador

Sacks, Oliver, 2015, 'Urge', *New York Review of Books*, 24 September 2015

Saunders, George, 2015, *On Story* (film by Sarah Klein and Tom Mason), Redglass Pictures

Schulz, Bruno, (date unknown), 'Republic of Dreams' in *The Street of Crocodiles and Other Stories*, Penguin (2008)

Schwartz, Barry, 2004, *The Paradox of Choice: Why More Is Less*, HarperCollins; see also 'The Paradox of Choice (TED talk), July 2005, https://www.ted.com/talks/barry_schwartz_on_the_paradox_of_choice

Sheridan, Margaret A., 2012, 'Variation in Neural Development as a Result of Exposure to Institutionalisation Early in Childhood', *Proceedings of the National Academy of Sciences*, vol. 109, no. 32
Sloboda, John, 2005, *Exploring the Musical Mind*, Oxford University Press; and personal communication with the author in 2015
Sloterdijk, Peter, 2013, *You Must Change Your Life*, Polity
Sophocles, 406 BC, *Oedipus at Colonus* in *The Three Theban Plays*, Penguin Classics (1984)
Stevenson, Bryan, 2014, *Just Mercy*, Spiegel & Grau
Tallis, Raymond, 2012, *In Defence of Wonder*, Acumen
Traherne, Thomas, *c.*1637–74, *Centuries of Meditations* (first published 1910)
Unger, Roberto Mangabeira, 2014, *The Religion of the Future*, Harvard University Press; see also openDemocracy, 31 March 2014, https://www.opendemocracy.net/transformation/roberto-mangabeira-unger/religion-of-future
Veloso, Caetano, 1998, *Livro*, Nonesuch
Ward, Colin, 1990, *The Child in the City*, Bedford Square Press
Weisberg, Jacob, 'We are Hopelessly Hooked', *New York Review of Books*, 25 February 2015
Whaley, John, Sloboda, John, and Gabrielsson, Alf, 2008, 'Peak Experiences with Music' in Hallam, Susan, et al. (eds.), *Oxford Handbook of Music Psychology*, Oxford University Press (2008)
White, Jon, 2015, 'Does Music Strike a Chord with Everyone?', *New Scientist*, 15 April 2015
Williams, Rowan, 2014, 'The Physicality of Prayer', *New Statesman*, 8 July 2014
Zweibel, Herman T., et al., 1999, *Our Dumb Century: The Onion Presents 100 Years of Headlines*, Crown Publishing

6. Of Maps and Dreams: World

Adams, Douglas, 1980, *The Restaurant at the End of the Universe*, Pan
Adams, John, 1995, *Harmonielehre*, 'Part III: Meister Eckhardt and Quackie', performed by Simon Rattle and CBSO, Associated Music, G. Schirmer Inc.
al-Koni, Ibrahim, 2014, 'Visiting Death' in Parker, David, *Myth and Landscape*, Kehrer Verlag (2014)
Aristotle, *Eudemian Ethics*, Oxford University Press (2011)
Barras, Colin, 'Mystery human species Homo naledi had tiny but advanced brain', *New Scientist*, 24 April 2017
Barthe, Christine, and Barral, Xavier (eds.), 2015, *The Lost Tribes of Tierra del Fuego*, with photographs by Martin Gusinde, Thames and Hudson
Beck, Ulrich, 2015, *The Metamorphosis of the World: How Climate Change is Transforming Our Concept of the World*, Polity
Benson, Michael, 2014, *Cosmigraphics: Picturing Space Through Time*, Abrams
Billings, Lee, 2013, *Five Billion Years of Solitude*, Current
Borges, Jorge Luis, 1946, 'On Exactitude in Science' in *A Universal History of Iniquity*, Penguin (2004)
Brothers, J. Roger, 2015, 'Evidence for Geomagnetic Imprinting and Magnetic

Navigation in the Natal Homing of Sea Turtles', *Current Biology*, vol. 25, issue 3
Brotton, Jerry, 2012, *A History of the World in Twelve Maps*, Allen Lane
Calvino, Italo, 1974, *Invisible Cities*, Harcourt Brace Jovanovich
Chandra X-Ray Observatory, 2016, 'Tycho's Supernova Remnant' (photographs), chandra.harvard.edu
Cox, Brian, and Cohen, Andrew, 2014, *The Human Universe*, William Collins
Cyranoski, David, 'Discovery of Long-Sought Biological Compass Claimed', *Nature*, 16 November 2015
Dillard, Annie, 1999, *For the Time Being*, Vintage
Doctorow, Cory, 2015, 'Skynet Ascendant', Locus Online (website), 2 July 2015, http://www.locusmag.com/Perspectives/2015/07/cory-doctorow-skynet-ascendant/
Ellis, George, 2014, interview with John Horgan, *Scientific American*, 22 July 2014
Finlay, Alec, 2015, 'The Princess Forest', *Forest Fables* (podcast), *Guardian*, 5 November 2015
Gaffney, Owen, and Steffen, Will, 2017, 'The Anthropocene Equation', *The Anthropocene Review*, 10 February 2017
Ghosh, Amitav, 2016, *The Great Derangement: Climate Change and the Unthinkable*, University of Chicago Press
Gooley, Tristan, 2011, *The Natural Navigator*, Virgin Books
Graham, W. S., 1954, 'As Brilliance Fell' in *Selected Poems*, Faber & Faber (1996)
Grinspoon, David, 2016, *Earth in Human Hands*, Grand Central Publishing
Guzmán, Patricio, 2010, *Nostalgia for the Light*, Icarus Films
Guzmán, Patricio, 2015, *The Pearl Button*, Atacama Productions et al.
Hoare, Philip, 2014, *The Sea Inside*, Fourth Estate
Hoffman, Donald, and Gefter, Amanda, 2016, 'The Evolutionary Argument Against Reality', Quanta Magazine (website), 21 April 2016, https://www.quantamagazine.org/20160421-the-evolutionary-argument-against-reality/
Hoggett, Paul, and Randall, Rosemary, 2016, 'Socially Constructed Silence? Protecting Policymakers from the Unthinkable', openDemocracy (website), 6 June 2016, https://www.opendemocracy.net/transformation/paul-hoggett-rosemary-randall/socially-constructed-silence-protecting-policymakers-fr
Jaubert, J., et al., 2016, 'Early Neanderthal Constructions Deep in Bruniquel Cave in Southwestern France', *Nature*, vol. 534
Jefferies, Richard, no date, *The Old House at Coate and Other Unpublished Essays*, Harvard University Press (1948)
Julian of Norwich, 1395, *Revelations of Divine Love*, Oxford University Press (2015)
Kalman, Maira, and Meyerowitz, Rick, 2001, cover of *The New Yorker*, 10 December 2001
Kant, Immanuel, 1755, *General History of Nature and Theory of the Heavens* in *Kant: Natural Science*, Cambridge University Press (2012)
Keneally, Thomas, 1988, 'Dreamscapes', *New York Times*, 13 November 1988
Kimmerer, Robin Wall, 2015, *Gathering Moss*, Oregon State University Press
Koberlin, Brian, 2013, 'Distant Star', One Universe at a Time (website), 22 December 2013, https://briankoberlein.com/2013/12/22/distant-star/

Kopenawa, Davi, and Albert, Bruce, 2013, *The Falling Sky*, Harvard University Press
Lawrence, D. H., 1928, 'Introduction to *Chariot of the Sun*' ('Chaos in Poetry') in *Phoenix: The Posthumous Papers of D. H. Lawrence*, Penguin (1978)
Leopold, Aldo, 1949, 'The Land Ethic' in *A Sand County Almanac*, Library of America (2013)
Levi, Primo, 'The First Atlas', Collected Poems, Faber & Faber (1988)
Lewis-Williams, David, 2011, *San Rock Art*, Jacana Media
Linklater, Andro, 2014, *Owning the Earth*, Bloomsbury
Lopez, Barry, 1986, *Arctic Dreams*, Charles Scribner & Sons
Lopez, Barry, 2015, 'The Invitation' in *Granta* 133: What Have We Done, autumn 2015
McCarthy, Michael J., 2015, *The Moth Snowstorm: Nature and Joy*, John Murray
Morphy, Howard, 1988, *Aboriginal Art*, Phaeton
Ornes, Stephen, 2016, 'Inside the Lost Cave World of the Amazon's Tepui Mountains', *New Scientist*, 26 April 2016
Parshley, Lois, 2017, 'Faultlines, Black Holes and Glaciers: Mapping Uncharted Territories', *Guardian*, 7 February 2017
Platoni, Kara, 2016 'Stanford's Virtual Reality Lab Turned Me into a Cow, Then Sent Me to the Slaughterhouse', KQED Science (website), 22 April 2016, https://ww2.kqed.org/futureofyou/2016/04/22/stanfords-virtual-reality-lab-turned-me-into-a-cow-then-sent-me-to-the-slaughterhouse/
Ramsey, Frank, 1931, *The Foundations of Mathematics: And Other Logical Essays*, Routledge
Ritz, Thorsten, 2009, 'Magnetic Compass of Birds is Based on a Molecule with Optimal Directional Sensitivity', *Biophysical Journal*, vol. 96, issue 8
Robinson, Kim Stanley, 2015, *Aurora*, Orbit
Robinson, Tim, 1986, *Stones of Aran: Pilgrimage*, Penguin
Schalansky, Judith, 2010, *Atlas of Remote Islands*, Particular Books
Schiffman, Richard, 'Four Countries are Acting as Safe Havens for African Elephants', *New Scientist*, 5 June 2015
Stapleton, Olaf, 1937, *Star Maker*, Methuen
Thompson, Nainoa, no date, 'On Wayfinding', Hawaiian Voyaging Traditions (website), http://pvs.kcc.hawaii.edu/ike/hookele/on_wayfinding.html
Tollefson, Jeff, 2016, 'Antarctic Model Raises Prospect of Unstoppable Ice Collapse', *Nature*, 30 March 2016
Tracy, Gene, 2015, 'Sky Readers', Aeon (website), 23 December 2015, https://aeon.co/essays/what-have-we-lost-now-we-can-no-longer-read-the-sky
Tyler, Dominik, 2015, *Common Ground: A World-Lover's Guide to the British Landscape*, Guardian Books/Faber & Faber
Vanz, Thomas, 2016, *Novae: An Aesthetic and Scientific Representation of a Supernova*, Vimeo
Vitebsky, Pier, 2005, *The Reindeer People: Living with Animals and Spirits in Siberia*, HarperCollins
Wang, J., and Marois, C., 2017, 'Four Planets Orbiting Star HR 8799' (video), NExSS (NASA)/Keck Observatory, https://apod.nasa.gov/apod/ap170201.html
Wernquist, Erik, 2014, *Wanderers*, https://vimeo.com/108650530
Winchester, Simon, 2001, *The Map That Changed the World: William Smith and the*

Birth of Modern Geology, HarperCollins
Wright, Thomas, 1750, *An Original Theory or New Hypothesis of the Universe*, Cambridge University Press (2014)
Wulf, Andrea, 2015, *The Invention of Nature: The Adventures of Alexander von Humboldt*, John Murray

7. Future Wonders: Adventures with Perhapsatron

Allen, Paul, 2011, 'The Singularity Isn't Near', *MIT Technology Review*, 12 October 2011
Andersen, Ross, 2014, 'Exodus: Elon Musk Puts His Case for a Multi-Planet Civilisation', Aeon (website), 30 September 2014, https://aeon.co/essays/elon-musk-puts-his-case-for-a-multi-planet-civilisation
Armstrong, D. J., et al., 2016, 'Variability in the Atmosphere of the Hot Giant Planet HAT-P-7 b', *Nature Astronomy*, vol. 1
Barrat, James, 2015, *Our Final Invention: Artificial Intelligence and the End of the Human Era*, St Martin's Griffin
Benford, James, and Benford, Gregory (eds.), 2013, *Starship Century: Toward the Grandest Horizon*, Microwave Sciences
Berger, John, undated, quoted by Ali Smith in 'John Berger Remembered', *Guardian*, 6 January 2017
Bernal, John Desmond, 1929, *The World, The Flesh and the Devil: An Enquiry into the Future of the Three Enemies of the Rational Soul*, https://www.marxists.org/archive/bernal/works/1920s/soul/
Bhorat, Ziyaad, 2016, 'Lost in Space: Silicon Valley and the Future of Democracy', openDemocracy (website), 13 July 2016, https://www.opendemocracy.net/transformation/ziyaad-bhorat/lost-in-space-silicon-valley-and-future-of-democracy
Bishop, Elizabeth, 'At the Fishhouses' in *The Complete Poems 1927–1979*, Farrar Straus and Giroux (1983)
Boden, Margaret, 2016, *AI: Its Nature and Future*, Oxford University Press
Borges, Jorge Luis, 1946, 'A New Refutation of Time' in *Other Inquisitions, 1937–1952*, University of Texas Press
Boston Dynamics, 2016, 'Atlas: The Next Generation' (video), https://www.youtube.com/watch?v=rVlhMGQgDkY
Bostrom, Nick, 2008, 'Why I Want To Be a Posthuman When I Grow Up' in Gordijn, Bert, and Chadwick, Ruth (eds.), *Medical Enhancement and Posthumanity*, Springer
Bostrom, Nick, 2014, *Superintelligence*, Oxford University Press
BP Statistical Review of World Energy, June 2015
Brodbeck, Luzius, et al., 2015, 'Morphological Evolution of Physical Robots through Model-Free Phenotype Development', *PLOS One*, 19 June 2015
Brooks, Michael, 2014, 'Turing's Oracle: The Computer That Goes Beyond Logic', *New Scientist*, 16 July 2014
Burton, Robert, 1621–39, *The Anatomy of Melancholy*, NYRB Classics (2001)
Butler, Declan, 2016, 'A World Where Everyone Has a Robot: Why 2040 Could Blow Your Mind', *Nature*, vol. 530, issue 7591
Butler, Samuel, 1863, 'Darwin Among the Machines', letter to *The Press*

(Christchurch, New Zealand), http://nzetc.victoria.ac.nz/tm/scholarly/tei-ButFir-t1-g1-t1-g1-t4-body.html
Carr, Nicholas, 2016, *Utopia is Creepy: And Other Provocations*, W. W. Norton & Co.
Carter, Mike, 2016, 'I Walked from Liverpool to London. Brexit was No Surprise', *Guardian*, 27 June 2016
Cave, Stephen, 2015, 'Rise of the Machines: Is There Anything to Fear?', *Financial Times*, 20 March 2015
Ceglowski, Maciej, 2016, 'Superintelligence: The Idea That Eats Smart People' (talk at Webcamp Zagreb), 26 October 2016, http://idlewords.com/talks/superintelligence.htm
Chang, Ha-Joon, 2010, *23 Things They Don't Tell You About Capitalism*, Allen Lane
Clancy, Kelly, 2017, 'A Computer to Rival the Brain', *The New Yorker*, 15 February 2017
Cole, K. C., 2015, 'Why You Didn't See It Coming', *Nautilus*, 15 October 2015
Cotton-Barrat, Owen, et al., 2016, 'Global Catastrophic Risks 2016', Global Challenges Foundation
Curry, Andrew, 2015, 'On Futures', Five Books (website), 19 June 2015, http://fivebooks.com/interview/andrew-curry-on-futures/
Davies, William, 2016, 'The New Neoliberalism', *New Left Review*, vol. 101 (Sep/Oct 2016)
de Mul, Jos, 2011, 'The Technological Sublime', Next Nature (website), 17 July 2011, https://www.nextnature.net/2011/07/the-technological-sublime/
Deutsch, David, 'Philosophy Will Be the Key That Unlocks Artificial Intelligence', *Guardian*, 3 October 2012
Drda, Darrin, 2014, 'The Selective Awareness of Wisdom 2.0', openDemocracy (website), 16 April 2014, https://www.opendemocracy.net/transformation/darrin-drda/selective-awareness-of-wisdom-20
Drexler, Eric K., 2013, *Radical Abundance: How a Revolution in Nanotechnology Will Change Civilization*, Public Affairs
Dyson, Freeman, 2016, 'The Green Universe: A Vision', *The New York Review of Books*, 13 October 2016
Dyson, George, 2013, *Turing's Cathedral: The Origins of the Digital Universe*, Penguin
Edin, Kathryn J., and Shaefer, H. Luke, 2015, *$2.00 a Day: Living on Almost Nothing in America*, Houghton Mifflin Harcourt
Enriquez, Juan and Gullans, Steve, 2016, *Evolving Ourselves: Redesigning the Future of Humanity – One Gene at a Time*, Current
Ford, Martin, 2015, *Rise of the Robots*, Basic Books
Frase, Peter, 2016, *Four Futures: Life After Capitalism*, Verso
Gaiman, Neil, 2015, 'How Stories Last' (seminar given for the Long Now Foundation), 9 June 2015, http://longnow.org/seminars/02015/jun/09/how-stories-last/
Garland, Alex, 2015, *Ex Machina*, Film4/DNA Films
Gillon, Michaël and Amaury Triaud, 2017 'Dwarf Planetary Systems will Transform the Hunt for Alien Life', *Aeon*, 2 May 2017
Gillon, Michaël, et al., 2017, 'Seven Temperate Terrestrial Planets around the Nearby Ultracool Dwarf Star TRAPPIST-1', *Nature*, vol. 542, issue 2762

Goodall, Chris, 2016, *The Switch*, Profile Books
Harari, Yuval Noah, 2015, *Homo Deus: A Brief History of Tomorrow*, Penguin Books
Harford, Tim, 2011, *Adapt*, Abacus
Harford, Tim, 2016, 'Brexit and the Power of Wishful Thinking', timharford.com (website), 19 July 2016, http://timharford.com/2016/07/brexit-and-the-power-of-wishful-thinking/
Helbing, Dirk, et al., 2017, 'Will Democracy Survive Big Data and Artificial Intelligence?', *Scientific American*, 25 February 2017
Herzog, Werner, 2016, *Lo and Behold: Reveries of the Connected World*, Magnolia Pictures
Hopkins, Gerard Manley, 1877, 'Pied Beauty' in *Poems and Prose*, Penguin Classics (2008)
Humphrey, Nicholas, 2015, 'The Colossus is a BFG', Edge (website), http://edge.org/response-detail/26063
Huxley, Aldous, 1959, 'Tomorrow and Tomorrow and Tomorrow' in *Collected Essays*, Harper
IEEE Spectrum, 2015, 'Robots Falling Down at the DARPA Robotics Challenge' (video), 5 June 2015, https://www.youtube.com/watch?v=goTaYhjpOfo
Impey, Chris, 2015, *Beyond: Our Future in Space*, W. W. Norton & Co.
Ings, Simon, 2016, 'Stanisław Lem: The Man with the Future Inside Him', *New Scientist*, 16 November 2016
International Energy Agency, 2016, *World Energy Outlook*
Johnson, Steven, 2014, *How We Got to Now: Six Innovations That Made the Modern World*, Particular Books
Jones, Jonathan, 2016, 'Out of This World: Why the Most Important Art Today is Made in Space', *Guardian*, 12 June 2016
Jones, Richard A. L., 2016, *Against Transhumanism* (self-published e-book), Soft Machines (website), http://www.softmachines.org/wordpress/?p=1772
Kelly, Kevin, 2016, *The Inevitable: Understanding the 12 Technological Forces That Will Shape Our Future* , Viking Press, https://backchannel.com/the-seven-steps-toward-making-humans-obsolete-3a5c4e24a19b - .3xeg1n8jg
Kelly, Kevin, 2016, 'The Next 30 Digital Years' (seminar given for the Long Now Foundation), 14 July 2016, http://longnow.org/seminars/02016/jul/14/next-30-digital-years/
Keynes, John Maynard, 1930, 'Economic Possibilities for Our Grandchildren' in *Essays in Persuasion*, New York: W. W. Norton & Co. (1963)
Keynes, John Maynard, 1937, 'Some Economic Consequences of a Declining Population', *The Eugenics Review*, vol. 29, no. 1
Khatchadourian, Raffi, 2014, 'A Star in a Bottle', *The New Yorker*, 3 March 2014
Kleeman, Jenny, 2017, 'The Race to Build the World's First Sex Robot', *Guardian*, 27 April 2017
KPMG India, 2015, *The Rising Sun: Disruption on the Horizon*, https://assets.kpmg.com/content/dam/kpmg/pdf/2016/01/ENRich2015.pdf
Kruze, Ethan, 2015, *Kepler Orrery IV*, University of Washington
Kurzweil, Ray, 2005, *The Singularity is Near: When Humans Transcend Biology*, Viking
LaFrance, Adrienne, 'When Robots Hallucinate', *The Atlantic*, 3 September 2015

Lanchester, John, 'The Robots Are Coming', *London Review of Books*, 5 March 2015
LeCun, Yann, et al., 2015, 'Deep Learning', *Nature*, vol. 521, no. 7553
Le Guin, Ursula, 2014, Speech upon receipt of the US National Book Foundation's Medal for Distinguished Contribution to American Letters, http://www.nationalbook.org/amerletters_2014_uleguin.html#.WMXzABKLRLU
Lem, Stanisław, 1961, *Solaris*, Faber & Faber (2003)
Lem, Stanisław, 1964, *Summa Technologiae*, Minneapolis: University of Minnesota Press (English edition, 2013)
Liu Cixin, 2008, *Remembrance of Earth's Past*, Head of Zeus (2015)
Louden, Tom, and Wheatley, Peter J., 2015, 'Spatially Resolved Eastward Winds and Rotation of HD 189733b', *The Astrophysical Journal Letters*, vol. 84, no. 2
Louwen, Atse, et al., 2016, 'Re-Assessment of Net Energy Production and Greenhouse Gas Emissions Avoidance after 40 Years of Photovoltaics Development', *Nature Communications* 7
Mailer, Norman, 1970, *A Fire on the Moon*, Penguin Classics (2014)
Marcus, Gary, 2013, 'Why We Should Think About the Threat of Artificial Intelligence', *The New Yorker*, 24 October 2013
Margolin, Madison, 2016, 'War Algorithms Will Save Lives Unless They Kill Us All', Inverse (website), 16 December 2016, https://www.inverse.com/article/25140-war-algorithms
Martin, Ursula, 2015, 'Thinking Saltmarshes', Edge (website), https://www.edge.org/response-detail/26120
Marx, Karl, 1845, *The German Ideology*, Prometheus Books (1998)
Mason, Michael, et al., 2015, 'The Potential of CAM Crops as a Globally Significant Bioenergy Resource', *Energy and Environmental Science*, vol. 8
McKay, Chris, 2016, 'Make Mars Great Again', *Nautilus*, 15 December 2016
Mendelson, Edward, 2016, 'In the Depths of the Digital Age', *New York Review of Books*, 23 June 2016
Mordvintsev, Alexander, et al., 2015, 'Inceptionism: Going Deeper into Neural Networks' (Google research blog), 17 June 2015, https://research.googleblog.com/2015/06/inceptionism-going-deeper-into-neural.html
Morris, William, 1890, *News from Nowhere*, Oxford World's Classics (2009)
Moskvitch, Katia, 2015, 'Artificial Photosynthesis: Mimicking Nature's Most Powerful Trick', *Engineering and Technology Magazine*, 9 November 2015
Mukherjee, Siddhartha, 2016, *The Gene: An Intimate History*, Scribner
New Scientist magazine, 'The Rapid Rise of Neural Networks and Why They'll Rule Our World', 8 July 2015
Nielsen, Michael, 2016, 'Is AlphaGo Really Such a Big Deal?', Quanta Magazine (website), 29 March 2016, https://www.quantamagazine.org/20160329-why-alphago-is-really-such-a-big-deal/
Nocera, Daniel, et al., 2016, 'Water Splitting–Biosynthetic System with CO2 Reduction Efficiencies Exceeding Photosynthesis', *Science*, vol. 352, issue 6290
Nordhaus, William D., 1994, 'Do Real Output and Real Wage Measures Capture Reality? The History of Lighting Suggests Not' in Bresnahan, Timothy F., and Gordon, Robert J. (eds.), *The Economics of New Goods*, University of Chicago Press (1996)

Nyquist, Eric, 2016, 'Running is Always Blind: How Your Brain Keeps You from Falling On Your Face', Nautilus (website), 7 July 2016, http://nautil.us/issue/38/noise/running-is-always-blind
Office of the Secretary of Defence Public Affairs, 2016, 'Perdix Swarm Demo Oct 2016' (video), https://www.dvidshub.net/video/504622/perdix-swarm-demo-oct-2016
Oreskes, Naomi, and Conway, Erik M., 2014, *The Collapse of Western Civilization: A View from the Future*, Columbia University Press
Osnos, Evan, 2017, 'Doomsday Prep for the Super-Rich', *The New Yorker*, 30 January 2017
Pamlin, Dennis, and Armstrong, Stuart, 2015, *12 Risks That Threaten Civilisation*, Oxford Martin School
Parfit, Derek, 2017, *On What Matters*, vol. 3, Oxford University Press
Pecchi, Lorenzo, and Piga, Gustavo, 2008, *Revisiting Keynes*, MIT Press
Radiohead, 2016, 'Daydreaming' (video, directed by Paul Thomas Anderson), https://www.youtube.com/watch?v=TTAU7lLDZYU
Raworth, Kate, 2017, *Doughnut Economics: Seven Ways to Think Like a 21st-Century Economist*, Random House Business
Regis, Ed, 2015, 'Let's Not Move to Mars', *New York Times*, 21 September 2015
Reid, Nicholas, et al., 2014, 'Indigenous Australian Stories and Sea-Level Change' in Heinrich, Patrick, and Ostler, Nicholas (eds.), *Indigenous Languages and their Value to the Community* (Proceedings of the 18th Foundation for Endangered Languages Conference, Okinawa, Japan)
REN21, 2016, *Renewables 2016 Global Status Report*, http://www.ren21.net/status-of-renewables/global-status-report/
Romesberg, Floyd E., et al., 2016, 'A Semisynthetic Organism Engineered for the Stable Expansion of the Genetic Alphabet', *Proceedings of the National Academy of Sceinces*, vol. 114, no. 6
Russell, Andrew, and Vinsel, Lee, 2017, 'Whitey on Mars', Aeon (website), 1 February 2017, https://aeon.co/essays/is-a-mission-to-mars-morally-defensible-given-todays-real-needs
Shanahan, Murray, 2015, *The Technological Singularity*, MIT Press
Shanahan, Murray, 2016, 'Conscious Exotica', Aeon (website), 19 October 2016, https://aeon.co/essays/beyond-humans-what-other-kinds-of-minds-might-be-out-there
Simonite, Tom, 'Google's Quantum Dream Machine', *MIT Technology Review*, 18 December 2015
Smith, Zadie, 2016, 'On Optimism and Despair', *New York Review of Books*, 22 December 2016
Sophocles, 441 BC, *Antigone* in *The Three Theban Plays*, Penguin Classics (1994)
Sparrow, Robert, 2016, 'The Argument Against Terraforming Mars', Nautilus (website), http://cosmos.nautil.us/short/85/the-argument-against-terraforming-mars
Streeck, Wolfgang, 2014, 'How Will Capitalism End?', *New Left Review*, vol. 81 (May–June 2014)
Sussams, Luke, et al., 2017, *Expect the Unexpected: The Disruptive Power of Low-Carbon Technologies*, Carbon Tracker/Grantham Institute, Imperial College London, http://www.carbontracker.org/report/expect-the-unexpected-disruptive-power-low-carbon-technology-solar-electric-vehicles-grantham-imperial/

Upton, John, 2015, 'Ancient Sea Rise Tale Told Accurately for 10,000 Years', *Scientific American*, 26 January 2015
Urban, Tim, 2015, 'How and Why Spacex will Colonise Mars', Wait But Why (website), 16 August 2015, http://waitbutwhy.com/2015/08/how-and-why-spacex-will-colonize-mars.html
Various contributors, 2015, responses to the 2015 Edge question 'What Do You Think about Machines That Think?', Edge (website), https://www.edge.org/annual-question/what-do-you-think-about-machines-that-think
Webster, C. R., et al., 2015, 'Mars Methane Detection and Variability at Gale Crater' *Science*, vol. 347, issue 6220
Wenger, Albert, 2016, *World After Capital*, http://worldaftercapital.org
Wheeler, Michael, 2015, 'Martin Heidegger' in Zalta, Edward N. (ed.), *The Stanford Encyclopedia of Philosophy*, https://plato.stanford.edu/
Winfield, Alan, and Vanderelst, Dieter, 2016, 'The Dark Side of Ethical Robots', arXiv:1606.02583v1
Wood, David, et al., 2014, *Anticipating 2025: A Guide to the Radical Changes That May Lie Ahead*, London Futurists
World Nuclear Association, 2016, 'Plans for New Reactors Worldwide', http://www.world-nuclear.org/info/current-and-future-generation/plans-for-new-reactors-worldwide/
Yong, Ed, 2015, 'What Can You Actually Do with Your Fancy Gene-Editing Technology? Wading through the Hype about CRISPR', *The Atlantic*, 2 December 2015
Yunipingu, Galarrwuy, 2016, 'Rom Watangu: An Indigenous Leader Reflects on a Lifetime Following the Law of the Land', *The Monthly*, July 2016
Zhang, Xiaochun, and Caldeira, Ken, 2015, 'Time Scales and Ratios of Climate Forcing Due to Thermal versus Carbon Dioxide Emissions from Fuels', *Geophysical Research Letters*, vol. 42, issue 11

Afterword: The Wonderer and his Shadow

Alexander, Stephon, 2016, *The Jazz of Physics: The Secret Link Between Music and the Structure of the Universe*, Basic Civitas Books
Augustine of Hippo, St, *c.* AD 400, *Confessions*, Penguin Classics (2002)
Benjamin, Walter, 1940, 'On the Concept of History' in *Selected Writings*, vol. 4, Belknap Harvard (2003)
Brent Tully, R., et al., 2014, 'The Laniakea Supercluster Galaxies', *Nature*, vol. 513, issue 7516; see also *Nature* video 'Laniakea: Our Home Supercluster', 3 September 2014, https://www.youtube.com/watch?v=rENyyRwxpHo
Browne, Thomas, 1658, *Hydrotaphia, or Urne Buriall*, New York Review Books (2012)
Cooper, Alan, et al., 2017, 'Aboriginal Mitogenomes Reveal 50,000 Years of Regionalism in Australia', *Nature*, 8 March 2017
González-Jiménez, M., et al., 2016, 'Observation of Coherent Delocalized Phonon-Like Modes in DNA under Physiological Conditions', *Nature Communications*, vol. 7, article 11799
Hopkins, Gerard Manley, 1879, 'Binsey Poplars' in *Poems and Prose*, Penguin Classics (2008)

Jaspers, Karl, 1966/7, radio talk quoted in Bakewell, Sarah, *At the Existentialist Cafe*, Chatto & Windus (2016)
Jefferies, Richard, 1883, *The Story of My Heart*, http://www.gutenberg.org/ebooks/2317
Johnson, Ronald, 1967, 'What the Leaf Told Me' in *The Book of the Green Man*, Uniform Books (2015)
Krause, Bernie, 2012, *The Great Animal Orchestra*, Little, Brown
Larkin, Philip, 1962, 'Here' in *The Whitsun Weddings*, Faber & Faber
Leopold, Aldo, *c.*1940–8, *Round River: From the Journals of Aldo Leopold*, Oxford University Press (1993)
Lucretius, 56 BC, *On the Nature of Things*, Penguin Classics (2007)
Macfarlane, Robert, 2012, *The Old Ways: A Journey on Foot*, Hamish Hamilton
MacNeice, Louis, 'Entirely' in *Selected Poems*, Faber & Faber (1964)
Messiaen, Olivier, 1974, 'Les ressuscités et le chant de l'étoile Aldebaran', from *Des canyons aux étoiles . . .*, Roger Muraro, Orchestre Philharmonique de Radio France, Myung-Whun Chung, Deutsche Gramaphon (2003)
Mishra, Pankaj, 2017, *The Age of Anger: A History of the Present*, Farrar, Straus and Giroux
Nietzsche, Friedrich, 1880, 'The Wanderer and His Shadow', Part 3 of *Human All Too Human*, Penguin Classics (1994)
Nussbaum, Martha, 2013, *Political Emotions*, Harvard University Press
Nussbaum, Martha, 2016, *Anger and Forgiveness: Resentment, Generosity, Justice*, Oxford University Press
Purdy, Jedediah, 2015, *After Nature*, Harvard University Press
Ruff, Willie, and Rogers, John, no date, *The Harmony of the Word: A Realisation for the Ear of Johannes Kepler's Astronomical Data from Harmonices Mundi 1619*, willlieruff.com
Saint-Exupéry, Antoine de, 1939, *Wind, Sand and Stars* (originally published in France as *Terre des hommes*), Penguin Classics (2000)
Schulz, Kathryn, 2017, 'When Things Go Missing', *The New Yorker*, 13 and 20 February 2017
Shriver, Lionel, 'Lionel Shriver Reads T. C. Boyle' (podcast, a reading and discussion of *Chicxulub* by T. C. Boyle), *The New Yorker* Fiction Podcast, 31 August 2015
Solnit, Rebecca, 2014, 'Coyote', *The New Yorker*, 22 December 2014
Solnit, Rebecca, 2016, *Hope in the Dark: Untold Histories, Wild Possibilities*, Canongate
Waits, Tom, 2002, 'Misery is the River of the World', *Blood Money*, ANTI-
Wheeler, John, no date, 'Quantum Ideas, Quantum Foam' (video), Web of Stories (website), https://www.webofstories.com/play/john.wheeler/77
Wittgenstein, Ludwig, 1958, *Philosophical Investigations*, Basil Blackwell

Thanks

Thanks to my family. Thanks to Phil Bloomer and George Monbiot who, over a pint, encouraged me to go for it. Thanks to my agent James Macdonald Lockhart and to my editor Laura Barber. Thanks to all at Granta and Chicago University Press who worked to make this book happen.

Thanks to Lucy Conway, Eddy Scott and The Bothy Project. Thanks to the staff at the Manuscripts and Rare Books Department of the Bodleian Library. Thanks to staff at Christchurch College Library for letting me see Richard Burton's own books.

Thanks to the Society of Authors for a grant for work in progress. It came at an especially helpful time. Thanks to all those who supported *A New Map of Wonders* in ways big and small, including Meg Berlin, Robert Butler, Ian Christie, Sheila Dillon, Chris Goodall, David Hough, Roman Krznaric, Sarah Laird, Pedro Moura Costa, Kate Raworth, Paul Stiga and Marina Warner.

Thanks to, *inter alia*, Peter Abrahams, Rebecca Abrams, Caspar Addyman, Phil Agland, Mark Anthony, Sonia Antoranz Contera, Anthony Barnett, Richard Berry, Matthew Bevis, James Bradley, Mark Brickman, Harvey Brown, Jorge Cabadas, Thomas Cabot, Susan Canney, Jo Cartmell, Barbara Casadei, Melanie Challenger, Nic Compton, Nick Cope, Tan Copsey, Jon Corbett, Julian Cottee, Miriam Darling, Sally Davies, Ken Dixon, Athene Donald, Danny Dorling, Steve Drummond, Jules Evans, Martyn Evans, Nick Forrester, Charles Foster, Gavin Francis, Darwin Franks, Marshall Ganz, Sam Geall, Ed Gillespie, Tom Goreau, Loren Griffith, Sam Guglani, Jon Halperin, Thomas Harding, Tim Harford, Roger Harrabin, Dennis Harrison, Ben Hennig, Judith Herrin, Steve Hicks, Laurence Hill, Isabel Hilton, Roland Hodson, Richard Holloway, Sara Holloway, Finn Jackson, Paul Kingsnorth, Justine Kolata, Charlie Kronick, Sally Lane, Antonia Layard, Rosemary Lee, John Letts, Mark Lynas, Richard Mabey, Robert Macfarlane, Alastair McIntosh, Richard Madden, Lambros Malafouris, Alberto Manguel, Pippa Marland, Harry Marshall, Maury Martin, Susan Miller, Fran Monks, Hugh Osborn, Maria Padget, John Parham, Stuart Paterson, Jan Pedersen, Ben Pope, Max Porter, Matt Prescott, Philip Pullman, David Pyle, Karthik Ramanna Adam Ramsay, Andrew Ray, Philip Read and Manchoir, Adam Ritchie,

Callum Roberts, Nick Robins, Oliver Robinson, Yazaan Romahi, Mike Rossi, Jonathan Rowson, Noemi Roy, Caleb Scharf, Louis Schwizgebel, Michael Segal, Natalie Shaw, John Shea, Iain Sinclair, Joe Smith, Amia Srinivasan, Veronica Strang, John Sutherland, Balasz Szendroi, Hanne Thoen, Sam Thompson, Alison Tickell, Oliver Tickell, Peter Trawny, Maya Tudor, Dominick Tyler, Renata Tyszczuk, Fran van Dijk, Zool Verjee, Tilman Vogt, John Walker, Martin Ward, Hugh Warwick, Christopher Whalen, Ben Wilmore, Alan Winfield, Chris Wood, Andrea Wulf and Xiao Xiao Zhu.

Aún aprendo

Picture Credits

The author and the publisher have made every effort to trace copyright holders. Please contact the publisher if you are aware of any omissions.

viii Henry Holiday, illustration in *The Hunting of the Snark* by Lewis Carroll. In the public domain via Wikimedia Commons.

6 Cueva de las Manos, Río Pinturas. UNESCO Creative Commons Attribution-ShareAlike 3.0 IGO http://whc.unesco.org/en/list/936.

17 Albrecht Dürer, *Melencolia I*. In the public domain via Wikimedia Commons.

31 Olaf Roemer, speed of light diagram. A diagram of Jupiter (B) eclipsing its moon Io (DC) as viewed from different points in earth's orbit around the sun. Olaf (Ole) Roemer, 'Demonstration tovchant le mouvement de la lumiere trouvé par M. Römer de l'Academie Royale des Sciences', December 7, 1676. In the public domain via Wikimedia Commons.

33 Philip Ronan and Gringer, visible light as proportion of electromagnetic spectrum. Wikimedia Commons. Copyright © Philip Ronan, Gringer/ Wikimedia Commons 2013. Used under license: CC-BY-SA-3.0.

35 Roland Deschain, rhodopsin. Copyright © 2006. In the public domain via Wikimedia Commons.

37 Ben Conrad, front view of Ben Conrad wearing pinhole suit and helmet with 135 cameras, 1994. Courtesy Palace of the Governors Photo Archives (NMHM/ DCA), HP.2012.15.1398.

44 Marin Cureau de la Chambre, rainbow. From *The Rainbow: From Myth to Mathematics* by Carl B. Boyer, Princeton University Press, 1987.

44 Rene Descartes, rainbow. In the public domain via Wikimedia Commons.

45 Thomas Young, double-slit experiment. In the public domain via Wikimedia Commons. Copyright © 2016.

47 Isaac Newton, Newton colour circle. In the public domain via Wikimedia Commons. Copyright © 2016.

50 Leonardo da Vinci, Sketch of Earthshine. Illustrations including sketch of earthshine from Leonardo da Vinci's *Codex Leicester*.

51 Galileo, sunspots. Images provided by The Galileo Project and taken from Professor Owen Gingerich's copy of the first edition of *Istoria e Dimostrazioni*.

53 Athanasius Kircher, The Sun. The Bodleian Libraries, The University of Oxford, Douce K 148, illustration between pp. 64 – 65 (Tomus I.180).

61 Drawn by the publisher to author's specifications.

65 Black Hole Outflow from Centaurus A. Centaurus A. Copyright ESO/WFI (Optical); MPIfR/ESO/APEX/A.Weiss et al. (Submillimetre); NASA/CXC/ CfA/R.Kraft et al. (X-ray).

67 Jean-Pierre Luminet, Black Hole (1978) Reproduced by kind permission of Jean-Pierre Luminet.

76 Redrawn by the publisher based on this original: NASA (Jennifer Johnson), The Origin of the Solar System Elements, Copyright © ESA/NASA/ AASNova. Graphic created by Jennifer Johnson.
77 HL Tauri, HL Tauri. Copyright ALMA (ESO/NAOJ/NRAO).
78 Theia. Copyright © NASA/JPL-Caltech.
85 Athanasius Kircher, Earth (fire). The Bodleian Libraries, The University of Oxford, Douce K 148, illustration between pp. 114–115 (Tomus I v.15).
86 Athanasius Kircher, Earth (water). The Bodleian Libraries, The University of Oxford, Douce K 148, illustration between pp.180–181 (Tomus I 64).
90 Major pathways in a cell. Copyright © 2017 F. Hoffman-La Roche Ltd.
92 Wilson Bentley, Snowflakes. Photos: Wilson Bentley Digital Archives of the Jericho Historical Society/snowflakebentley.com.
93 NASA, Barchan dunes on Mars Dunes. Copyright NASA/JPL/University of Arizona. Reprinted with their kind permission.
94 Soundwaves on Chladni plate. https://physics.stackexchange.com/ questions/90021/theory-behind-patterns-formed-on-chladni-plates.
96 Hand and arm graphic for deep time. Redrawn by the publisher with author's alterations, based on based on http://nautil.us/issue/17/big-bangs/ the-greatest-animal-war.
99 Graphics of RNA and DNA. Sponk / Wikimedia Commons / CC BY-SA 3.0.
105 Bacterial ribosome. Copyright © David S. Goodsell and the RCSB Protein Data Bank. Reproduced by kind permission of David Goodsell.
108 ATP synthase. Copyright © David S. Goodsell and the RCSB Protein Data Bank. Reproduced by kind permission of David Goodsell.
117 Leonardo da Vinci, RCIN 919073 Verso: Studies of the coronary vessels and valves of the heart, (on spreadsheet as Recto: the Cranium sectioned), Royal Collection Trust / © HM Queen Elizabeth II 2017.
120 Simplified diagram of the heart. Wikimedia Commons. CC BY-SA 3.0. Copyright © 2000, 2001, 2002, 2007, 2008 Free Software Foundation, Inc. Permission granted under the terms of the GNU Free Documentation License, Version 1.3, or any later version published by the Free Software Foundation.
122 Laurence Jackson, rainbow 'bird's nest' MRI of heart fibres. Copyright © Laurence Jackson.
127 Richard Wheeler, haemoglobin molecule. In the public domain via Wikimedia Commons.
130 Henry Vandyke Carter, embryological development of the heart. From *Gray's Anatomy.*
136 Robert Fludd, The Divine Monochord. SLUB Dresden / Deutsche Fotothek
139 Old maps of ocean circulation. Reproduced from The Department of Rare Books and Special Collections, Princeton.
140 NASA Perpetual. Copyright © NASA/SVS (Scientific Visualisation Studio)
145 Starling murmuration. http://www.murmuration.community.
153 Santiago Ramon y Caja, Dendrite and the Forest of Cells. Three drawings by Santiago Ramon y Cajal, taken from the book 'Comparative study of the sensory areas of the human cortex', pages 314, 361, and 363. Left: Nissl-stained visual cortex of a human adult. Middle: Nissl-stained motor cortex of a human adult. Right: Golgi-stained cortex of a 1 1/2 month old infant.
156 Lobes of the cerebral cortex. Illustrations from the University of Queensland Brain Institute website.

157 Midbrain. Illustration from Anatomy & Physiology, Connexions website. © Wikimedia Commons/OpenStax College.
158 Henry Vandyke Carter, Hindbrain. From *Gray's Anatomy*.
159 Cortical Homunculus. https://i.pinimg.com/736x/e3/a5/98/e3a598678b4b8b6d0f4418e865fdb0d8--homunculus-brain-emergency-medicine.jpg.
161 Memory palace. Illustration from Giulio Camillo's *L'idea del Theatro*.
171 Leonardo da Vinci, RCIN 919058 Recto: The cranium sectioned. Royal Collection Trust / © HM Queen Elizabeth II 2017
172 Thomas Willis, Anatomy of the Brain, 1664. Provided by the Museum of the History of Science, Oxford.
173 Thomas Willis, Map of Nerves. From Thomas Willis. Cerebri Anatome: Cui Accessit Nervorum Descriptio et Usus. London: Typis T. Roycroft, Impensis J. Martyn & J. Allestry, 1664. By permission of the Lilly Library, Indiana University Bloomington.
175 Brainwaves. Redrawn by the publisher from http://www.makeuseof.com/tag/tapping-brainwaves-will-brains-soon-hackable/.
176 Santiago Ramon y Caja, Microscopic Brain Structure, 1904. Science Source/Science Photo Library.
197 Abraham Ortelius, Islandia. In the public domain via Wikimedia Commons.
206 Art Wolfe, Japanese cranes dancing. Copyright © Art Wolfe/artwolfe.com.
210 Robin Carhart-Harris, psilocybin increased brain activity. Image taken from Homological scaffolds of brain functional networks, G. Petri, P. Expert, F. Turkheimer, R. Carhart-Harris, D. Nutt, P. J. Hellyer, F. Vaccarino. Published in Proceedings of the Royal Society Interface., 6 December 2014, Volume 11, issue 101.
224 Francisco Goya, *Aun aprendo (I Am Still Learning)*. Copyright © Museo Nacional del Prado
231 Roger Penrose, Three Worlds, by Sir Roger Penrose.
231 Roger Penrose, Penrose's Triangle by Sir Roger Penrose. Wikimedia Commons. Copyright © 2007. Shared under the CC BY-SA 2.5 licence.
232 Details from Hereford Mappa Mundi (*c.*1300), Hereford Cathedral.
233 Hereford Mappa Mundi (*c.*1300), Hereford Cathedral.
234 [*inverted*] al-Idrisi's Tabula Rogeriana (1154). Tabula Rogeriana, 1154. In the public domain via Wikimedia Commons. Copyright © 2013.
235 Charles Minard, 1869 chart showing the number of men in Napoleon's 1812 Russian campaign army, their movements, as well as the temperature they encountered on the return path. In the public domain via Wikimedia Commons. Copyright © 2008.
236 Benjamin Hennig, A Modern Mappa Mundi. Copyright © Benjamin Hennig, Worldmapper Project (image published on www.viewsoftheworld.net).
241 Thomas Dawson, line drawing of the Linton Panel. Copyright © Thomas Dawson, 1987. https://archaeology-travel.com/friday-find/linton-panel/
244 Hōkūle'a star compass, Polynesian Voyaging Society (PVS).
246 Bibi Saint-Pol, Anaximander's World Map. Image by Bibi Saint-Pol, based on an image found in *An Introduction to Early Greek Philosophy* by John Mansley Robinson, Houghton and Mifflin, 1968. In the public domain via Wikimedia Commons. Copyright © 2006.
249 EA1978.2573 The angel Ruh holding the celestial spheres. Page from a

manuscript of the Aja'ib al-Makhluqat (Wonders of Creation) by Zakariya ibn Muhammad al-Qazwini. Image © Ashmolean Museum, University of Oxford

252 Thomas Wright, Galaxy Models. The Bodleian Libraries, The University of Oxford, Rigaud d. 170, Plate XXVII

257 A bridge between two galaxies. By NASA, ESA, and the Hubble Heritage Team (STScI/AURA). In the public domain via Wikimedia Commons.

258 Sequence from Roundhay Garden Scene. National Science and Media Museum, Bradford.

261 David Parker, *Oracle 1*. Reproduced with the kind permission of David Parker.

265 Alexander von Humboldt, Physical Geography. Humboldt's Distribution of Plants in Equinoctial America, According to Elevation Above the Level of the Sea. Drawn by George Aikman. Copyright © Geographic/Wikimedia Commons.

266 Galileo, Drawings of the Moon. In the public domain via Wikimedia Commons. Copyright © 2016.

267 Martin Gusinde, Telil, Shenu. Martin Gusinde, Télen, northern Shoort, Sénu, western Shoort. Copyright © 1923 Martin Gusinde / Anthropos Institut / Éditions Xavier Barral

269 Redrawn by the publisher from a detail of the Hereford Mappa Mundi. Hereford Cathedral.

279 Comprehensive Space plan. Copyright © Rockwell International, Space Transportations Systems Division.

286 Wendelstein 7-X Stellarator. Image of Wendelstein 7 X Stellerator with the kind permission of Max Planck Institute for Plasma Physics.

288 Reuben Wu, solar panels. Crescent Dunes Solar Energy Project © Reuben Wu 2017.

291 Darius Siwek, artifical leaf. Copyright © Californian Institute of Technology. Artwork by Darius Siwek.

293 Extreme Deep Field (2012), Hubble Goes to the eXtreme to Assemble Farthest-Ever View of the Universe. Copyright © NASA; ESA; G. Illingworth, D. Magee, and P. Oesch, University of California, Santa Cruz; R. Bouwens, Leiden University; and the HUDF09 Team

308 Murray Shanahan, Spectrum of All Possible Minds. Copyright © Shanahan, M. Conscious Exotica. *Aeon* magazine, October 2016.

312 Graph of US manufacturing real output and employment 198 –2015. Redrawn by the publisher. Data taken from The Brookings Institute.

321 Pole on Maplin Sands. Photograph by author.

325 Paul Klee, *In Engelshut* (*In Angel's Care*). Provided to Wikimedia Commons by the Solomon R. Guggenheim Museum as part of a cooperation project. Solomon R. Guggenheim Museum, New York Estate of Karl Nierendorf, by purchase.

Text Credits

The author and the publisher have made every effort to trace copyright holders. Please contact the publisher if you are aware of any omissions.

Lines from 'Heavy Thought' in *Collected Poems* by W.H. Auden. Reproduced with the kind permission of Penguin Random House.

Lines from 'The Meadow Across the Creek' in *The Great Work: Our Way Into the Future* by Thomas Berry. Copyright © 1999 by Thomas Berry.

Lines from 'The Moose' from *Complete Poems* by Elizabeth Bishop published by Chatto & Windus. Copyright © 1980 by Elizabeth Bishop. Reprinted by permission of The Random House Group Limited.

Lines from 'A New Refutation of Time' in *Labyrinths* by Jorges Luis Borges. Reprinted by permission of Pollinger Limited (www.pollingerltd.com) on behalf of the Estate of Jorges Luis Borges.

Lines from *An American Childhood* by Annie Dillard. Copyright © 1987 Annie Dillard. Reprinted by the permission of Russell & Volkening as agents for the author.

Lines from 'Total Eclipse' in *Teaching a Stone to Talk* by Annie Dillard. Copyright © 1982 Annie Dillard. Reprinted by the permission of Russell & Volkening as agents for the author.

Lines from 'The Princess Forest' by Alec Finlay reproduced with the kind permission of Alec Finlay. 'The Princess Forest' was written and read by Alec Finlay and produced by with original recordings and sound design by Chris Watson

Lines from *The Epic of Gilgamesh* translated by Andrew George. Reprinted with permission from Professor Andrew George.

Lines from 'As Brilliance Fell' in *Selected Poems* by W.S. Graham. Copyright © The Esate of W. S. Graham. Reproduced by permission of Rosalind Mudaliar, the Estate of W.S. Graham.

Lines from 'Postscript' in *The Spirit Level* by Seamus Heaney. Copyright © The Estate of Seamus Heaney. Reprinted with permission from Faber & Faber Ltd.

Lines from *A Phone Call From Paul: A Conversation with W.S. Merwin*, 12 January 2017, by Paul Holdengräber. Published by Literary Hub. Copyright © 2017 by Paul Holdengräber. All rights reserved.

Lines from *When Breath Becomes Air* by Paul Kalanithi © 2016 by Paul Kalanithi. Reproduced by permission of The Random House Group Ltd.

Lines from *A Death in the Family: My Struggle Book 1* by Karl Ove Knausgaard. Copyright © Forlaget Oktober 2009. English translation copyright © Don Bartlett 2012. Reproduced by permission of The Random House Group Ltd.

Lines from *Some Rain Must Fall: My Struggle Book 5* by Karl Ove Knausgaard.

Copyright © Karl Ove Knausgaard and Forlaget Oktober 2010. English translation copyright © Don Bartlett 2016. Reproduced by permission of The Random House Group Ltd.

Lines from 'The Snow Path' by Ko Un, translated by Chae-Pyong Song and Anne Rashid, *Azalea: Journal of Korean Culture and Literature*, volume 5, 2012.

Lines from 'Water' in *The Complete Poems* by Philip Larkin. Copyright © The Estate of Philip Larkin. Reprinted with permission from Faber & Faber Ltd.

Lines from *Arctic Dreams* by Barry Lopez. Copyright © 1986 Barry Lopez. Reproduced with the kind permission of Sterling Lord Literistic.

Lines from 'An Encounter, Wilno Lithuania 1936' from *New and Old Collected Poems: 1931–2001* by Czesław Miłosz. Copyright © 1988, 1991, 1995, 2001 by Czesław Miłosz Royalties, Inc. Reprinted by permission of HarperCollins Publishers.

Lines from 'A Defence of Poetry' by Les Murray. Permission is given by Margaret Connolly & Associates on behalf of Les Murray.

Lines from *The Living Mountain* by Nan Shepherd. Copyright © Nan Shepherd 2008. Reprinted with permission from Canongate Books.

Lines from *Seven Brief Lessons on Physics* by Carlo Rovelli, Simon Carnell and Erica Segre (Penguin Books, 2014). Copyright © Carlo Rovelli, 2014. Translation copyright © Simon Carnell and Erica Segre, 2015. The moral right of the author and translators has been asserted.

Lines from *The Republic of Dreams* by Bruno Schulz, translated by Nina Ogrodowczyk. Reprinted with the permission of Abner Stein on behalf of the Estate of Bruno Schulz.

Lines from *Braiding Sweetgrass: Indigenous Wisdom, Scientific Knowledge and the Teachings of Plants* by Robin Wall Kimmerer. Copyright © 2013 by Robin Wall Kimmerer. Reprinted with permission from Milkweed Editions. www.milkweed.org

Lines from 'Rom Watangu: The Law of the Land' by Galarrwuy Yunupingu, from *The Monthly*, July 2016, pp.18–29.

Index